Potato Crop: Current Developments

Potato Crop: Current Developments

Editor: Patrick Garter

MURPHY & MOORE
www.murphy-moorepublishing.com

Murphy & Moore Publishing,
1 Rockefeller Plaza,
New York City, NY 10020, USA

Visit us on the World Wide Web at:
www.murphy-moorepublishing.com

ISBN: 978-1-63987-443-9 (Hardback)

Cataloging-in-Publication Data

Potato crop : current developments / edited by Patrick Garter.
p. cm.
Includes bibliographical references and index.
ISBN 978-1-63987-443-9
1. Potatoes. 2. Crop science. 3. Potatoes--Breeding. 4. Potatoes--Diseases and pests. 5. Potatoes--Varieties. 6. Potatoes--Harvesting. I. Garter, Patrick.

SB211.P8 P683 2022
635.21--dc23

Table of Contents

Preface

Potato is one of the world's main food crops which is grown for its edible tubers. It is a temperate crop which is a perennial in the nightshade family. It is also known as Solanum tuberosum. Potatoes are a rich source of vitamin C, vitamin B1, starch and carbohydrates. Essential amino acids such as tryptophane, leucine, and isoleucine are found within it in large quantities. Depending upon the variety, potato plants can grow upto 24 inches in height. The leaves of potato plants die after the formation of fruits, flowers and tuber. There are around 4000 varieties of potatoes which are often differentiated on the basis of their starch content into baking and boiling potatoes. They can grow in almost every type of soil except saline and alkaline soil. From theories to research to practical applications, case studies related to all contemporary topics of relevance to this field have been included in this book. It is a vital tool for all researching or studying potato crops as it gives incredible insights into emerging trends and concepts. Those in search of information to further their knowledge will be greatly assisted by this book.

All of the data presented henceforth, was collaborated in the wake of recent advancements in the field. The aim of this book is to present the diversified developments from across the globe in a comprehensible manner. The opinions expressed in each chapter belong solely to the contributing authors. Their interpretations of the topics are the integral part of this book, which I have carefully compiled for a better understanding of the readers.

At the end, I would like to thank all those who dedicated their time and efforts for the successful completion of this book. I also wish to convey my gratitude towards my friends and family who supported me at every step.

Editor

1

Genetics and Cytogenetics of the Potato

Rodomiro Ortiz and Elisa Mihovilovich

Abstract Tetraploid potato (*Solanum tuberosum* L.) is a genetically complex, polysomic tetraploid ($2n = 4x = 48$), highly heterozygous crop, which makes genetic research and utilization of potato wild relatives in breeding difficult. Notwithstanding, the potato reference genome, transcriptome, resequencing, and single nucleotide polymorphism (SNP) genotyping analysis provide new means for increasing the understanding of potato genetics and cytogenetics. An alternative approach based on the use of haploids ($2n = 2x = 24$) produced from tetraploid *S. tuberosum* along with available genomic tools have also provided means to get insights into natural mechanisms that take place within the genetic load and chromosomal architecture of tetraploid potatoes. This chapter gives an overview of potato genetic and cytogenetic research relevant to germplasm enhancement and breeding. The reader will encounter findings that open new doors to explore inbred line breeding in potato and strategic roads to access the diversity across the polyploid series of this crop's genetic resources. The text includes classical concepts and explains the foundations of potato genetics and mechanisms underlying natural cytogenetics phenomena as well as their breeding applications. Hopefully, this chapter will encourage further research that will lead to successfully develop broad-based potato breeding populations and derive highly heterozygous cultivars that meet the demands of having a resilient crop addressing the threats brought by climate change.

Introduction

The most grown potatoes are tetraploid ($2n = 4x = 48$), but farmers in the Andes grow diploid ($2n = 2x = 24$), triploid ($2n = 3x = 36$), and pentaploid ($2n = 5x = 60$) cultivars (Watanabe 2015). The basic chromosome number of these tuber-bearing

R. Ortiz (✉)
Swedish University of Agricultural Sciences (SLU), Alnarp, Sweden
e-mail: rodomiro.ortiz@slu.se

E. Mihovilovich
Independent Consultant, Lima, Peru

Solanum species is 12. Diploid cultivars along with tetraploid cultivars are used in potato breeding through ploidy manipulations with haploids and $2n$ gametes (or gametes with the sporophytic chromosome number), and chromosome engineering using aneuploidy (Ortiz 1998). Improvement of cultivated potato is challenged by its high heterozygosity (Bradshaw et al. 2006; Hirsch et al. 2013) and complex polysomic tetraploid inheritance (Howard 1970; Ortiz and Peloquin 1994; Ortiz and Watanabe 2004). The tetraploid *Solanum tuberosum* has four homologues, which include 12 unique chromosomes each, thus showing tetrasomic inheritance (Bradshaw 2007). Epistasis and heterozygosity are key for succeeding in $4x$ potato breeding because multi-allelic quantitative trait loci showing high-order genic interactions, while additivity also contributes to quantitative traits with high heritability.

The genetics of tetrasomic potato depends on four sets of homologous chromosomes (instead of two as in diploid potato). Three genotypes (*AA*, *Aa*, *aa*) are expected after selfing a heterozygous diploid (*Aa*), while the selfing of a comparable tetraploid (*AAaa*) gives five different genotypic classes in its offspring: *AAAA* (quadruplex), *AAAa* (triplex), *AAaa* (duplex), *Aaaa* (simplex), and *aaaa* (nulliplex). Double reduction—related to tetrasomic inheritance—occurs when two chromosomes in a gamete derive from two sister chromatids; i.e., the sister chromatids end in same gamete. Quadrivalent formation, a single crossing over between the centromere and the locus to allow sister chromatids to attach to two different centromeres, that these centromeres with sister chromatids move to the same pole in anaphase I, and sister chromatids go the same pole in anaphase II are necessary for double reduction. The probability of double reduction to occur is noted as α, which is equalto $\frac{qea}{2}$, where q is the quadrivalent frequency, e is the frequency of equational separation that depends on the gene–centromere map distance, and a is the frequency of non-disjunction (often $\frac{1}{3}$). Chromosome segregation arises when $\alpha \neq 0$,thus indicating that a locus of interest lies close to the centromere; while if $\alpha = \frac{1}{7}$ or $\frac{1}{6}$ then chromatid segregation or maximal equational division (MED), respectively, occurs. MED is rarely found because of the requirements for its occurrence (Burham 1984). DNA-aided marker analysis confirmed the occurrence of double reduction and that it increases with distance from the centromeres (Bourke et al. 2015). As noted by Gálvez et al. (2017), the potato reference genome and transcriptome, research on both gene expression and regulatory motif, plus resequencing and SNP genotyping analyses, provide new means for increasing the understanding of potato genetics. For example, DNA resequencing allows assembling a genome reference for each cultivar or landrace, which may provide useful knowledge regarding structural differences between the various potato groups. In this regard, the resequencing of diversity panel including wild species, landraces, and cultivars demonstrated that a limited gene set accounts for early improvement of the potato cultigen, while distinct loci seems to be involved on the adaptation *S. tuberosum* group Andigenum (upland potato) and *S. tuberosum* groups Chilotanum and Tuberosum (lowland potato) populations (Hardigan et al. 2017). Signatures of selection in genes regulating pollen development/gametogenesis reduced fertility. Introgression of truncated alleles of wild species, particularly *S. microdontum*, was noted in long day cultivars, thus showing how wild tuber-bearing *Solanum* species are key sources of variation for breeding.

Haploids and Disomic Inheritance

Tetraploid potato shows significant inbreeding depression (De Jong and Rowe 1971) because polyploidy and heterozygosity mask deleterious recessive mutations and buffer genomic imbalance (Comai 2005; Henry et al. 2010; Tsai et al. 2013). These characteristics led to an alternative breeding approach based on the use of haploids ($2n = 2x = 24$) produced from tetraploid *S. tuberosum*. The homozygosity/heterozygosity of methylated DNA may be, however, involved in inbreeding depression/heterosis in self-compatible diploid potatoes because DNA methylation may suppress gene expression (Nakamura and Hosaka 2010).

The induction of haploid plants is generally referred to as "haploidization." There are two main pathways by which haploid formation can be induced in potato: androgenesis and gynogenesis. Androgenesis is through in vitro culturing of whole anthers or free microspores on a nutrient rich medium to induce plantlet regeneration from single gametic cells or haploid calli (Veilleux 1996), while gynogenesis is haploidization via the "maternal" or seed parent's genome. Potato haploids are routinely obtained by gynogenesis, a process in which specific *S. tuberosum* Group Phureja ($2n = 2x = 24$) selections, known as "haploid inducers," contribute the paternal gametes for pollination of the desired haploid progenitor. The formation of a haploid embryo begins when the egg is either induced into parthenogenesis (Hermsen and Verdenius 1973) or when the zygote experiences spontaneous abortion of the pollen donor's set of chromosomes (Clulow et al. 1991). Evidence for the latter has been the identification of Phureja-specific molecular markers in aneuploids ($2n = 2x + 1 = 25$) among the offspring of some haploid induction crosses (Clulow et al. 1991; Clulow and Rousselle-Bourgeois 1997; Samitsu and Hosaka 2002; Straadt and Rasmussen 2003; Ercolano et al. 2004). Ortiz et al. (1993a) suggested that the genetics of the ability to induce haploids is relatively simple.

Putative haploids are identified firstly by the lack of a dominant morphological marker for anthocyanin pigmentation on developing embryos (embryo spots) or on seedling shoots (nodal bands). This marker that allows early haploid selection is present in homozygosis in certain pure *S. tuberosum* Group Phureja clones or has been bred to homozygosity in its derived hybrids. The ploidy of the resultant seedlings is confirmed by counting chromosomes in mitotically dividing root cells (Sopory 1977), counting chloroplasts in stomatal guard cell pairs (Singsit and Veilleux 1991) or through flow cytometric analysis (Owen et al. 1988). The haploid-inducing clones often used owing to their relatively superior haploid-induction frequency and homozygosity for the seed marker "embryo-spot" are the following

Table Haploid induction ability of "IVP 35," "IVP 101," and "PL-4"

Character[a]	"IVP 35"	"IVP 101"	"PL-4"
Number of haploids per 100 berries	69	62	96
Number of haploids per 1000 seeds	52	85	103

[a]Pooled data per haploid inducer from 13 seed parents

Group Phureja clones: "IVP 35," "IVP 48", and "IVP 101" (Ross 1986). "IVP 101" has been derived from the cross [(G609 × "IVP 48") × ("IVP 10" × "IVP 1")]. G609 is a haploid from Group Tuberosum cultivar "Gineke" that combines its own haploid induction ability with a high degree of male fertility, profuse flowering, and vigor.

The efficiency of haploid production is determined by both, production ability of the tetraploid seed parent and induction ability of the diploid pollinator (Hougas et al. 1964; Frandsen 1967; Hermsen and Verdenius 1973). However, interaction between seed parents and pollinators were also noted (Frandsen 1967). Despite this interaction, haploid induction ability of "IVP 101" has proved to be higher than "IVP 35" and "IVP 48" (Hutten et al. 1993). By the end of the 1990s a promising haploid inducer named "PL-4" (CIP596131.4) was selected at the International Potato Center (CIP, Lima, Perú) as a transgressive genotype from the cross between "IVP 35" × "IVP 101" due to its highest haploid inducer ability, degree of flowering, shedding and pollen viability relative to its parents (M. Upadhya and R. Cabello, CIP, unpublished data). Historical data accumulated from 2001 to 2009 from haploid induction crosses between 37, 4*x* breeding lines with both "IVP 101" and "PL-4" showed that the latter produced twice the amount of seeds without embryo spot (putative haploids) of "IVP 101." A more comprehensive study to determine the haploid inducer ability of "PL-4" relative to their parents was performed during 2015 and 2016 at CIP. This involved haploid induction of 13, 4*x* breeding clones with the three haploid inducers. Haploid confirmation was made by counting chloroplasts in stomatal guard cell pairs and flow cytometric analysis in seedlings grown from seeds without embryo spot. "PL-4" produced a significantly higher number of haploids than its parents. Meanwhile, "IVP 101" outperformed "IVP 35" in number of haploid per 1000 seeds. There were also differences in haploid production ability between seed parents. Two breeding clones, CIP 300056.33 and CIP 392820.1, showed the highest number of haploids.

Further Research and New Evidence on Haploid Origin

Previous research in potato haploids originated by gynogenesis detected aneuploids ($2n = 2x + 1 = 25$ and $2n = 2x + 2 = 26$) instead of the expected 24-chromosome karyotypes with concurrent appearance of Group Phureja-specific molecular markers (Clulow et al. 1991, 1993; Waugh et al. 1992; Wilkinson et al. 1995; Clulow and Rousselle-Bourgeois 1997; Ercolano et al. 2004). Moreover, there was one case in which translocation of a Group Phureja chromosomal segment to the Group Tuberosum genome was detected by genomic in situ hybridization (GISH; Wilkinson

Table Haploid production ability of 13, 4*x* breeding clones from CIP potato breeding program

CIP-number[a]	Breeding code	Number of haploids per 100 berries	Number of haploids per 1000 seeds
CIP 300056.33	LR00.014	141	169
CIP 300072.1	LR00.022	68	101
CIP 300093.15	LR00.027	82	58
CIP 301023.15	C01.020	40	63
CIP 388615.22	C91.640	75	141
CIP 388676.1	Y84.027	40	32
CIP 390478.9	C90.170	2	2
CIP 390637.1	93	76	107
CIP 391931.1	458	11	17
CIP 392780.1	C92.172	9	21
CIP 392820.1	C93.154	258	175
CIP 397073.16	WA.104	39	30
CIP 397077.16	WA.077	100	68
Average		72	76

[a]Pooled data per breeding clone from three haploid inducers

et al. 1995). The cause, frequency, and nature of these introgressions remain unknown, though they may affect performance of the haploids (Allainguillaume et al. 1997). A similar phenomenon has been observed in CenH3-based haploid induction in *Arabidopsis thaliana* (Ravi and Chan 2010; Ravi et al. 2014; Tan et al. 2015). In this system, mis-segregation of the haploid inducer chromosomes leads to genome elimination. In a fraction of the haploid progeny, one or few of the haploid inducer chromosomes were retained, resulting in aneuploid progeny. This DNA introgression was identified readily by low-pass sequencing and single nucleotide polymorphism (SNP) analysis (Tan et al. 2015). These findings added evidence that DNA introgression from a haploid inducer is expected to involve large contiguous segments, and often whole chromosomes. Lately, K.R. Amundson et al. (unpublished) surveyed a haploid segregating population for aneuploids by low-pass sequencing. The population was previously developed at CIP for tetraploid genetic mapping of a major gene controlling *Potato leaf roll virus* (PLRV) resistance in the Group Andigena cultivar "Alca Tarma" (Velásquez et al. 2007). They identified 19 haploids (11.4%) that displayed elevated relative sequence read coverage of a single chromosome consistent with 25-chromosome karyotypes in root tip metaphase spreads in these putatively aneuploid clones. By sequencing parental genotypes to higher depth (40–66×) and identifying homozygous SNP between "Alca Tarma" and either of the two haploid inducers, "IvP-101" or "PL-4," plus assuming to have sired each haploid (Velásquez et al. 2007), they found nearly 0% haploid inducer

SNP for all chromosomes of these aneuploids. Thus, the additional chromosomes observed likely did not originate from the haploid inducer genome, but were maternally inherited. This lack of Group Phureja SNP in haploids was previously reported in another study that employed DNA markers (Samitsu and Hosaka 2002). The production of aneuploid gametes is a common property of polysomic polyploids (Comai 2005), and "Alca Tarma" was not an exception. Admusson et al. (unpublished) concluded that for haploid inducers "IvP-101" and "PL-4," either the mechanism of haploid induction does not involve egg fertilization or genome elimination in "Alca Tarma" was very efficient.

Relevance of Haploids in Plant Breeding and Genetics

Haploids showed disomic inheritance, which means that each chromosome paired with its homolog, thus providing means for simplifying genetic research in potato. They can also be efficiently used for research on chromosome pairing and natural mutation accumulated at the tetraploid level. Initially, potato haploids were envisioned as a tool to simplify the breeding of *S. tuberosum* cultivar production by reducing tetraploid germplasm to a diploid breeding level (Chase 1963). A second early reason for the production of haploids was to acquire a "genetic bridge" between the various genomes of *Solanum* species. Ploidy barriers between the cultivated and wild *Solanum* species could be circumvented by crossing haploids to the wild diploid *Solanum spp.* and novel hybrid germplasm incorporated back into tetraploid breeding programs through $4x \times 2x$ crosses using $2n$ gamete formation, or by colchicine-doubling of the novel diploid hybrid (Ross 1986). In addition to their use in breeding, diploid potato hybrids represent a powerful tool for genetic analysis due to its much simpler segregation ratios compared to tetraploid cultivated potatoes (Ortiz and Peloquin 1994). Thus, diploid potato has been used to determine the inheritance of economically important traits such as tuber shape (De Jong and Burns 1993; Van Eck et al. 1994b), tuber flesh and skin pigmentation (De Jong 1987; Van Eck et al. 1994a), and tuber skin texture (De Jong 1981). The genetic basis of some physiological mutants has also been analyzed with the use of diploids (De Jong et al. 1998, 2001). Haploids have been convenient for trait mapping (Kotch et al. 1992; Pineda et al. 1993; Freyre et al. 1994; Simko et al. 1999; Naess et al. 2000; Velásquez et al. 2007) and in the development of an online catalogue of amplified fragment length polymorphisms (AFLP) covering the potato genome (Rouppe van der Voort et al. 1998). On the other hand, many breeders have extracted haploids from superior parents and also maintain a diploid gene pool composed of hybrids between haploids and diploid wild species carrying specific quality and host plant resistance genes not found in cultivars (Carputo and Barone 2005; Ortiz et al. 2009).

Currently, there are some ongoing efforts towards genetically restructuring potato as a diploid inbred line-based crop (Jansky et al., 2016; Lindhout et al. 2011). Here, the vision is a diploid potato crop composed of a series of inbred lines that capture the favorable genetic diversity available in the potato cultigen. This diploid genepool with a broad suite of traits represent a valuable stock for fixing desirable gene combinations and realized breeding gains. Last but not least, haploids have been regarded as a tractable ploidy state for copy number variation (CNV) analysis in tetraploid potatoes as these variants are meiotically transmissible and diploid gametes can likely shield deficiencies such as recessive lethal and dosage sensitive loci (Comai 2005; Lovene et al. 2013; Henry et al. 2015). CNV is defined as stretches of DNA from 1 kilobase (kb) to several megabases (Mb) that display different copy numbers in populations (Feuk et al. 2006). Analyses of multiple genotypes in *Arabidopsis* and maize suggest that CNV may play a significant role in phenotypic diversity and hybrid heterosis in plant species (Swanson-Wagner et al. 2010; Cao et al. 2011). Moreover, they can affect phenotype impacting important agronomic and host plant resistance traits (Maron et al. 2013; Díaz et al. 2012; Zhu et al. 2014; Cook et al. 2012). Direct CNV detection in tetraploid potato is challenging. The dosage increase associated with a duplication is subtler in tetraploids (25% increase) than in diploids (50% increase), and high heterozygosity impedes haplotype assembly (Potato Genome Sequencing Consortium 2011). As a consequence, CNV analysis is often limited to few loci of interest and is costly to be practical in breeding programs. Hence, sampling the gametophyte genome of tetraploids by ploidy reduction through haploidy is an alternative approach.

2*n* Gametes

Gametes with the sporophytic chromosome number should be named as 2*n* gametes and not as "unreduced" gametes as wrongly dubbed. They result from pre-meiotic, meiotic, or post-meiotic abnormalities during gametogenesis. The modes of formation are pre-meiotic doubling, first division restitution (FDR), chromosome replication during meiotic interphase, second division restitution (SDR), post-meiotic doubling, and apospory (diploid sac formed from nucellus or integument cell). FDR and SDR mechanisms are the most common for 2*n* pollen and 2*n* egg formation in potato (Ortiz 1998). Heterozygous 2*x* parents transmit 80% and 40% of their heterozygosity to their 4*x* hybrid offspring after sexual polyploidization with FDR or SDR 2*n* gametes, respectively.

The parallel orientation of the spindles in the second meiotic division accounts frequently for FDR 2*n* pollen, while omission of the second division after a normal first division seems to be often involved in SDR 2*n* eggs. The abnormal meiosis leading to these 2*n* gametes are under the genetic control of recessive mutants: *ps* for 2*n* pollen and *os* for 2*n* egg, both of which appear to be ubiquitous in *Solanum* species. The finding of genes whose mutations led to a high frequency of 2*n* gametes in the model plant species *Arabidopsis thaliana* provided further means for understanding their formation in plants (Brownfield and Kölher 2011). It appears to be very likely that a mechanism related to a loss of protein function leads to the formation of 2*n* gametes.

The frequency of $2n$ gametes may be affected by incomplete penetrance and variable expressivity, which is under minor modifier genes and influenced by plant age and the environment. Phenotypic recurrent selection could be effective for increasing the frequency of FDR $2n$ pollen, while recurrent selection with progeny testing may raise SDR $2n$ egg expressivity.

There are synaptic mutants affecting gametogenesis in haploids, *Solanum* species and haploid-species hybrids. They may cause poor pairing, reduced chiasma, or both, thus reducing recombination. For example, the synaptic mutant sy_3—found in Group Phureja–haploid hybrids—along with *ps* produces FDR $2n$ pollen without crossing over (FDR-NCO), while the desynaptic mutant *ds-1* generates sterile *n* eggs and fertile FDR $2n$ eggs owing to a direct equational division of univalent chromosomes at anaphase I, i.e., pseudohomotypic division. Desynaptic gametes may transfer about 95% of the $2x$ genotype to their $4x$ hybrid offspring.

Cytoplasm Diversity and Male Sterility

There are six distinct cytoplasmic genome types in potato, namely, M, P, A, W, T, and D. Many clones bred worldwide show a genetic bottleneck in cytoplasmic diversity due to the continuous use of cytoplasmic-based male sterility "lineages" derived from *S. demissum* or *S. stoloniferum* that are often used as sources of host plant resistance. For example, T (45%), D (38%), and W (11%) are the most frequent types in CIP bred germplasm (Mihovilovich et al. 2015); while the most popular among EU cultivars and breeding clones are T (59%), D (27%), and W (12%) (Sanetomo and Gebhardt 2015), and cultivars and breeding lines from Japan plus a sample of landraces and foreign cultivars show 73.9% T, 17.4% D, and 2.4% W (Hosaka and Sanetomo 2012).

Cytoplasmic factors and nuclear alleles are involved in indehiscence, shriveled microspores, sporad formation, anther-style fusion, ventral-styled anthers, and thin anthers (Grun et al. 1977). Group Andigena and its ancestors, Group Stenotomum and Phureja, share most plasmon factors, of which many differ from those found in Group Tuberosum. Hence, cytoplasmic-genetic male sterility is often noted in hybrid offspring among some tuber-bearing *Solanum* species because interactions between sensitive factors in the cytoplasm of one species and nuclear genes from the other species. Hybrids derived from crossing Group Tuberosum haploids as female and Group Phureja or Stenotomun as males are very often male sterile, but the reciprocal cross show male fertile offspring. Male sterility ensues from the interaction of a dominant gene (*Ms*) from the Group Phureja or Stenotomun with Group Tuberosum sensitive cytoplasm. Diploid (2EBN) wild species such as *S. chacoense, S. berthaultii,* and *S. tarijense* do not carry genes that interact with the Group Tuberosum cytoplasm as shown by the high male fertile hybrid offspring between them.

The frequency of male fertile offspring in hybrids between Group Tuberosum and groups Stenotomum or Phureja may vary because some tetraploid cultivars bear a dominant male fertility restorer (*Rt*) gene (Iwanaga et al. 1991b). The *Ms* and *Rt* genes, which are independently inherited, are very distal from the centromere, thus showing both loci chromatid segregation (Ortiz et al. 1993b). The *Rt* gene allows to partially bypass male sterility using crossbreeding, e.g. by crossing Group Tuberosum haploids bearing *Rt* with 2× species carrying *Ms* because ½ (*rt*/*rt* × *Ms*/*ms*) ¾ (*Rt*/*rt* × *Ms*/*ms*) or 100% (*Rt*/*Rt* × *Ms*/*Ms* or *Rt*/*Rt* × *Ms*/*ms*) of the resulting hybrid offspring will be male fertile.

Self-Incompatibility and *s* Locus Inhibitor Mechanism

Diploid potatoes and their related wild tuber-bearing *Solanum* species are self-incompatible due to a gametophytic self-incompatibility (GSI) system controlled by the interaction of a pollen *S* gene with pistil *S* gene(s). GSI inhibits fertilization by self-pollen or pollen from closely related (sibling) plants (Hanneman Jr 1999). In this system, compatibility is controlled by the *S* locus, which consists of two closely linked genes: *S-RNase* and *S-Locus F-box* (*SLF/SFB*) that control the female and male specificity, respectively. S-RNase-based self-incompatibility systems are widespread mechanisms for controlling selfing (Hancock et al. 2003). *S* locus variants, currently known as S-haplotypes, determine self-incompatible pollen rejection when there is a match between the single S-haplotype in the haploid pollen and either of the two haplotypes in the diploid system. S-RNases are the determinants of *S*-specificity in the pistil and act in pollen recognition as well as in direct pollen growth inhibition by degrading pollen RNA in incompatible pollinations (McClure et al. 1990). A distinct S-RNase protein is expressed from each functional S-haplotype and upon recognition the protein enters intact pollen tubes retaining its potentially cytotoxic enzyme activity (Gray et al. 1991). Conversely, pollen RNA is stable in compatible pollinations as interaction between S-RNase and *SLF* confers resistance to the cytotoxic effects of S-RNase (Fig.; Golz et al. 2001). The *SLF/SFB* is a family of F-box protein genes whose most well-known role is ubiquitin-mediated protein degradation. Pollen modifier genes encoding proteins that form complexes with SLF provides for ubiquitylation and degradation of non-self S-RNase as a necessary step to overcome the cytotoxicity of S-RNase. On the other hand, self S-RNases fail to bind productively and thus escape degradation (Hua et al. 2008; Zhang et al. 2009).

Modifier genes encoding putative pistil self-incompatibility factors were found in potato wild relatives. *HT-A* and *HT-B* are two similar genes expressed in the genus *Solanum*, but only *HT-B* has shown to be strongly suppressed in self-incompatibility breakdown in *S. chacoense*. HT-B proteins appear to be degraded in pollen tubes after compatible pollination, while this is not the case in incompatible pollen tubes where substantial amounts of HT-B reactive protein were found (Fig. ; Goldraij et al. 2006).

Self-compatibility can be obtained by converting a self-incompatibility diploid (e.g., S_1S_2) to a tetraploid (de Nettancourt 1977). Here the defect occurs only in the pollen due to the so-called heteroallelic pollen (HAP) effect. Thus, $S_1S_1S_2S_2$ pistils

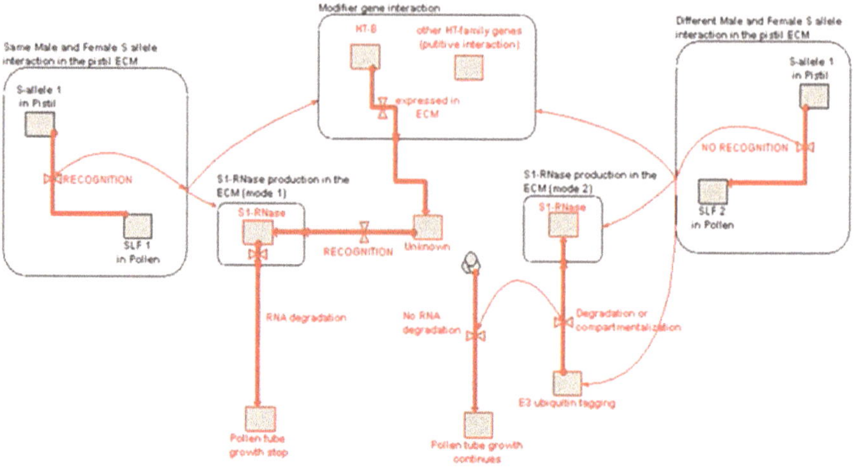

Fig. A model for S-RNase-based self-incompatibility. When the pollen tube enters the extra-cellular matrix (ECM) the proteins HT-B, S-RNase, and 120K (not represented) are expressed. On a positive recognition of self-pollen (or a self *SLF* gene) the pollen tube RNA is broken down by the production of S-RNase and fertilization is unlikely. In the opposite reaction, when an unrecognizable match is made, S-RNase is degraded after being tagged by an E3 ubiquitin ligase complex. (Courtesy: Dr. Philippe Kear, International Potato Center)

reject S_1- and S_2-pollen normally, but diploid pollen is not rejected. Self-incompatibility breakdown only occurs in the HAP case, S_1S_2. Pollen *S* functions to provide resistance to S-RNase. Self-compatible variants have often been described among genotypes of self-incompatible potato species, such as, *inter alia*, *S. chacoense*, *S. kurtzianum*, *S. neohawkesii*, *S. pinnatisectum*, *S. raphanifolium*, *S. sanctaerosae*, *S. tuberosum* Groups Phureja and Stenotomum (Cipar et al. 1964), and *S. verrucosum* (Eijlander 1998). SC variants were also noted in haploids ($2n = 2x = 24$ chromosomes) from Group Tuberosum (De Jong and Rowe 1971; Olsder and Hermsen 1976) and in hybrids between Group Phureja and haploids ($2n = 2x = 24$ chromosomes) of Group Andigenum (Cipar 1964).

Self-compatible variants in *S. chacoense* show a single dominant gene "*Sli*" that is expressed in a sporophytic fashion (Hanneman 1985; Hosaka and Hanneman 1998a). The existence of self-incompatible progeny segregating from *S. chacoense* selfed plants having *Sli* gene in a heterozygous condition shows that fully functional S-haplotypes are transmitted through pollen even when the *Sli* factor is not. *Sli* was mapped to the end of chromosome 12 (Hosaka and Hanneman 1998b). Since the *S* locus has been localized on chromosome 1 (Gebhardt et al. 1991; Jacobs et al. 1995; Rivard et al. 1996), it is evident that the *Sli* gene is independent of the *S* locus. *Sli* can be regarded as a dominant gain-of-function (GOF) pollen-part mutant (PPM) that interacts in some way with the GIS system in pollen and results in self-compatible plants. This pollen side gene may inhibit S-RNase uptake, break down the pollen-stigma recognition system, overcome the cytotoxic activity of S-RNase independent of its pollen *S*-genotype or through the action on a non-S-specific factor (Hosaka and Hanneman 1998b; McClure et al. 2011). Likewise, self-compatible variants in *S. verrucosum* (*ver*) has been assumed to be a pistil side nonfunctional S-RNase haplotype (*Sv* allele) that allows *ver* plants be pollinated by its own pollen as well by the pollen of other self-incompatible potato species (Eijlander 1998). Absence of pistillate S-RNases seems to be a characteristic feature of this species (Makhan'ko 2011).

Self-Compatibility in Breeding

Most cultivated tetrasomic polyploid or self-incompatible diploid potatoes have not realized breeding gains due to low recombination, long generation cycles, polyploidy, inbreeding depression, and poor adaptation of wild potato germplasm (Visser et al. 2009; Lindhout et al. 2011). New breeding methods that involve the development of diploid inbred lines in potato were proposed as a strategy to address many perceived limitations faced by potato breeders (Birhman and Hosaka 2000; Phumichai et al. 2005). In the last few years a trend emerged in a group of potato breeders to reconsider the crop as a diploid species composed of a series of inbred lines that capture the favorable genetic diversity available in cultivated and wild potatoes. Inbreeding due to selfing may be efficient for organizing the whole gene pool into various favorably interacting and stable epistatic systems (Allard 1999).

The self-incompatibility inhibitor (*Sli*) gene opens new doors to explore inbred line breeding in potato (Lindhout et al. 2011). Highly inbred *S. chacoense* lines such as M6, which has been self-pollinated for seven generations, are vigorous and fertile (Jansky et al. 2014). CIP has incorporated *Sli* into diploid cultivars of *S. tuberosum* groups Stenotomum and Phureja and selected a panel of 20 *Sli* bearing self-compatible hybrids to provide a more desirable self-compatibility source than wild *S. chacoense* for the development of inbred lines. These self-compatible hybrids denoted BSLi, are being used to incorporate novel diversity from wild species and take full advantage of modern genetics and genomics tools to generate inbred genetic resources such as recombinant inbred lines (RILs), for fundamental gene discovery and gene mapping.

Self-compatibility has been identified in five diploid cultivated potatoes of *S. tuberosum* Phureja (phu) and Stenotomum (stn) Groups held in CIP's genebank. Selfing three of these self-compatible diploid cultivars; i.e., CIP705468 (goniocalyx), CIP703320 (stn), and CIP701165 (stn) yielded progenies that segregated for self-compatible and self-incompatible individuals. The segregation ratio 2 self-compatible: 1 self-incompatible was significantly skewed from the expected ratio of 3 self-compatible: 1 self-incompatible for a mutant factor in heterozygosis because of the small population size ($N < 60$) analyzed in each selfed cultivar. Self-compatible plants due to pistil side mutations that compromised S-RNase or HT factors produce only self-compatible plants. The same is true for GOF mutations of the *SLF* gene

(McClure et al. 2011). On the other hand, self-compatible plants with a *Sli* mutation in heterozygocity produce 3 self-compatible: 1 self-incompatible plants regardless of its haplotype condition in the *SLF* locus; i.e., *SxSx*, *SxSy*, *SySy*. Hence, the presence of self-compatible plants in the offspring of self-compatible 2*x* cultivars suggests a dominant pollen-side mutation similar to the *Sli* gene since this is the only scenario that yields self-incompatible offspring. Further research will be required to elucidate whether this *Sli*-like phenotype is a novel pollen-side GOF factor or *Sli* gene variant. Whatever its nature, self-compatible 2*x* cultivars would provide a more desirable self-compatible source than *S. chacoense* as they will avoid the undesirable linkage drag associated with the use of a wild species in the development of 2*x* inbred lines.

Interspecific Crosses and Incompatibility

Interspecific reproductive barriers (IRBs) complicate using wild germplasm species for crop improvement (Zamir 2001; Jansky et al. 2013). Therefore, the rich trait diversity available in CIP's extensive collection of tuber-bearing *Solanum* accessions requires overcoming IRBs to be introgressed into cultivated *Solanum* taxa. These barriers include incompatibility between pollen and pistil, male sterility resulting from interactions between nuclear and cytoplasmic genes, and endosperm failure (Camadro et al. 2004). Post-zygotic IRBs are due to EBN incompatibilities that lead to endosperm failure, whereas pre-zygotic IRBs are typically associated with pollen tube growth inhibition (Camadro and Peloquin 1981; Fritz and Hanneman 1989; Novy and Hanneman 1991; Camadro et al. 1998; Erazzú et al. 1999; Hayes et al. 2005).

Genetic and molecular research shows that some self-incompatibility factors also function in prezygotic IRBs. However, self-incompatibility and IRBs differ in terms of specificity and the precise factor requirements. IRBs show broad specificity, and a single S-RNase can cause rejection of pollen from species or groups of species (Murfett et al. 1996; Tovar-Méndez et al. 2014). *S-RNase* and *HT* genes dual roles in self-incompatibility and interspecific pollen rejection points out pleiotropic effects and hence linkage between these two mechanisms.

IRB mechanisms' complexity is such that multiple redundant mechanisms can contribute to interspecific incompatibility, even between a single pair of species (Murfett et al. 1996; McClure et al. 2000). This may complicate experiments because defects in one rejection mechanism do not necessarily result in compatibility. For example, HT-proteins previously found implicated only in S-RNase-dependent self-incompatibility and IRBs have also been involved in S-RNase-independent pollen rejection in tomato (McClure et al. 2011).

Selective pressures like reproductive assurance make self-incompatible to self-compatible mating system transitions (MSTs) common in nature (Barrett 2002; Goldberg et al. 2010; Goldberg and Igic 2012). Self-compatibility has not been extensively investigated in the potato clade and few self-compatible species have been recognized. Loss of S-RNase function is a common route to self-compatibility. This is the case of self-compatible variants of *S. verrucosum* in which pollen tubes of other potato species, including those having 1EBN can grow without inhibition in their pistils reaching ovules in great quantity (Hermsen and Ramanna 1976; Makhan'ko 2011). Furthermore, dominant gain-of-function (GOF) pollen-part mutants such as *Sli*, which results in self-compatible variants in *S. chacoense*, increase seed set in interspecific crosses to *S. pinnatisectum* in addition to suppressing self-incompatibility (Sanetomo et al. 2014).

Phumichai et al. (2006) were able to introduce after crossing the *S*-locus inhibitor gene (*Sli*), which can inhibit gametophytic self-incompatibility, in diploid potatoes and alter self-incompatible to self-compatible plants, into 32 diploid genotypes. *Sli* has been also successfully introduced to diploid cultivars from *S. tuberosum* Phureja and Stenotomum Groups using SC *S. chacoense* variants as male parents at CIP. Assuming that this pollen side mutant acts either breaking down pollen-stigma recognition system or overcoming S-RNAse cytotoxic activity (Hosaka and Hanneman 1998b), then a change in interspecific compatibility may occur after crossing *Sli*-bearing self-compatible hybrids as pollen parents and self-incompatible sources of late blight resistance from wild *S. piurae* and *S. chiquidenum* species. Previous attempts at CIP to cross these wild species with 2*x* *S. tuberosum* cultivars produced very few seeds with *S. chiquidenum* and no seed set at all with *S. piurae*. Embryo rescue was often required to save the few hybrids produced from *S. chiquidenum*. Linkage between self-incompatibility and IRBs results in MSTs with significant implications in germplasm enhancement programs particularly when decisions between different crop wild relatives have to be made in interspecific crosses.

Unilateral Compatibility

This is a very common IRB pattern that refers to crosses that are compatible in only one direction (Lewis and Crowe 1958). Most IRBs conform to the self-incompatible × self-compatible rule, in which pollen from the self-compatible species is rejected on pistils of related self-incompatible species but the reciprocal pollination is compatible (Nathan Hancock et al. 2003; Bedinger et al. 2011). In potato only some IRBs conform to this rule (Hermsen and Ramanna 1976; Eijlander et al. 2000). For example, crosses between the self-compatible 2*x* (1EBN) species *S. pinnatisectum* and the 2*x* (1EBN) self-incompatible species *S. cardiophyllum* were successful only when self-compatible *S. pinnatisectum* was used as the female parent (Chen et al. 2004). A similar pattern was observed in crosses between the self-compatible 2*x* (1EBN) species *S. commersonii* and the self-incompatible 2*x* (2EBN) species *S. chacoense* (Summers and Grun 1981). Although it is not absolute, the consistency of the "self-incompatible × self-compatible" rule suggests a link between inter- and intraspecific pollen rejection. The suggested linkage is that the *S*-locus controls unilateral incompatibility as well as self-incompatibility. Conversely, there are many exceptions to the self-incompatible × self-compatible rule in potato such as the occurrence of both unilateral and bilateral self-incompatible × self-incompatible conflicts (Camadro et al. 1998, 2004; Kuhl et al. 2002; Raimondi et al. 2003). Genetic systems entirely independent of the *S*-locus have been proposed to explain cross incompatibility, which is a term used to emphasize potato clade cross distinctive complexity. Cross incompatibility includes both post- and pre-zygotic mechanisms.

An important phenomenon worth mentioning is compatibility encountered in crosses between self-compatible *S. verrucosum* plants (2*x*, 2EBN) and a wide range of accessions of various 1EBN potato species. Success on producing these novel sexual hybrids was achieved with several 1EBN diploid species, such as *inter alia*, *S. bulbocastanum*, *S. pinnatisectum*, *S. polyadenium*, *S. commersonii*, and *S. circaeifolium* (Yermishin et al. 2014). Absence of S-RNases in pistils in self-compatible *S. verrucosum* allowed growth of pollen tubes from these 1EBN wild species and fertilization of egg cells. In addition, "rescue pollination" was used to improve hybridization effectiveness. "Rescue pollination" also known as "double pollination" is a technique that reduces premature fruit drop and involves the application of pollen from the incompatible species, followed a day or 2 later by that of a compatible species, denoted as "mentor pollen" (Singsit and Hanneman Jr 1990). The "mentor pollen" fertilizes several ovules, stimulating the development of fruit and pollen tubes from the incompatible parent to reach the ovules and effect fertilization (Singsit and Hanneman Jr 1990; Yermishin et al. 2014). Pollen of Group Phureja pollinators are used as "mentor pollen" because of their typical dominant seed spot marker, so its offspring can be visually identified and eliminated (Brown and Adiwilaga 1991; Iwanaga et al. 1991a).

A feature of inter-EBN interspecific hybridization using self-compatible *S. verrucosum* as "bridge species" is male sterility in most resulting hybrids (Abdalla and Hermsen 1972–Abdalla and Hermsen 1973; Yermishin et al. 2014). Cytoplasmic male sterility factors (CMS) from *S. verrucosum* have been assumed to interact with dominant nuclear genes from the 1EBN male parents, resulting in male sterility of hybrids. CMS has also been suggested to account for male sterility of hybrids produced when 2*x* (2EBN) Group Tuberosum haploids were used as male parents in crosses with cultivated 2*x* (2EBN) Phureja and Stenotomum Groups (Grun et al. 1962; Ross et al. 1964; Carroll 1975) as well as with a wide range of wild species (Hermundstad and Peloquin 1985; Tucci et al. 1996; Santini et al. 2000). Male fertile hybrids obtained when the haploids were the male parent corroborated this assumption (Tucci et al. 1996; Novy and Hanneman 1991). Consequently, this type of barrier can be overcome by carrying out reciprocal crosses.

Male sterility of hybrids from *S. verrucosum* did not represent a drawback since crosses with Tuberosum haploids were successful when the (*S. verrucosum* × 1EBN) hybrids were used as females (Yermishin et al. 2014). However, male fertile Tuberosum haploids are not widely available (Makhan'ko 2008). Breeders may overcome this drawback by extracting haploids from 4*x* *S. tuberosum* cultivars or breeding clone bearing the male fertility dominant restorer gene *Rt* that gives fertility to plants that contain the dominant male sterility gene *Ms* in the presence of sensitive cytoplasm (Iwanaga et al. 1991b). Selection of Group Tuberosum haploids carrying a restorer of fertility (*Rt*) gene can be used to pollinate (*S. verrucosum* × 1EBN) hybrids and produce male fertile hybrids for further backcrossing with cultivated potato.

An alternative appealing approach for creating opportunities to solve the problem of the prezygotic interspecific incompatibility with a number of 1EBN wild species was proposed by Polyukhovich et al. (2010). These investigators transferred the nonfunctional S-RNase *Sv* haplotype from self-compatible *S verrucosum* directly to Group Tuberosum haploids using some rare pollen receptive haploids. Further development of homozygote *SvSv* hybrids were achieved after selfing or sib-mating of F_1 self-compatible hybrids with high functional pollen fertility. These *SvSv* hybrids were identified based on their good penetration of pollen tubes in the styles of 1EBN *S. bulbocastanum* and *S. pinnatisectum* species. These *SvSv* Tuberosum hybrids may produce fertile hybrids in direct crosses with 1EBN wild diploid potato species.

According to available knowledge, 2*x* (2EBN) self-compatible *S. verrucosum*, which is used as a "bridge species", provides potato breeders an ideal route in germplasm enhancement aimed at introgressing valuable genes from 1EBN wild diploid *Solanum* species into breeding populations. Efficiency of hybrid production using self-compatible *S. verrucosum* is greater than other methods based on ploidy manipulations such as somatic hybridization or embryo rescue, which require considerable experience and expenditure of time and resources (Jansky 2006).

Group Tuberosum haploids are the most promising recipients for introgression of novel genes for traits of interest from diploid wild species germplasm (Peloquin et al. 1989; Jansky et al. 1990). Smaller population size in comparison with tetraploids is needed for selecting recombinants that meet breeding requirements due to their disomic inheritance. Accumulation of desirable genes and elimination of undesirable ones flows faster at this ploidy level. In addition, the presence of naturally occurring meiotic mutations in the potato gene pool, which leads to the production of 2*n* gametes, allows chromosome doubling for the transfer of valuable gene combinations to the tetraploid level (Carputo et al. 2000; Yermishin et al. 2014).

Endosperm Balance Number (EBN) and Interspecific Reproductive Barriers

Various genetic mechanisms account for the success or failure of seed development in flowering plants, which undergo double fertilization during sexual reproduction (Kinoshita 2007). Failure for producing triploids after tetraploid × diploid crosses—also known as "triploid block"—led to studying as a reproductive barrier the endosperm, which provides nourishment to the seed embryo. A first concept was to consider a 2:3:2 ploidy balance between maternal tissue, endosperm and embryo but further research demonstrated that normal endosperm development depends on having a 2:1 maternal to paternal genome dosage in the endosperm (Ehlenfeldt and

Ortiz 1995 and references therein). This endosperm dosage system for both intra- and inter-specific crossing seems to be multigenic in *Solanum* species, in which is known as the endosperm balance number (EBN) that also explains some aspects of species evolution therein. For example, Ortiz and Ehlenfeldt (1992) indicated the role of EBN in the origin of both diploid and polyploid potato species or how it becomes a hybridization barrier for speciation among sympatric *Solanum* species with same ploidy.

The EBN is a unifying concept that may predict endosperm function in intraspecific, interploidy, and interspecific crosses in potato and wild crop relatives (Johnston et al. 1980). Each species carries an EBN value that is constant across interspecific crossing, thus determining the effective ploidy in the endosperm, which must be in a 2 maternal:1 paternal ratio that is a necessary for successful endosperm development. The EBN, which is in itself an arbitrary value, is given to a species based on its crossing behavior with known EBN standards. The EBN does not reflect directly the ploidy of a species (Hanneman Jr 1999). For example, there are 2*x* (1EBN), 2*x* (2EBN), 3*x* (2EBN), 4*x* (2EBN), 4*x* (4EBN), 5*x* (4EBN), and 6*x* (4EBN) *Solanum* species.

The EBN concept was useful to elucidate the nature of the pollinator effect in haploid extraction (Peloquin et al. 1996). Haploid embryos appear to be associated with hexaploid endosperms as a result of having the union of 2-chromosome sets from the pollinator with the polar nuclei and lack of fertilization of the egg, thus having a 2 maternal: 1 paternal EBN in the endosperm that normal seed development requires.

The "triploid block" is a reproductive barrier resulting from endosperm malfunction due to the epigenetic phenomenon of genomic imprinting (Ehlenfeldt and Ortiz 1995; Köhler et al. 2009). Evidence shows that the endosperm dosage systems are imprinted within the gametes, thus the same gene being functionally different in maternal and paternal chromosomes. The maternally and paternally imprinted genes often carry a DNA methylation or histone modification in their vicinity (Kinoshita 2007). The position of these epigenetic modifications determines how the gene will be expressed; i.e., either from the maternal or paternal inherited allele (Köhler et al. 2012). Maternally imprinted genes appear to repress endosperm proliferation, which seems to be promoted by paternally imprinted genes. Abnormal endosperm development results from the imbalance of these imprinted genes. This "parental conflict" is consistent with the differential maternal and paternal genome effects in interploidy mating (Köhler et al. 2009). Products of imprinted PcG genes such as *Medea* (*MEA*) and *Fertilization Independent Seed2* (*FIS2*) could be limiting factors affecting endosperm and seed development. For example, FIS PcG proteins repress fertilization-independent seed formation and restrict endosperm proliferation (Köhler and Makaverich 2006).

The EBN provides useful prior knowledge for germplasm transfer from 2*x* (1EBN), 2*x* (2EBN), 4*x* (2EBN) and 6*x* (4EBN) species into 4*x* (4EBN) potato (Ortiz and Ehlenfeldt 1992). For example, chromosome engineering using 4*x* (2EBN) and 2*x* (2EBN) addition lines will allow to introduce specific chromosomes bearing the target allele into the 4*x* cultigen pool for further use in breeding.

Likewise, 2*x* chromosome addition lines from may result using 2*x* (1EBN) and 2*x* (2EBN) *Solanum* species. The first step will be to get a 3*x* (2EBN) interspecific hybrids after crossing both 2*x* species and thereafter using these hybrids by crossing them with 2*x* (2EBN) species to obtain 2*x* (2EBN) chromosome addition lines, which can be further used for breeding at 2*x* level. Haploids from 5*x* (4EBN) clones may be another method for producing 2*x* (2EBN) offspring with traits from the 2*x* (1EBN) parent because haploid extraction will allow screening for the right balance in the endosperm.

Trait Genetic Research: A Summary Prior to DNA Markers

Swaminathan and Howard (1953) provided the first summary of the inheritance of most important traits in potato, which was further updated by Howard (1960, Howard 1970), Bradshaw and Mackay (1994 and chapters therein), and Tiemens-Hulscher et al. (2013). Table give some details about trait genetics and gene symbols as noted in some of these publications.

Further research using the candidate gene approach led to identifying diagnostic DNA-based markers for genes involved, *inter alia*, in quantitative resistance to late blight (*R1* gene family in chromosome V) or cyst nematode (major QTL in same resistance "hot spot" on potato chromosome V), and both chip color (e.g., co-localized with a cold-sweetening QTL in chromosome IX and other sugar QTL in chromosome X) and tuber starch content in tetraploid potato cultivars (Gebhardt et al. 2007). Other annotated loci in the potato genetic map are *Y* (yellow flesh color) in chromosome III, *H3* of *Gpa4* in chromosome IV, Rx_2 (extreme resistance to *Potato Virus X*) and Nx_{tbr} (hypersensitivity to *Potato Virus X*) in chromosome V, Nx_{phu} in chromosome IX, *Ro* (tuber shape) and anthocyanin (including flower and skin color) in chromosome X, Ry_{sto}, and *Ry-adg* in chromosome XI, and both *Gpa 2* (resistance to potato cyst nematode) plus Rx_1 in chromosome XII.

Cytogenetics for Crossing, Scaling Up and Down Ploidy, and Chromosome Engineering

Germplasm enhancement (mistakenly replaced by some as pre-breeding, Ortiz 2002) is the early component of sustainable plant breeding that includes identifying a useful character, "capturing" its genetic diversity, and putting those genes into a "usable" form (Peloquin et al. 1989). Potato provides an interesting example of using crop wild relatives in germplasm enhancement. The wild species *S. demissum* (6*x*, 4EBN), *S. stoloniferum* (4*x*, 2EBN), and *S. vernei* contributed host plant resistance genes to late blight, *Potato Virus Y*, and cyst nematode, respectively; while

Table Gene symbols and their genetics of key traits for potato breeding

Trait	Gene symbol	Genetics (disomic inheritance unless indicated otherwise)
Plant growth type		Upright dominant to prostrate, and intermediate procumbent recessive to upright but unknown to prostrate
Dwarfism		Recessive phenotype producing compact, dark green, rosette plant
Flower color	*D* red	*F* involved in anthocyanin expression in flowers
	P blue	Purple = *D_P_F_*, Red to rose *D_ppF_*, Blue = *ddP_F_*, White = *D_P_ff* or *ddppF_*
Skin color	*D* red	*I* engaged in anthocyanin expression in tuber skin
	P blue	Purple = *D_P_I_*, Red *D_ppI_*, Blue = *ddP_I_*, White to yellow to brown = *D_P_ff* or *ddppF_*
Flesh color	*Y* yellow, *y* white	Yellow caused by carotenoids dominant to white
	Or allele at *Y* locus	Orange depends on two genes: one determining production and other (recessive) accounting for accumulation of zeazhantin. Orange dominant to yellow
	D red	Purple or red due to anthocyanins regulated by *B*.
	P blue	Purple = *D_ P_ B_*, Red = *D_ pp B_*, Blue = *dd P_ B_* (2*x*)
Tuber color pattern		Splashed and spotted only in eyes while speckled (dominant) reverses; i.e., eyes are either white or yellow
Tuber shape	*Ro* round	Round (*Ro__*) dominant to long (*roro*)
	ro long	
Russet skin		Three independent loci having an additive effect to each other; i.e., *AABBCC* show more russet skin that *AaBbCc*
Stem pubescence		Single gene, being pubescent dominant to glabrous stem
Early maturity		Earliness due to dominant allele with additive effects; i.e., duplex (*AAaa*) earlier than simplex (*Aaaa*) in 4*x*
Tuber dormancy		Dominant early sprouting with short dormancy
Tuberization under long days		High heritability but influenced by day length, light intensity, and temperature

(continued)

Table (continued)

Trait	Gene symbol	Genetics (disomic inheritance unless indicated otherwise)
Chip color		3-Gene hypothesis for both reversion resistance (ability to produced light-colored chips after harvest or short storage) and reconditioning (controlled tuber warming after storing fall crop for 4–6 months) to eliminate reducing sugars: A dominant allele in each of three loci for good chipping, being one or two loci common to both traits
Late blight resistance	12 known *R* genes derived from *S. demissum*, *Rpi-blb3* from *S. bulbocastanum*, and *Rpi-abpt* derived from quadruple hybrid involving *S. acaule*, *S. bulbocastanum*, plus groups Phureja and Tuberosum	Dominant *R* for race-specific host resistance corresponding to virulence genes in oomycete *Phytophthora infestans* (*R8* and *R9* seem to be "durable") Many genes are involved in partial host plant resistance that slow down the development of all *Phytophthora infestans* races, thus being likely more durable than race-specific resistance
Early blight resistance		High heritability for partial resistance, thus additivity being the most important gene action
Potato leaf roll virus resistance	N_L for hypersensitivity	Dominance of resistance but major gene does not provide enough durable resistance to a cultivar
Potato virus X resistance	Rx_i for extreme resistance, while Nx_i controls hypersensitivity	Dominant inheritance
Potato virus Y resistance	Ny_{chc}, Ny_{dms}, ${Ry_{sto}}^{n1}$ and ${Ry_{sto}}^{n2}$ gives hipersensitivity, whereas *Ry*-adg, *Ry-chc*, *Ry-hou* and *Ry-sto* are for extreme resistance	Dominant extreme resistance from *S. stoloniferum*, *S. hougasii*, and Group Andigena, while multigenic for Group Phureja
Potato virus A	*Na* protects for infection through hypersensitivity	
Wart resistance		Dominant alleles in two loci necessary for resistance
Verticillium wilt		2 genes
Bacterial wilt resistance		3–4 dominant genes required
Black leg and bacterial soft rot resistance		Minor genes with additive effects increase resistance
Potato tuber moth resistance		Simple inheritance due to additivity

(continued)

Table (continued)

Trait	Gene symbol	Genetics (disomic inheritance unless indicated otherwise)
Root-knot nematode resistance		High narrow-sense heritability for resistance in Group Phureja (susceptible)—*S. sparsipilum* (likely three major complementary genes in heterozygous state) segregating population, but reciprocal effects suggest maternal effects of *S. sparsipilum* or cytoplasmic-nuclear gene interaction
Potato cyst nematode resistance	*H1* (from Group Andigena), *H2* (from CPC2602), *Fa* and *Fb* (from *S. spegazzinii*), plus *B* and *C* from *S. verneii* provide host plant resistance against *Globodera rostochiensis*, while *Gpa2*, *Gpa5*, *GPa1*, *H2*, and *H3* of *Gpa4* give resistance against *G. pallida*	*H1* gives host plant resistance against Ro-1 and Ro-4 pathotypes

processing quality was derived from S. *chacoense* (2*x*, 2EBN) in various cultivars (Jansky et al. 2013).

Introgression and incorporation are the two approaches for using wild species in plant breeding (Simmonds 1993). Introgression refers to transferring one or a few alleles from wild germplasm to breeding populations that lack them, while a large-scale program for developing breeding populations using wild germplasm to broaden its genetic base is known as incorporation. Figure illustrates both in potato breeding. "Bridge" species, double pollination, embryo rescue, 2*n* gametes, and the EBN knowledge allows using chromosome engineering, while haploids, wild 2*x* (2EBN) species, and 2*n* gametes are used in ploidy manipulations.

CIP released in the mid-1990s diploid bred-germplasm generated from using haploids from tetraploid cultivars and breeding clones, plus diploid landraces and wild species (Watanabe et al. 1994). These genetic resources are of value for potato breeding due to their genetic diversity, crossability with the 4*x* cultigen pool facilitated by 2*n* gametes (mostly FDR 2*n* pollen), and high host plant resistance to pathogens and pests derived mostly from wild relatives. Ploidy manipulations at CIP led to the transmission of host plant resistance to cyst and root-knot nematodes, bacterial wilt, early blight, and potato tuber moth, as well as producing high-yielding 4*x* (4EBN) breeding clones also having yield stability over environments (Ortiz et al. 1994). Burundi released in the 1990s the tetraploid potato cultivar "Nemared," which derived from this diploid breeding population , because of both its host plant resistance to root knot nematode and desired agronomic traits.

Potato, as shown above, is the model crop species for germplasm enhancement of polysomic polyploids. Crop wild relatives and landraces are the diversity sources, while haploid derived from the tetraploid cultigen pool "capture" this diversity after crossing them with the former. The haploid-species hybrids producing 2*n* gametes

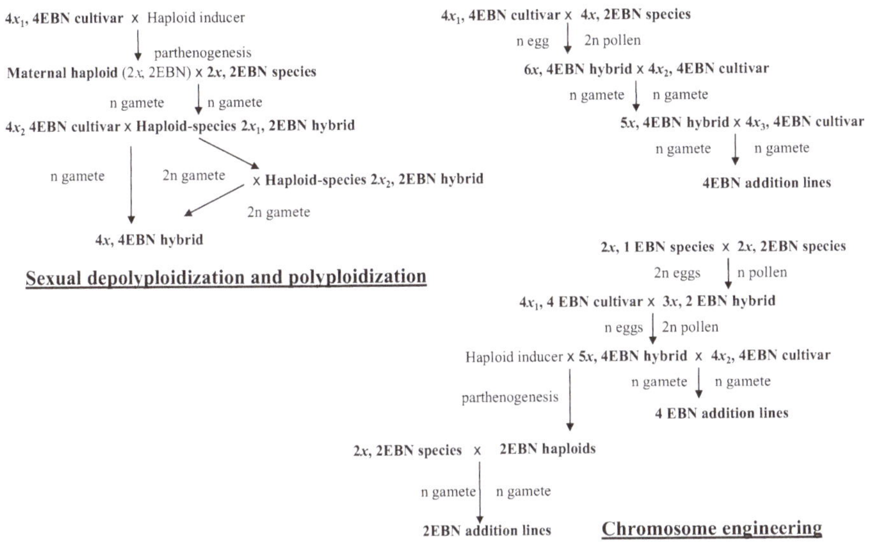

Fig. Germplasm enhancement approaches in potato breeding: introgression through chromosome engineering and incorporation using sexual depolyploidization and polyploidization

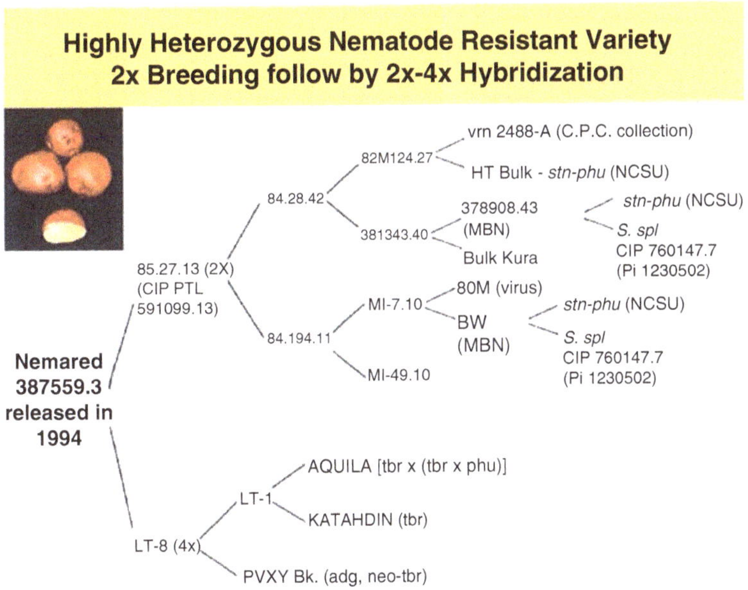

Fig. Pedigree of tetraploid cultivar potato "Nemared" resulting from ploidy manipulations

transfer this diversity to the tetraploid breeding pool through sexual polyploidization in which the EBN ensures the resulting ploidy of the hybrid offspring.

Concluding Remarks

Despite its the biological characteristics which made genetic improvement more complex in potato than in other crops, potato breeders have at their disposal such a powerful and effective approach as the unique ability to conduct across ploidy and species-wide crosses in order to introgress relevant genetic variation from its genetic resources into potato breeding programs. Progress in the field of cytogenetics of potato enables a more effective transfer of relevant genetic variation from wild *Solanum* accessions kept in genebanks or through in situ approaches. The increasing demands to develop resilient potato varieties able to withstand the threats brought by climate change, as well as the substantial progress recently achieved with the development of potato hybrid cultivars at the $2x$ level, underpin the increasing contribution of cytogenetics to the genetic improvement of the potato crop. The increasing availability of vast amounts of genomic information, as well as the continuous reduction of expenses associated with the sequencing of whole genomes are expected to further increase the contribution of potato cytogenetics, in terms of accelerating the development of varieties which provide not only superior adaption to changing production environments but also increased food and nutrient security to the millions of people to whom potato represents a staple crop.

References

Abdalla MMF, Hermsen JGT (1972) Plasmons and male sterility types in *Solanum verrucosum* and its interspecific hybrid derivatives. Euphytica 21:209–220

Abdalla MMF, Hermsen JGT (1973) An evaluation of *Solanum verrucosum* Schlechtd. For its possible use in potato breeding. Euphytica 22:19–27

Allainguillaume J, Wilkinson MJ, Clulow SA et al (1997) Evidence that genes from the male parent may influence the morphology of potato dihaploids. Theor Appl Genet 94:241–248

Allard RW (1999) History of plant population genetics. Ann Rev Genet 33:1–27

Barrett SCH (2002) The evolution of plant sexual diversity. Nat Rev Genet 3:274–284

Bedinger PA, Chetelat RT, McClure B et al (2011) Interspecific reproductive barriers in the tomato clade: opportunities to decipher mechanisms of reproductive isolation. Sex Plant Reprod 24:171–187

Birhman RK, Hosaka K (2000) Production of inbred progenies of diploid potatoes using an S-locus inhibitor (Sli) gene and their characterization. Genome 43:495–502

Bourke PM, Voorrips RE, Visser RGF, Maliepaard C (2015) The double-reduction landscape in tetraploid potato as revealed by a high-density linkage map. Genetics 201:853–863

Bradshaw JE (2007) The canon of potato science: 4 tetrasomic inheritance. Potato Res 50:219–222

Bradshaw JE, Mackay GR (eds) (1994) Potato genetics. CAB International, Wallinford

Bradshaw JE, Bryan GJ, Ramsay G (2006) Genetic resources (including wild and cultivated *Solanum* species) and progress in their utilization in potato breeding. Potato Res 49:49–65

Brown CR, Adiwilaga KD (1991) Use of rescue pollination to make a complex interspecific cross in potato. Am J Potato Res 68:813–820
Brownfield L, Kölher C (2011) Unreduced gamete formation in plants: mechanisms and prospects. J Exp Bot 62:1659–1668
Burham C (1984) Discussion in cytogenetics. Burgess Publishing, Minneapolis
Camadro EL, Peloquin SJ (1981) Cross-incompatibility between two sympatric polyploid *Solanum* species. Theor Appl Genet 60:65–70
Camadro EL, Verde LA, Marcellán ON (1998) Pollen-pistil incompatibility in a diploid hybrid potato population with cultivated and wild germplasm. Am J Potato Res 75:81–85
Camadro EL, Carputo D, Peloquin SJ (2004) Substitutes for genome differentiation in tuber-bearing Solanum: interspecific pollen-pistil incompatibility, nuclear-cytoplasmic male sterility, and endosperm. Theor Appl Genet 109:1369–1376
Cao J, Schneeberger K, Ossowski S et al (2011) Whole-genome sequencing of multiple *Arabidopsis thaliana* populations. Nat Genet 43:956–963
Carputo D, Barone A (2005) Ploidy level manipulations in potato through sexual hybridization. Ann Appl Biol 146:71–79
Carputo D, Barone A, Frusciante L (2000) 2n gametes in the potato: essential ingredients for breeding and germplasm transfer. Theor Appl Genet 101:805–813
Carroll CP (1975) The inheritance and expression of sterility in hybrids of dihaploid and cultivated diploid potatoes. Genetica 45:149–162
Chase SS (1963) Analytic breeding in *Solanum tuberosum* L.—a scheme utilizing parthenotes and other diploid stocks. Can J Genet Cytol 5:359–363
Chen Q, Lynch D, Platt HW et al (2004) Interspecific crossability and cytogenetic analysis of sexual progenies of Mexican wild diploid 1EBN species *Solanum pinnatisectum* and *S. cardiophyllum*. Am J Potato Res 81:159–169
Cipar MS (1964) Self compatibility in hybrids between Phureja and haploid Andigena clones of *Solanum tuberosum*. Eur Potato J 7:152–160
Cipar MS, Peloquin SJ, Hougas R (1964) Variability in the expression of self-incompatibility in tuber-bearing diploid *Solanum* species. Amer Potato J 41:155–162
Clulow SA, Rousselle-Bourgeois F (1997) Widespread introgression of *Solanum phureja* DNA in potato (*S. tuberosum*) dihaploids. Plant Breed 116:347–351
Clulow SA, Wilkinson MJ, Waugh R et al (1991) Cytological and molecular observations on Solanum phureja-induced dihaploid potatoes. Theor Appl Genet 82:545–551
Clulow SA, Wilkinson MJ, Burch LR (1993) *Solanum phureja* genes are expressed in the leaves and tubers of aneusomatic potato dihaploids. Euphytica 69:1–6
Comai L (2005) The advantages and disadvantages of being polyploid. Nat Rev Genet 6:836–846
Cook DE, Lee TG, Guo X et al (2012) Copy number variation of multiple genes at *Rhg1* mediates nematode resistance in soybean. Science 338:1206–1209
De Jong H (1981) Inheritance of russeting in cultivated diploid potatoes. Potato Res 24:309–313
De Jong H (1987) Inheritance of pigmented tuber flesh in cultivated diploid potatoes. Amer Potato J 64:337–343
De Jong H, Burns VJ (1993) Inheritance of tuber shape in cultivated diploid potatoes. Am Potato J 70:267–283
De Jong H, Rowe PR (1971) Inbreeding in cultivated diploid potatoes. Potato Res 14:74–83
De Jong H, Kawchuk LM, Burns VJ (1998) Inheritance and mapping of a light green mutant in cultivated diploid potatoes. Euphytica 103:83–88
De Jong H, Kawchuk LM, Coleman WK et al (2001) Development and characterization of an adapted form of droopy, a diploid potato mutant deficient in abscisic acid. Am J Potato Res 78:279–290
de Nettancourt D (1977) The genetic basis of self-incompatibility. In: Incompatibility in angiosperms. Monographs on theoretical and applied genetics, vol 3. Springer, Berlin

Díaz A, Zikhali M, Turner AS et al (2012) Copy number variation affecting the Photoperiod-B1 and Vernalization-A1 genes is associated with altered flowering time in wheat (*Triticum aestivum*). PLoS One 7:e33234
Ehlenfeldt MK, Ortiz R (1995) On the origins of endosperm dosage requirements in *Solanum* and other angiosperma genera. Sex Plant Reprod 8:189–196
Eijlander R (1998) Mechanisms of self–incompatibility and unilateral incompatibility in diploid potato (*Solanum tuberosum* L.). Dissertation, Wageningen University
Eijlander R, Ter Laak W, Hermsen JGT et al (2000) Occurrence of self-compatibility, self-incompatibility and unilateral incompatibility after crossing diploid *S. tuberosum* (SI) with *S. verrucosum* (SC): I. Expression and inheritance of self-compatibility. Euphytica 115:127–139
Erazzú LE, Camadro EL, Clausen AM (1999) Pollen-style compatibility relations in natural populations of the wild diploid potato species *Solanum spegazzinii* Bitt. Euphytica 105:219–227
Ercolano MR, Carputo D, Li J, Monti L et al (2004) Assessment of genetic variability of haploids extracted from tetraploid ($2n = 4x = 48$) *Solanum tuberosum*. Genome 47:633–638
Feuk L, Carson AR, Scherer SW (2006) Structural variation in the human genome. Nat Rev Genet 7:85–97
Frandsen NO (1967) Haploidproduktion aus einem Kartoffelzuchtmaterial mit intensiver Wildarteinkreuzung. Der Zuchter 37:120–134
Freyre R, Warnke S, Sosinski B et al (1994) Quantitative trait locus analysis of tuber dormancy in diploid potato (*Solanum* spp.). Theor Appl Genet 89:474–480
Fritz NK, Hanneman RE (1989) Interspecific incompatibility due to stylar barriers in tuber-bearing and closely related non-tuber-bearing Solanums. Sex Plant Reprod 2:184–192
Gálvez JH, Tai HH, Barkley NA et al (2017) Understanding potato with the help of genomics. AIMS Agric Food 2:16–39
Gebhardt C, Ritter E, Barone A et al (1991) RFLP maps of potato and their alignment with the homeologous tomato genome. Theor Appl Genet 83:49–57
Gebhardt C, Li L, Pajerowska-Mukthar K et al (2007) Candidate gene approach to identify genes underlying quantitative traits and develop diagnostic markers in potato. Crop Sci 47:S106–S111
Goldberg EE, Igic B (2012) Tempo and mode in plant breeding system evolution. Evolution 66:3701–3709
Goldberg EE, Kohn JR, Lande R et al (2010) Species selection maintains self-incompatibility. Science 330:493–495
Goldraij A, Kondo K, Lee CB et al (2006) Compartmentalization of S-RNase and HT-B degradation in self-incompatible Nicotiana. Nature 439(7078):805–810
Golz JF, Oh HY, Su V et al (2001) Genetic analysis of Nicotiana pollen-part mutants is consistent with the presence of an S-ribonuclease inhibitor at the S-locus. Proc Natl Acad Sci U S A 98:15372–15376
Gray JE, McClure BA, Bonig I et al (1991) Action of the style product of the self-incompatibility gene of Nicotiana alata (S-RNase) on in vitro-grown pollen tubes. Plant Cell 3:271–283
Grun P, Aubertin M, Radlow A (1962) Multiple differentiation of plasmons of diploid species of *Solanum*. Genetics 47:1321–1333
Grun P, Ochoa C, Capage D (1977) Evolution of cytoplasmic factors in tetraploid cultivated potato (Solanaceae). Am J Bot 64:412–420
Hancock CN, Kondo K, Beecher B, McClure B (2003) The S-locus and unilateral incompatibility. Phil Trans R Soc Lond B 358:1133–1140
Hanneman RE Jr (1999) The reproductive biology of the potato and its implication for breeding. Potato Res 42:283–312
Hanneman RE (1985) Self-fertility in *Solanum chacoense*. Am Potato J 62:428–429. (abstract)
Hardigan MA, Parker F, Laimbeer E et al (2017) Genome diversity of tuber-bearing Solanum uncovers complex evolutionary history and targets of domestication in the cultivated potato. Proc Natl Acad Sci USA. https://doi.org/10.1073/pnas.1714380114

Hayes RJ, Dinu II, Thill CA (2005) Unilateral and bilateral hybridization barriers in interseries crosses of 4x 2EBN *Solanum stoloniferum*, *S. pinnatisectum*, *S. cardiophyllum*, and 2x 2EBN S. tuberosum haploids and haploid-species hybrids. Sex Plant Reprod 17:303–311
Henry IM, Dilkes BP, Miller ES, Burkart-Waco D et al (2010) Phenotypic consequences of aneuploidy in *Arabidopsis thaliana*. Genetics 186:1231–1245
Henry IM, Zinkgraf MS, Groover AT et al (2015) A system for dosage-based functional genomics in poplar. Plant Cell 27:2370–2383
Hermsen JGT, Ramanna MS (1976) Barriers to hybridization of *Solanum bulbocastanum* Dun. and *S. verrucosum* Schlechtd. and structural hybridity in their F1 plants. Euphytica 25:1–10
Hermsen JT, Verdenius J (1973) Selection from *Solanum tuberosum* group Phureja of genotypes combining high-frequency haploid induction with homozygosity for embryo-spot. Euphytica 22:244–259
Hermundstad SA, Peloquin SJ (1985) Male fertility and 2n pollen production in haploid-wild species hybrids. Am Potato J 62:479–487
Hirsch CN, Hirsch CD, Felcher K et al (2013) Retrospective view of North American potato (*Solanum tuberosum* L.) breeding in the 20th and 21st centuries. G3 3:1003–1013
Hosaka K, Hanneman RE (1998a) Genetics of selfcompatibility in a self-incompatible wild diploid potato species *Solanum chacoense*. I. Detection of an S locus inhibitor (*Sli*) gene. Euphytica 99:191–197
Hosaka K, Hanneman RE (1998b) Genetics of self-compatibility in a selfcompatible wild diploid potato species *Solanum chacoense*. 2. Localization of an S-locus inhibitor (Sli) gene on the potato genome using DNA markers. Euphytica 103:265–271
Hosaka K, Sanetomo R (2012) Development of a rapid identification method for potato cytoplasm and its use for evaluating Japanese collections. Theor Appl Genet 125:1237–1251
Hougas RW, Peloquin SJ, Gabert AC (1964) Effect of seed-parent and pollinator on frequency of haploids in *Solanum tuberosum*. Crop Sci 4:593–595
Howard HW (1960) Potato cytology and genetics, 1952-59. Biblphia Genet 19:87–216
Howard HW (1970) Genetics of the potato *Solanum tuberosum*. Logos Press, London
Hua ZH, Fields A, Kao TH (2008) Biochemical models for S-RNase-based self-incompatibility. Mol Plant 1:575–585
Hutten RCB, Scholberg EJMM, Huigen DJ et al (1993) Analysis of dihaploid induction and production ability and seed parent x pollinator interaction in potato. Euphytica 72:61–64
Iwanaga M, Freyre R, Watanabe K (1991a) Breaking the crossability barriers between disomic tetraploid *Solanum acaule* and tetrasomic tetraploid *S. tuberosum*. Euphytica 52:183–191
Iwanaga M, Ortiz R, Cipar M, Peloquin SJ (1991b) A restorer gene for genetic-cytoplasmic male sterility in cultivated potatoes. Am Potato J 68:19–28
Jacobs JME, Van Eck HJ, Arens P et al (1995) A genetic map of potato (*Solanum tuberosum*) integrating molecular markers, including transposons, and classical markers. Theor Appl Genet 91:289–300
Jansky S (2006) Overcoming hybridization barriers in potato. Plant Breeding 125:1–12
Jansky S, Peloquin SJ, Yerk GL (1990) Use of potato haploids to put 2x wild species germplasm in usable form. Plant Breed 104:290–294
Jansky S, Dempewolf H, Camadro EL, Simon R, Zimnoch-Guzowska E, Bisognin DA, Bonierbale M (2013) A case for crop wild relative preservation and use in potato. Crop Sci 53:746–754
Jansky S, Chung YS, Kittipadukal P (2014) M6: a diploid potato inbred line for use in breeding and genetics research. J Plant Reg 8:195–199
Jansky SH, Charkowski AO, Douches DS, Gusmini G, Richael C, Bethke PC, Spooner DM, Novy RG, De Jong H, De Jong WS, Bamberg JB, Thompson AL, Bizimungu B, Holm DG, Brown CR, Haynes KG, Sathuvalli VR, Veilleux RE, Miller JC, Bradeen JM, Jiang J (2016) Reinventing potato as a diploid inbred line–based crop. Crop Sci 56:1412–1422
Johnston SA, den Nijs TP, Peloquin SJ, Hanneman RE Jr (1980) The significance of genic balance to endosperm development in interspecific crosses. Theor Appl Genet. 57:5–9

Kinoshita T (2007) Reproductive barrier and genomic imprinting in the endosperm of flowering plants. Genes Genet Syst 82:177–186
Köhler C, Makaverich G (2006) Epigenetic mechanisms governing seed development in plants. EMBO Rep 7:1223–1227
Köhler C, Scheid OM, Erilova A (2009) The impact of the triploid block on the origin and evolution of polyplpid plants. Trends Genet 26:142–148
Köhler C, Wolff P, Spillane C (2012) Epigenetic mechanisms underlying genomic imprinting in plants. Annu Rev Plant Biol 63:331–352
Kotch GP, Ortiz R, Peloquin SJ (1992) Genetic analysis by use of potato haploid populations. Genome 35:103–108
Kuhl JC, Havey MJ, Hanneman RE (2002) A genetic study of unilateral incompatibility between diploid (1EBN) Mexican species *Solanum pinnatisectum* and *S. cardiophyllum* subsp. *cardiophyllum*. Sex Plant Reprod 14:305–313
Lewis D, Crowe LK (1958) Unilateral interspecific incompatibility in flowering plants. Heredity 12:233–256
Lindhout P, Meijer D, Schotte T et al (2011) Towards F_1 hybrid seed potato breeding. Pot Res 54:301–312
Lovene M, Zhang T, Lou Q, Buell CR et al (2013) Copy number variation in potato—an asexually propagated autotetraploid species. Plant J 75:80–89
Makhan'ko OV (2008) Interspecific incompatibility in diploid potato breeding. Zemlyarobstva i Akhova Raslin 1:11–14
Makhan'ko OV (2011) Incompatibility in interspecific and intraspecific hybridization of diploid potato and ways of its overcoming. Dissertation, Institute of Genetics and Cytology Belarusian National Academy of Sciences
Maron LG, Guimarães CT, Kirst M et al (2013) Aluminum tolerance in maize is associated with higher *MATE1* gene copy number. Proc Natl Acad Sci USA 110:5241–5246
McClure B, Cruz-Garcia F, Romero C (2011) Compatibility and incompatibility in S-RNase based systems. Ann Bot 108:647–658
McClure BA, Gray JE, Anderson MA et al (1990) Self-incompatibility in Nicotiana alata involves degradation of pollen rRNA. Nature 347:757–760
McClure BA, Cruz-Garcia F, Beecher BS, Sulaman W (2000) Factors affecting inter- and intraspecific pollen rejection in Nicotiana. Ann Bot 85:113–123
Mihovilovich E, Sanetomo R, Hosaka K et al (2015) Cytoplasmic diversity in potato breeding: case study from the International Potato Center. Mol Breed 35:137. https://doi.org/10.1007/s11032-015-0326-1
Murfett J, Strabala T, Zurek D et al (1996) S RNase and interspecific pollen rejection in the genus Nicotiana: multiple pollen-rejection pathways contribute to unilateral incompatibility between self-incompatible and self-compatible species. Plant Cell 8:943–958
Naess SK, Bradeen JM, Wielgus SM et al (2000) Resistance to late blight in *Solanum bulbocastanum* is mapped to chromosome 8. Theor Appl Genet 101:697–704
Nakamura S, Hosaka K (2010) DNA methylation in diploid inbred lines of potatoes and its possible role in the regulation of heterosis. Theor Appl Genet 120:205–214
Nathan Hancock C, Kondo K, Beecher B et al (2003) The S locus and unilateral incompatibility. Phil Trans R Soc Lond B 358:1133–1140
Novy RG, Hanneman RE (1991) Hybridization between Gp. Tuberosum haploids and 1EBN wild potato species. Am Potato J 68:151–169
Olsder J, Hermsen JT (1976) Genetics of selfcompatibility in dihaploids of Solanum tuberosum L. 1. Breeding behaviour of two selfcompatible dihaploids. Euphytica 25:597–607
Ortiz R (1998) Potato breeding via ploidy manipulations. Plant Breed Rev 16:15–86
Ortiz R (2002) Germplasm enhancement. In: Engels JMM, Ramanatha Rao V, Brown AHD, Jackson MT (eds) Managing plant genetic diversity. CAB International, Wallingford—International Plant Genetic Resources Institute, Rome, pp 275–290

Ortiz R, Ehlenfeldt M (1992) The importance of endosperm balance number in potato breeding and the evolution of tuber bearing Solanums. Euphytica 60:105–113
Ortiz R, Peloquin SJ (1994) Use of 24-chromosome potatoes (diploids and dihaploids) for genetic analysis. In: Bradshaw JE, Mackay GR (eds) Potato genetics. CAB International, Wallinford, pp 133–153
Ortiz R, Watanabe KN (2004) Genetic contributions to breeding polyploid crops. Recent Res Devel Genet Breed 1:269–286
Ortiz R, Iwanaga M, Camadro EL (1993a) Utilización potencial de progenie autofecundada de IvP-35 como inductor de haploides en papa por cruzamientos 4*x*-2*x*. Revista Latinoamericana de Papa 5/6:46–53
Ortiz R, Iwanaga M, Peloquin SJ (1993b) Male sterility and 2*n* pollen in 4*x* progenies derived from 4*x* × 2*x* and 4*x* × 4*x* crosses in potatoes. Potato Res 36:227–236
Ortiz R, Iwanaga M, Peloquin SJ (1994) Breeding potatoes for the developing countries using wild tuber bearing *Solanum* species and ploidy manipulations. J Genet Breed 48:89–98
Ortiz R, Simon P, Jansky S et al (2009) Ploidy manipulation of the gametophyte, endosperm and sporophyte: a tribute to Professor Stanley J. Peloquin (1921–2008). Ann Bot 104:795–807
Owen HR, Veilleux RE, Haynes FL et al (1988) Photoperiod effects on 2*n* pollen production, response to anther culture, and net photosynthesis of a diplandrous clone of *Solanum phureja*. Am Potato J 65:131–139
Peloquin SJ, Yerk GL, Werner JE, Darmo E (1989) Potato breeding with haploids and 2n gametes. Genome 31:1000–1004
Peloquin SJ, Gabert AC, Ortiz R (1996) Nature of "pollinator" effect in potato haploid production. Ann Bot 77:539–542
Phumichai C, Mori M, Kobayashi A et al (2005) Toward the development of highly homozygous diploid potato lines using the self-compatibility controlling Sli gene. Genome 48:977–984
Phumichai C, YIkeguchi-Samitsu Y, Fujimatsu M et al (2006) Expression of S-locus inhibitor gene (Sli) in various diploid potatoes. Euphytica 148:227–234
Pineda O, Bonierbale MW, Plaisted RL et al (1993) Identification of RFLP markers linked to the H_1 gene conferring resistance to the potato cyst nematode *Globodera rostochiensis*. Genome 36:152–156
Polyukhovich YV, Makhan'ko OV, Savchuk AV et al (2010) Production of bridge lines for overcoming interspecific incompatibility in potato. Vestsi Natsiyanalnai Akademii Navuk Belarusi. Seryia Biyalagichnykh Navuk 2:51–58
Potato Genome Sequencing Consortium (2011) Genome sequence and analysis of the tuber crop potato. Nature 475:189–195
Raimondi JP, Sala RG, Camadro EL (2003) Crossability relationships among the wild diploid potato species *Solanum kurtzianum*, *S. chacoense* and *S. ruiz-lealii* from Argentina. Euphytica 132:287–295
Ravi M, Chan SW (2010) Haploid plants produced by centromere-mediated genome elimination. Nature 464:615–618
Ravi M, Marimuthu MPA, Tan EH et al (2014) A haploid genetics toolbox for *Arabidopsis thaliana*. Nat Commun 5:5334
Rivard SR, Cappadocia M, Landry BS (1996) A comparison of RFLP maps based on anther culture derived, selfed, and hybrid progenies of *Solanum chacoense*. Genome 39:611–621
Ross H (1986) Potato breeding-problems and perspectives. Verlag Paul Parey, Berlin
Ross RW, Peloquin SJ, Hougas RW (1964) Fertility of hybrids from *Solanum phureja* and haploid *S. tuberosum* matings. Eur Potato J 7:81–89
Rouppe van der Voort JR, Van Eck HJ, Draaistra J et al (1998) An online catalogue of AFLP markers covering the potato genome. Mol Breed 4:73–77
Samitsu Y, Hosaka K (2002) Molecular marker analysis of 24-and 25-chromosome plants obtained from *Solanum tuberosum* L. subsp. *andigena* (2n = 4x = 48) pollinated with a *Solanum phureja* haploid inducer. Genome 45:577–583

Sanetomo R, Gebhardt C (2015) Cytoplasmic genome types of European potatoes and their effects on complex agronomic traits. BMC Plant Biol 15:162. https://doi.org/10.1186/s12870-015-0545-y

Sanetomo R, Akino S, Suzuki N et al (2014) Breakdown of a hybridization barrier between *Solanum pinnatisectum* Dunal and potato using the S locus inhibitor gene (Sli). Euphytica 197:119–132

Santini M, Camadro EL, Marcellán ON et al (2000) Agronomic characterization of diploid hybrid families derived from crosses between haploids of the common potato and three wild Argentinean tuber-bearing species. Am J Potato Res 77:211–218

Simko I, Vreugdenhil D, Jung CS et al (1999) Similarity of QTLs detected for in vitro and greenhouse development of potato plants. Mol Breed 5:417–428

Simmonds NW (1993) Introgression and incorporation: strategies for the use of crop genetic resources. Biol Rev 68:539–562

Singsit C, Hanneman RE Jr (1990) Isolation of viable inter-EBN hybrids among potato species using double pollinations and embryo culture. Am Potato J 67:578

Singsit C, Veilleux RE (1991) Chloroplast density in guard cells of leaves of anther-derived potato plants grown in vitro and in vivo. HortScience 26:592–594

Sopory SK (1977) Differentiation in callus from cultured anthers of dihaploid clones of *Solanum tuberosum*. Z Pflanzenphysiol 82:88–91

Straadt IK, Rasmussen OS (2003) AFLP analysis of *Solanum phureja* DNA introgressed into potato dihaploids. Plant Breed 122:352–356

Summers D, Grun P (1981) Reproductive isolation barriers to gene exchange between *Solanum chacoense* and *S. commersonii* (Solanaceae). Am J Bot 68:1240–1248

Swaminathan MS, Howard HW (1953) The cytology and genetics of the potato (*Solanum tuberosum*) and related species. Biblphia Genet 16:1–192

Swanson-Wagner RA, Eichten SR, Kumari S et al (2010) Pervasive gene content variation and copy number variation in maize and its undomesticated progenitor. Genome Res 20:1689–1699

Tan EH, Henry IM, Ravi M et al (2015) Catastrophic chromosomal restructuring during genome elimination in plants. Elife 4:e06516

Tiemens-Hulscher M, Delleman J, Eising J, Lammerts van Beuren ET (2013) Potato breeding. Aardappelwereld, The Hague

Tovar-Méndez A, Kumar A, Kondo K et al (2014) Restoring pistil-side self-incompatibility factors recapitulates an interspecific reproductive barrier between tomato species. Plant J 77:727–736

Tsai H, Missirian V, Ngo KJ et al (2013) Production of a high-efficiency TILLING population through polyploidization. Plant Physiol 161:1604–1614

Tucci M, Carputo D, Bile G et al (1996) Male fertility and freezing tolerance of hybrids *involving Solanum tuberosum* haploids and diploid Solanum species. Potato Res 39:345–353

Van Eck HJ, Jacobs JM, Van den Berg PM et al (1994a) The inheritance of anthocyanin pigmentation in potato (*Solanum tuberosum* L.) and mapping of tuber skin colour loci using RFLPs. Heredity 73:410–421

Van Eck HJ, Jacobs JM, Stam P et al (1994b) Multiple alleles for tuber shape in diploid potato detected by qualitative and quantitative genetic analysis using RFLPs. Genetics 137:303–309

Veilleux RE (1996) Haploidy in important crop plants—potato. In: Mohan Jain S, Sopory SK, Veilleux RE (eds) In vitro haploid production in higher plants, vol 3. Kluwer Academic, Dordrecht, pp 37–39

Velásquez AC, Mihovilovich E, Bonierbale M (2007) Genetic characterization and mapping of major gene resistance to potato leafroll virus in *Solanum tuberosum* ssp. *andigena*. Theor Appl Genet 114:1051–1058

Visser RGF, Bachem CWB, Boer JM et al (2009) Sequencing the potato genome: outline and first results to come from the elucidation of the sequence of the world's third most important food crop. Am J Potato Res 86:417–429

Watanabe K (2015) Potato genetics, genomics, and applications. Breed Sci 65:53–68

Watanabe K, Orrillo M, Iwanaga M et al (1994) Diploid potato germplasm derived from wild and landrace germplasm resources. Am Potato J 71:599–604

Waugh R, Baird E, Powell W (1992) The use of RAPD markers for the detection of gene introgression in potato. Plant Cell Rep 11:466–469

Wilkinson MJ, Bennett ST, Clulow SA et al (1995) Evidence for somatic translocation during potato dihaploid induction. Heredity 74:146–151

Yermishin AP, Polyukhovich YV, Voronkova EV et al (2014) Production of hybrids between the 2EBN bridge species *Solanum verrucosum* and 1EBN diploid potato species. Am J Potato Res 91:610–617

Zamir D (2001) Improving plant breeding with exotic genetic libraries. Nat Rev Genet 2:983–989

Zhang Y, Zhao Z, Xue Y (2009) Roles of proteolysis in plant self-incompatibility. Annu Rev Plant Biol 60:21–42

Zhu J, Pearce S, Burke A et al (2014) Copy number and haplotype variation at the VRN-A1 and central FR-A2 loci are associated with frost tolerance in hexaploid wheat. Theor Appl Genet 127:1183–1197

2

The Genes and Genomes of the Potato

Marc Ghislain and David S. Douches

Abstract During the last decade, genomics research has generated new insights into potato genetics and made possible new strategies for varietal improvement. The most commonly grown and eaten potato is an autotetraploid, highly heterozygote crop suffering from rapid inbreeding depression. The genetic improvement of the potato presents numerous challenges using conventional tetraploid breeding techniques. However, novel breeding technologies are now available to increase precision and gains for varietal improvement. The public availability of the first potato genome sequence has created new ways to identify the genetic determinants of key traits of the potato as well as ways to use this knowledge for speeding up variety development. Genomic selection applied to tetraploid breeding promises to increase prediction of progeny performance by a more efficient selection of parents. Diploid hybrid breeding is finally making its way two decades after discovering a suppressor gene of the self-incompatibility locus of diploid potatoes. Direct gene transfer into existing varieties of major genes for key traits has been successful but biotech potato development has been constrained by public perception and issues related to the regulation of the technology. Although genome or gene editing is still in its primary stage in potato, it has already been successful in modifying gene expression in a controlled way, and it might face a lower regulatory burden and easier adoption than biotech, transgenic potatoes. Concluding on an optimistic note, we have many reasons, and evidence is starting to mount, that potato crop improvement is finally benefiting from decades of investment in molecular genetics and that the future hold the promises of faster releases of more robust varieties to pest, disease, and climatic extremes, as well as nutritionally enhanced varieties to feed an ever-growing world population.

M. Ghislain (✉)
International Potato Center, Nairobi, Kenya
e-mail: m.ghislain@cgiar.org

D. S. Douches
Michigan State University, East Lansing, MI, USA
e-mail: douchesd@msu.edu

At the Crossroad of Potato Improvement

The Numerous Challenges of Tetraploid Potato Breeding

Potato breeding has a long and successful history of improved variety released after breeding from mostly advanced breeding clones and landraces of tetraploid nature from essentially two genepools, the short-day adapted upland Andigenum and the long-day adapted lowland Chilotanum (Bonierbale et al., this volume; Spooner et al. 2007). The latter group has given rise to the modern cultivar well-adapted to the Northern hemisphere referred to as the Tuberosum group (Gavrilenko et al. 2013). In the US and Canada, variety replacement has been particularly disappointing in potato since turnover is in the range of several decades, unlike grain crops (Walker 1994). In developing countries, there has been marked change in variety adoption as evidenced by a study on CIP-related varieties which grew from nothing to beyond one million hectares in 35 years (Thiele et al. 2008). A more recent study by Gatto et al. (2018) provides further insight on the impact of CIP's breeding efforts, and documents that in China, the world's largest potato producer, there are over one million hectares planted with varieties that trace back to CIP pedigrees. Furthermore, CIP's genetic footprint in China reaches above 35% of all varieties currently in use, be it through the registration of CIP advanced breeding lines as varieties or through using CIP's elite breeding lines as parents in crosses initiated by Chinese breeding programs. The main bottleneck explaining the difference in varietal turnover is market-driven mostly by processing industry in the US and Canada, unwilling to adjust their manufacturing processes to the cooking and frying attributes of the new varieties. Regardless, potato breeding has shown the ability to deliver varieties with market-demanded processing qualities and new traits for higher resilience against climatic extremes, pest and diseases threats, and with enhanced nutritional qualities.

Phenotypic recurrent selection has been the method of choice to select for improved potato lines starting usually with about 100,000 seedlings from 200 to 300 crosses followed by clonal selection over many years (Bradshaw 2009, 2017). A recent review of the history of conventional potato breeding revealed many examples of important varieties been released after 30 years or more of crossing and clonal selection when an optimal timeline should be 13–14 years (Bradshaw 2009; Jansky and Spooner 2017). The main reason for such long cycle for variety development is the quantitative nature of most important traits, the rapid inbreeding depression, and the low intensity of selection in early generations. The propagation through tubers adds on delays due to low multiplication rate and ease of contamination with pathogens which delay bulking enough quality seed tubers for multilocation field selection.

The genetic base of potato varieties grown in large commercial area is relatively narrow compared to the accessible gene pool for conventional breeding of the potato. This is likely due to the narrow genetic base of the original tetraploid sources from Andigenum and Tuberosum which were used as starting material for breeding.

Ellis et al. (this volume) provide a detailed current status of the gene pool and the germplasm of potato. Many wild potato species can be crossed with cultivated potatoes directly or via another wild species used as a bridge (Plaisted and Hoopes 1989). However, only a fraction of useful genes from wild species have been introgressed successfully into modern potato varieties. About 40% of the wild species carry interesting genetic traits value for pests, diseases and abiotic stresses (Sood et al. 2017). Major genes for disease resistance from wild relatives of the potato were introgressed into breeding lines for late blight (*S. demissum, S. bulbocastanum*), viruses (*S. stoloniferum*), and nematodes (*S. spegazzini, S. vernei*) resistance (Bradshaw and Ramsay 2005; Bradshaw 2009; Finkers-Tomczak et al. 2011).

The real contribution of wild relatives to modern potato varieties is likely underestimated due to uncertainty in pedigree information and quantitative nature of many important traits of the potato. *S. acaule,* which has been a source of disease resistance and abiotic stress tolerance but has been more used as a bridge between wild species and the cultivated potato (Watanabe et al. 1992). From the first cross between *S. bulbocastanum* bearing late blight-resistance genes and the bridging wild species *S. acaule*, 46 years of crossing and selection with cultivated potato, first diploid *S. Phureja* and then tetraploid *S. tuberosum*, were necessary to release the late blight-resistant varieties "Bionica" and "Toluca" (Haverkort et al. 2009). Introgression of wild species genomes into the cultivated groups has been facilitated by unreduced (*2n*) gametes in diploid potatoes and was shown to be highest in group Tuberosum because of intense breeding effort using a dozen of wild species (Plaisted and Hoopes 1989; Hardigan et al. 2017). Wild species carry genes for wild characteristics which are introduced as genetic drag with the disease-resistance genes lead to the notion that there could be a tradeoff between disease resistance and yield (Ning et al. 2017). This might have contributed to the limited use of wild species in potato breeding.

Assuming allelic combination has to consider one positive allele from a wild species and three neutral alleles from cultivated potato, the introgression of only ten positive alleles from wild species is only one in a million genotypes [$(1/4)^n$ where $n = 10$ genes]. This number becomes quickly without practical reach considering epistatic effects from the cultivated potato alleles, and that each cross redistributes the 20 or so quantitative trait loci which are priority traits of modern varieties (Bradshaw 2017). Potato being an auto-tetraploid clonally propagated crop has also accumulated rare mutations and epigenetic changes in alleles otherwise identical. This complicates further the straightforward exploitation of emerging molecular breeding approaches (Visser et al. 2014).

Increased selection intensity before clonal selection has been proposed by progeny tests and full-sib family selection (Bradshaw et al. 1995, 2000). Marker assisted selection can also screen at early stage major genes and Quantitative Trait Loci (QTL) with large effects (Gebhardt 2013; Sharma et al. 2014). In recent years, markers flanking major genes and QTL were developed for resistance to viruses (Mihovilovich et al. 2014; del Rosario et al. 2018), tuber starch and yield (Schönhals et al. 2016), and other important traits (Ramakrishnan et al. 2015). If applied at the early clonal generation stage and multiplexed, marker-assisted selection can be cost-effective (Slater et al. 2013). Estimated breeding value for traits which can be inferred from pedigree information was also proposed to accelerate the intensity of section and result in shorter breeding cycle (Slater et al. 2014a, b). Genomic selection is also proposed to improve combining unknown QTL at an early stage in the breeding process (Slater et al. 2016). New high density and high throughput polymorphic marker systems have been developed for potato (Hamilton et al. 2011; Uitdewilligen et al. 2013; Vos et al. 2015). Using a panel of 83 cultivars of mostly European origin, a high frequency of relatively rare variants and/or haplotypes, with 61% of the variants having a minor allele frequency below 5%, was found which can be explained by the limited number of meiosis separating these cultivars (Uitdewilligen et al. 2013). Recent estimates of linkage disequilibrium in modern potato cultivar populations confirmed the relatively limited number of meiosis separating modern cultivars and therefore limited power of Genome Wide Assisted Studies (GWAS) for allele/gene discovery (Vos et al. 2017; Sharma et al. 2018).

Hence, ways to improve conventional potato breeding exist and are under development but the fundamental inherent limitations of the narrow genetic base of advanced tetraploid potato germplasm used by breeders, the rapid inbreeding depression, and the low multiplication rate of seed tubers call for new methods and tools to improve the potato which will be complementary to conventional tetraploid breeding for some and an alternative for others.

New Potato Breeding Technologies

Potato genetic improvement has taken shortcuts many times to circumvent the limited genepool accessible by crossing and the tedious phenotypic recurrent selection.

Mutagenesis has long been used to improve yield, quality, biotic and abiotic stress resistance, and tolerance of many crops (Maluszynski et al. 1995). According to this review, more than 1,700 mutant varieties involving 154 plant species have been officially released. However, the tedious process of segregating out the rare positive mutation from the negative ones represent a bottleneck for potato crop improvement. Nevertheless, a novel form of doing mutagenesis is making a surprising come-back for potato crop improvement as mentioned below.

Somatic hybridization has been used in potato to bypass sexual incompatibilities between cultivated potato and wild species for about 40 years (Tiwari et al. 2018). These authors reported successful fusion products obtained from 23 *Solanum* species that were characterized for multiple traits. Numerous studies were generated from somatic hybrids to understand the genetic architecture of important traits including isolating important genes. However, no variety has apparently been released from breeding somatic hybrids with cultivated potato likely due to the limitation of tetraploid potato breeding to efficiently remove undesirable alleles.

Although direct gene transfer through transgenics has a much shorter history in potato crop improvement than conventional breeding approaches, it has been highly successful, though for a limited number of traits for which natural genetic variation was not readily available. Virus resistance was the first trait successfully engineered in potato in the late 1980s, and a commercial cultivar was first reported with the combined resistance to PVX and PVY (Lawson et al. 1990). Soon after, an insect resistance was engineered and led to the production of new commercial cultivars (Perlak et al. 1993). This first generation of biotech potatoes were commercialized under the name of NewLeaf™ from 1995 through 2001 in the United States and Canada, but potato processors and retailers realized soon that the NewLeaf potatoes were going to increase their costs without a share of their benefits which precipitated their decline (Thornton 2003). By 2004, none of them were commercialized anymore. For the next decade, no new biotech potato was released. Recently, a series of potato biotech varieties have been released with reduced bruising and browning first and then with late blight resistance, low acrylamide potential, reduced black spot, and lowered reducing sugars while others are near to be released (see below). The opportunities for engineering new traits that bring benefits to the producer and the consumer are numerous and applicable to potato (Halterman et al. 2016). However, the public acceptance of biotech crops remains volatile and unpredictable, and the current lack of science-based regulatory frameworks in many potato producing countries constrain the scope of genetic engineering to products with sizable benefits to both producers and consumers unachievable by other means.

Genome (gene) editing is the most recent and significant genetic engineering technique targeting specific DNA sequences in the crops' genome (Scheben et al. 2017; Yin et al. 2017). Targeted mutagenesis of specific genes for knock-out, deletion, or allelic changes are now possible with a final product free of foreign DNA. Potato has already been shown to be amenable to genome editing and even to develop novel useful products (Butler et al. 2015, 2016; Clasen et al. 2016; Nicolia et al. 2015; Wang et al. 2015; Andersson et al. 2017; Ma et al. 2017). The two editing tools, TALEN and CRISPR/Cas9, must access the genome without integration of foreign DNA since it cannot be eliminated by crossing without losing most of the qualities of the original commercial variety. Transient expression by PEG-mediated protoplast transfection or Agrobacterium-mediated leaf infiltration generated the intended mutation, but the absence of a selectable agent and somatic variation of plants regenerated from protoplasts can make these strategies labor-intensive. Delivery of the editing reagent may be mediated by virus vectors, but their spread and elimination pose additional difficulties. Hence, genome editing technology offers great opportunities in potato but is still at its first stage of development.

True hybrid potato is a new potato breeding strategy which is increasingly been regarded as the game-changing solution to many of the pitfalls of conventional tetraploid breeding. The sparking step goes two decades back with the discovery of a self-incompatibility inhibitor gene in the wild species *S. chacoense* (Hosaka and Hanneman 1998a, b). The *S* locus inhibitor gene (*Sli*) was introgressed into diploid cultivated potato and shown to confer self-compatibility

(Phumichai et al. 2005). Soon after, few breeding programs started to introgress the *Sli* gene into their diploid potato lines and obtained S3 diploid lines with 80% homozygosity and good agronomic performance including yield (Lindhout et al. 2011). In parallel, an inbred line of *S. chacoense* (M6) was developed to produce recombinant inbred line populations (Jansky et al. 2014). As stated in the title of an opinion paper by a large community of US potato geneticists and breeders, it is proposed to "reinvent the potato as a diploid inbred-based line crop" (Jansky et al. 2016). New sources of self-compatibility system are needed to circumvent the use of the wild species *S. chacoense*. Within the diploid cultivated potato germplasm, self-compatible landraces exist, though rare, but can be used to develop inbred lines from distinct gene pools such as the Stenotomum group and Phureja group (Haynes and Guedes 2018). Recently, an even more promising new system has been developed by knocking out the self-incompatibility gene *S-RNase* using the CRISPR–Cas9 gene editing system (Ye et al. 2018).

In addition to obtaining quick genetic gain by fixing major genes for disease resistance or other important traits and exploiting heterosis by hybridization, the hybrid variety propagation is via true seeds. The use of botanical seeds has long been known to be an extremely interesting alternative to tuber seeds because of it low weight, lower content of pathogens, good storability, option for beneficial coating, and high multiplication rate. Previous work by CIP and other potato research organizations on the concept of True Potato Seed (TPS) aimed to complement traditional seed systems by the use of botanical seed as a mean to propagate potato. However, its actual adoption by farmers has been much less than originally expected. Breeding for good parental clones from tetraploid breeding lines led to the development of several varieties but adoption remained conditioned to reduced or scarcity of seed tuber supply at affordable prices (Almekinders et al. 2009). Recently, a TPS variety, Oliver F1, was developed by the Dutch breeding company Bejo Zaden B.V. (http://www.bejo.com/magazine/bejo-introduces-its-first-true-potato-seed-variety) and is now under deployment in some African countries where quality seed availability is rare. Many years of breeding to develop superior parental inbred lines with disease-resistance genes adapted to the various agroecologies and markets are needed but the potential benefits that could be derived from true hybrid potato seeds are immense. The next decades will tell us whether reinventing the potato as hybrid varieties from diploid inbred parental lines will be adopted by small-holder farmers in developing countries who are the likely first adopters of this new technology.

The Genome of the Potato

Cultivated, Wild Potato Genome Sequences Towards a Pan-Genome

The identification of the first cultivated potato genome sequence is and will remain a turning point in the history of potato science. Prior to its discovery, genetic markers were associated to genetic determinants of traits breeders and geneticists had been working on. A fraction of the genes was known while transcriptomes were describing their expression in tissues, at different times, and under various environmental situations. Candidate genes were tested for association with these quantitative trait loci but for the most, genes underlying QTL remained unknown. The potato genome sequence brought together all this genetic knowledge into a physical perspective for the first time. Eighty-six percent of the 844 Mb genome was assembled into 12 pseudomolecules where 39,031 protein-coding genes were predicted (The Potato Genome Sequencing Consortium 2011). The potato whose genome was sequenced is a homozygote diploid plant obtained after chromosome doubling of a monoploid derived by anther culture of a heterozygous diploid potato from the *Solanum tuberosum* Group Phureja (Paz and Veilleux 1999). This cultivar groups are diploid short-day adapted cultivars producing tubers lacking dormancy. They occur throughout the eastern slope of the Andes from western Venezuela to central Bolivia at elevation between 2000 and 3400 masl (Ochoa 1990). Other genome sequence from cultivated potato, in particular from the Group Andigenum and Tuberosum including modern cultivars are still missing and expected to reveal insight into the domestication/wild species contributions to the various groups of cultivars. The difficulties lie in the presence of four genome sequences derived by auto-ploidization and multiple introgression of chromosome segments from wild species (Rodríguez et al. 2010; Spooner et al. 2014). This makes assembly and phasing particularly difficult and only recently claims of successful assembly of all four genome sequences from modern potato cultivars were made (*NRGene* at http://www.nrgene.com). The comparison of 99 Mb of genome from a potato of the group Tuberosum with the DM potato genome sequence revealed collinearity and high sequence identity (The Potato Genome Sequencing Consortium 2011). A year after the release of the potato genome sequence, the tomato genome sequence was published together with its closest wild relative and compared to the potato genome sequence (The Tomato Genome Consortium 2012). As known from previous cytological and comparative genetic mapping and synteny studies, the tomato genome presents very similar chromosomal organization but nine large and several smaller inversions. The euchromatic, gene rich regions diverge by 8.7%, whereas the intergenic and repeat-rich heterochromatic regions diverge by 30%. The potato genome sequence was greatly improved by ordering and reordering 93% of the previously assembled genome into 12 pseudomolecules representing the 12 chromosomes of the potato (Sharma et al. 2013). This genome sequence continues to be the sole publicly available genome sequence of a cultivated potato. It is accessible through a friendly

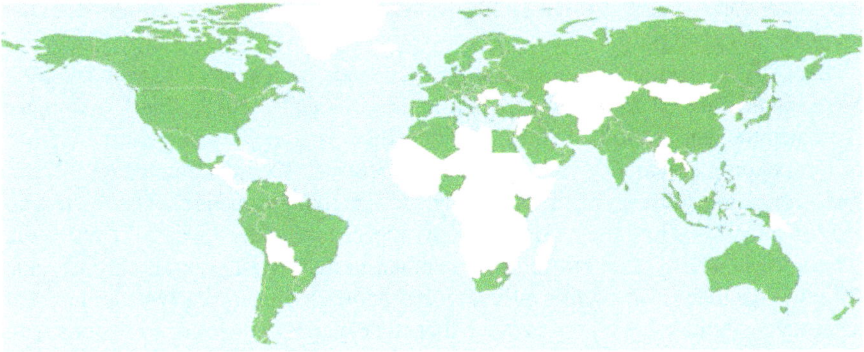

Fig. Geographic distribution of SpudDB users. Filled in countries were the source of ten or more unique visits to SpudDB (http://solanaceae.plantbiology.msu.edu/) from October 2017 to October 2018 (Courtesy from John Hamilton, Robin Buell Michigan State University)

web genome browser hosted and maintained by the Buell Lab at Michigan State University in United States and includes annotation datasets, phenotypic and genotypic data from a diversity panel of 250 potato clones (Hirsch et al. 2013). This genomic resource is actively used by potato scientists worldwide.

The initial efforts of the potato sequencing consortium were on resolving the two genome sequences of the dihaploid clone *S. tuberosum* Group Tuberosum RH89-039-16 (RH), but these could not be fully assembled in spite of the availability of the DM sequence. Higher level of heterogeneity was found among the two RH genomes than between RH and DM genomes (The Potato Genome Sequencing Consortium 2011). About 5% of the RH genome sequence (free of repetitive sequences) were aligned with the DM genome sequence and found to be mainly collinear with 97.5% sequence identity, whereas the two RH genome sequences presented 96.5% sequence identity. However, when larger RH genome sequences were obtained, loss of collinearity was frequently observed for the euchromatic region and the three highly diverged pericentric heterochromatin haplotypes of the chromosome 5 (de Boer et al. 2015). These findings stress the importance of sequencing other cultivated potato genome and perform de novo assembly.

Being a diploid and polyploid crop with frequent inbreeding depression and wild species introgression, genome sequence diversity is expected and has contributed to the difficulties of assembling more genome sequences from the cultivated potato.

The genome sequence of a wild species, *Solanum commersonii* was assembled using the potato genome sequence as reference (Aversano et al. 2015). This species has interesting sources of resistance to important diseases of the potato and is known for its freezing resistance and cold acclimation. The species has been used recently in breeding potato for resistance to bacterial wilt in potato and increased levels of resistance were observed (Carputo et al. 2009; Boschi et al. 2017). Flow cytometry estimated a total genome size of 830 Mb. The genome appears to be slightly smaller mainly due to differences in the intragenic regions, to have lower amount of repetitive DNA, and to have 126 cold-related genes not present in the *S. tuberosum* genome.

Recently, the genome sequence of another wild species was resolved using the M6 inbred clone of *Solanum chacoense* (Leisner et al. 2018). Flow cytometry estimated the genome to be 882 Mb. Using a de novo assembly procedure, 508 Mb of the genome assembly could be used to construct 12 pseudomolecules. These were compared to those of the first published genome sequence and shown with concordance for all of them. Interestingly, the genotype used was a 7-generation selfed *S. chacoense* plant but retained residual heterozygosity on all chromosome with three of them with significantly higher proportion. It is too early to assume that heterozygosity in some region is due either to deleterious alleles or to regions with reduced recombination. Genome annotation for gene-models revealed the presence of 37,740 genes. The *S. chacoense* genome sequence is a new resource for identifying important genes of key traits in population derived from M6.

The pan-genome of the cultivated potato covering traditional landraces (diploids to pentaploids), and modern potato cultivars of the Andigenum and Chilotanum gene pools represents today a huge endeavor due to its extraordinary diversity. When available, it would be a powerful resource for breeders to understand the genome structure of the cultivated potato between the core genome with genes present in all cultivars and the dispensable genome made of genes present only in some cultivar groups. The concept is not restricted to modern cultivars but can include wild relatives, or higher taxonomic level (Vernikos et al. 2015). Clearly, more genomes of wild species are also needed to be assembled to improve our understanding of the interspecific genome variation. Ten years after the beginning of sequencing the potato genome, it is worth noting that only one cultivated potato genome is publicly available unlike maize or rice. This highlights the complexity of resolving uneven heterozygosity of the two or four genomes present in wild and cultivated potato. New sequencing technologies and genome assembly software are about to deliver the genomes sequences from heterozygous potatoes. This is highly desirable due to the diversity of species that have contributed to the potato.

The Genome Plasticity of the Cultivated Potato

Comparative analysis of genome sequences in a small panel of closely related potatoes revealed extensive genome plasticity in potato (Hardigan et al. 2016). This study used a panel of doubled monoploid potatoes derived from *S. tuberosum* Group Phureja landraces with limited introgression from Group Stenotomum, Group Tuberosum, and *Solanum chacoense*. Large regions of the potato genome bearing stress-related gene families are duplicated or deleted revealing a possible evolutionary adaptation response to environmental stresses. Copy number variation (CNV) assessed with a minimum 100-bp size revealed that about 30% of the genes are duplicated or deleted in this panel of 12 closely related potatoes. The duplicated regions varied from 500 bp to 575 kb, with total CNV calls per individual varying from 2,978 to 10,532 located preferentially in intergenic sequences in pericentromeric region of the chromosomes. This genome plasticity concerns >7000 genes referred to as dispensable genes. A remarkably high level of genome heterogeneity is found in diploid potato, which is retained through clonal propagation.

Genome heterogeneity is responsible for differential gene expression observed among the genes of tetraploid cultivars (Pham et al. 2017). Genome-wide study of genomic variation and transcription in a panel of six North American tetraploid cultivars revealed the importance of preferential allele expression often associated with evolutionarily conserved genes. Additive allele expression genes in leaves and tubers were only slightly more abundant than preferred-allele expression genes. This can be due to the differential presence of regulatory sequences (promoters, enhancers) but also to structural differences (chromatin structure, epigenetic control). Copy number was frequent; about 40% of the genes from each cultivar were in variable copy number. Again here, copy number variation seemed to be more recent and concerning genes involved in response to biotic and abiotic stresses.

Resequencing of the genomes of a representative sample of cultivated potatoes revealed about 2622 genes under domestication selection, with only 14–16% shared by the North American modern potato cultivars and the Andigenum landraces (Hardigan et al. 2017). This relatively small original gene set suggests a relatively short original common domestication of cultivated potato which diverged into two geographically distinct and long-day adapted cultivar groups by the contribution of wild species. An equally plausible interpretation is two independent domestication events from distinct wild species. This hypothesis has been debated since the early days of potato taxonomy at the beginning of the twentieth century opposing the Russian and the English taxonomist schools advocating respectively multiple and single origin of the cultivated potato (reviewed in Spooner et al. 2014). The absence of extant wild species closely related to the ancestor species of the Southern domestication is the weakest support to this hypothesis (Spooner et al. 2012). The Hardigan study revealed the role of specific wild Solanum species in the evolution of the long-day adapted *S. tuberosum* cultivar group and adaptation to upland and lowland distinguishing the Andigenum and Chilotanum groups. However, both cultivated groups presented a significant contribution from the domestication progenitor *Solanum candolleanum* suggesting the differential contribution from wild species occurred after the domestication from the *S. candollearum* progenitor. Considering variants from the regions of introgression of wild species DNA, the nuclear phylogeny resolved the Chilotanum group and modern cultivars as deriving from the Andigenum group. This study brings closer to closure of a century-old controversy on the independent domestication event leading to the Chilotanum group of cultivars.

New Genomic Tools for Potato Improvement

Potato genomic resources are gradually expanding since the availability of the first potato genome sequence from an Andean potato landrace of the Phureja group (Hirsch et al. 2014). Partial genome sequences are available from dihaploid from modern cultivars and fully resolved haplotypes from tetraploid potato cultivars have been recently achieved. Transcriptomes corresponding to these genome sequences and similar ones have been produced under many important developmental and stress conditions.

The exploitation of potato genomic resources in modern cultivar development is mostly exemplified by the use of the Single Nucleotide Polymorphisms (SNP) arrays developed by the potato community (Douches et al. 2014). Several generations of SNP arrays were generated building on the original Infinium 8303 SNP array (Felcher et al. 2012). As listed by Hirsch et al. (2014), the SolCAP array was used to understand variation for glycoalkaloid biosynthesis in wild and cultivated potato, genotype several diversity panels for a retrospective view of North American potato breeding, for a taxonomic alignment, and for genetic structure of European potato cultivars. Since then, it has been used for genetic mapping in populations derived from a diploid inbred parent (Endelman and Jansky 2016; Peterson et al. 2016), genetic mapping of agronomic traits (Manrique-Carpintero et al. 2015), combined with other SNPs to extend its use to European potato breeding germplasm (Vos et al. 2015), assess linkage decay and testing GWAS models (Sharma et al. 2018), and test genetic identity of accessions in genebanks (Ellis et al. 2018).

From Genomes to the Genes of the Potato

Gene Discovery Facilitated by the Genome Sequence

Genomics-derived strategies for gene discovery have emerged with the availability of high density markers, decrease in sequencing costs, and the increasing power of bioinformatics.

GWAS has the potential to associate markers with regions, genes, underlying the phenotypic variation of trait of interest, and therefore to increase the effectiveness of potato breeding efforts. Unlike marker association studies based on biparental populations, GWAS is not constrained by the performance of one single genotype as the sole source of an allele of interest, and instead it exploits the power of large populations to identify marker-trait associations. A recent review of GWAS in potato highlighted the importance of understanding the structure (kinship) of the population under study (Sharma et al. 2018). Potato populations made of varieties and breeding lines have been studied to establish Linkage Disequilibrium (LD) between adjacent makers. This is an important parameter of the population under study because the shorter it is the higher is the significance of the association. LD decay in earlier studies were found to present large variation (1–10 cM until equilibrium) depending on the population, the locus, and the type of markers (reviewed in Spooner et al. 2014). The first whole-genome scan of LD decay on a large European potato cultivar population estimated LD decay to 5 cM (D'Hoop et al. 2010), concluding that association studies can be performed at moderate marker densities. Since the advent of SNP arrays, new GWAS have been conducted and revealed the power of this mapping approach over the biparental mapping (Stich et al. 2013). An extended SNP array of the 8303 SolCAP (SolSTW) was used to genotype 569 potato cultivars with 20k SNP markers (Vos et al. 2017). Although this study used a different estimator of LD decay than previous studies, it was found to be in the range of 1.5 Mb for old potato cultivars and 0.6 Mb for those of the second half of twentieth century, values which are compatible with the known limited number of meiosis (5–10) having taken place in the development of these European cultivar populations (Gebhardt et al. 2004; van Berloo et al. 2007). The most recent study using the SolCAP SNP array on a large European cultivar population of 351 tetraploid potatoes estimated LD decay in different regions (short and long arms, and pericentromeric heterochromatin) of each chromosome (Sharma et al. 2018). Again here, their estimates were in the range of 2.73 Mb for euchromatin and 3.27 Mb for whole chromosomal regions. Hence, most studies of LD decay report a modest decay of LD in European potato cultivar populations ranging from 0.6 to 20 Mb depending on the region and chromosome. Interestingly, smaller values of 0.3 Mb in chromosome 4 to 8 Mb in chromosome 8 were estimated for a population of 652 Andigenum cultivars (Berdugo-Cely et al. 2017). The lower distance for LD decay in these native cultivars is expected though a much lower distance could have been anticipated for a population from cultivar domesticated between 8,000 BC and 11,500 BC based on fossil evidence from the dry coast of Peru and south-central Chile (Spooner et al. 2014). It does appear that GWAS in potato can be successful at a modest marker density conferred by current SNP arrays in particular for traits with large QTL effects. However, GWAS alone will not be sufficient to associate markers directly to a specific gene contributing to the trait of interest in potato cultivar populations due to the limited number of meiotic recombination.

Annotation of the potato genome revealed the large family of plant resistance (R) genes discovered by motif sharing (nucleotide-binding site and leucine-rich repeat domain, NB-LRR) with an estimated number per haploid genome of 438 (Jupe et al. 2012). By rescreening the potato genome for NB-LRR target sequences, a total of 755 *R* gene homolog were identified (Jupe et al. 2013). This *R* gene enrichment and sequencing (RenSeq) method was applied to identify markers co-segregating with *R* genes for LB resistance and rapidly clone them (Jupe et al. 2013; Witek et al. 2016; Chen et al. 2018). A derived application of this genome-wide gene discovery is the diagnostic resistance gene enrichment sequencing (dRenSeq) which identifies full *R* genes and their homologs in breeding materials (Armstrong et al. 2018). Combined strategies to identify or clone, multiple resistance genes for diseases such as late blight, viruses, and nematodes will speed up the development of new cultivar with stacked resistance genes.

Progress Toward Next Generation of Potato Varieties

The exploitation of the pan-genome of the potato for varietal improvement will increase as more genomes are sequenced and traits phenotyped more accurately in broad germplasm. However, genetic gain will continue to be low in tetraploid breeding though faster and more predictable by the application of genomic selection. New breeding technologies and diploid hybrid breeding can generate unachievable genetic gains by tetraploid breeding.

Direct gene transfer in potato has been successful in generating disease resistance varieties since the early days of genetic engineering (Halterman et al. 2016). Existing widely grown varieties were genetically upgraded by addition of transgenes conferring resistance to pest and diseases, improved processing qualities, and consumer preferences.

These transgenes produced new pest and pathogen toxins, silenced incoming viruses or endo-genes, or new enzymes for metabolite engineering. After a short life, the first generation of biotech potatoes were withdrawn as reported above. However, a renewed interest of the industry lead to the release of new biotech potatoes in the US (Waltz 2015). A long awaited released came about the same time in Argentina with a PVY virus-resistant potato variety (Bravo-Almonacid and Segretin 2016). With the exception of the latter, all biotech potato released so far were developed by the private sector. Efforts towards future release of late blight-resistant varieties have increased in the last years. A 10-year research project in The Netherlands developed transgenic and cisgenic potatoes from four varieties with single and multiple *R* genes (Haverkort et al. 2016). *R* gene stacking was shown to confer high levels of resistance in the filed over several seasons (Zhu et al. 2012; Haesaert et al. 2015). One of these was even fully tested for regulatory approval (Storck et al. 2012). This biotech variety, Fortuna, was unfortunately withdrawn from regulatory approval because of the unfavorable European environment. In the US, a 5-year project aimed at the release in Indonesia and Bangladesh of late blight-resistant local varieties with three *R* genes (https://www.canr.msu.edu/biotechpp/index). These three-*R*-gene biotech potatoes are also the focus of a project aiming at release in sub-Saharan African countries potato varieties with an extremely high level of genetic tolerance to late blight, the most devastating disease in potato, caused by *Phytophtora infestans*, unrivalled by the genetic tolerance achieved to date through conventional breeding(Ghislain et al. 2018). Biotech potatoes have been field tested under natural infection for five seasons and have not shown any lesions caused by *P. infestans*.

The latter two projects are benefiting from the release in the US of the Innate potato with late blight resistance for which the regulatory dossier is publicly available (Clark et al. 2014). One of the important costs in regulatory dossier development is the toxicity assessment of the new proteins for which the 3*R* gene technology can build a weight of evidence instead of costly purification, stability, and gavage testing (Habig et al. 2018). Therefore, when both projects estimated their regulatory costs, these were found to be reasonable unlike those reported by

Table Traits of biotech potato from potatoes approved for food and cultivation (source ISAAA GM crop database)

Trait(s)	Trade name	Developer	First approved for food	First approved for cultivation
Colorado Potato Beetle resistance[a]	New Leaf™ Russet Burbank potato	Monsanto Co.	CAN USA (1995); AUS JPN NZL (2001); PHL (2003); KOR (2004)	USA (1994); CAN (1995)
	Atlantic NewLeaf™ potato	Monsanto Co.	CAN MEX USA (1996); AUS NZL (2001)	USA (1995); CAN (1997)
	Superior NewLeaf™ potato	Monsanto Co.	CAN (1995); USA (1996); MEX (1996); AUS JPN NZL (2001); PHL (2003); KOR (2004)	USA (1995)
Colorado Potato Beetle and PVY resistance[a]	New Leaf™ Y Russet Burbank potato	Monsanto Co.	USA (1998); CAN (1999); AUS JPN MEX NZL (2001); PHL KOR (2004)	USA (1997); CAN (1999)
	Shepody NewLeaf™ Y potato	Monsanto Co.	USA (1998); CAN (1999); AUS MEX NZL (2001); JPN PHL (2003); KOR (2004)	USA (1997); CAN (2001)
	Hi-Lite NewLeaf™ Y potato	Monsanto Co.	USA (1998)	
Modified starch (high amylose)[a]	Amflora™	BASF	EU (2010)	EU (2010)
	Starch Potato	BASF	USA (2014)	
Low asparagine (acrylamide), low black spot bruise	Innate® Cultivate	JR Simplot Co.	USA (2014); CAN (2016); AUS JPN MEX MYZ NZL (2017)	USA (2014); CAN (2016)
	Innate® Generate	JR Simplot Co.	USA (2014); CAN (2016); AUS MEX NZL (2017)	USA (2014); CAN (2016)
	Innate® Accelerate	JR Simplot Co.	USA (2014); CAN (2016); AUS MEX NZL (2017)	USA (2014); CAN (2016)
Low asparagine (acrylamide), low black spot bruise, late blight resistance	n/a (Russet Burbank)	JR Simplot Co.	USA (2015); AUS CAN NZL (2017)	USA (2015); CAN (2017)
	Innate® Acclimate	JR Simplot Co.	USA (2016); AUS CAN NZL (2017)	USA (2015); CAN (2017)
	Innate® Hibernate	JR Simplot Co.	USA (2016); AUS CAN NZL (2017)	USA (2015); CAN (2017)
PVY resistance	n/a (Spunta)	Technoplant Argentina	ARG (2018)	ARG (2018)

Countries are represented by three letter codes
[a]Refers to products phased out of the market

Fig. Confined field trial conducted in Uganda with transgenic potato with three *R* genes (dark green plots), developed as described by Ghislain et al. (2018), and nontransgenic potato varieties (severely damaged plots)

larger players for commodities like maize (Kalaitzandonakes et al. 2006; Schiek et al. 2016). However, the adoption of biotech potatoes remains challenging due to negative perception by a large part of the public unfamiliar with the challenges and potential solutions to improve agriculture production. The long-standing opposition to industrialization of agriculture, the concerns about multinational corporate dominance, the lack of trust in risk assessment of regulatory agencies, the growing conflict of interest of the organic industry, and the fear of unknown manipulations of our food, have delayed the approval and adoption of biotech crops. The release of biotech potatoes addressing a major long-lasting threat on its production which calls back bad memories to Europeans and North Americans, may well result in a perception change provided strong public education is developed (Hallerman and Grabau 2016).

Gene editing in potato has already passed the stage of proof-of-concept as reviewed above. There are yet no potato products on the market, but gene-edited varieties will soon be released with traits governed by known existing genes whose regulation and allele structure determine the trait value.

It is important to realize that gene editing is a complement to transgenesis, not replacement, because it is limited to the existing endogenous genes of the potato. Disruptive news came up recently when the European Court of Justice passed a judgment that genome edited crops should be regulated using the same regulatory framework as the transgenic crops (Callaway 2018). This decision is reminiscent of a previous one in 2012 when the European Food Safety Authority concluded that cisgenic crops should be regulated as transgenic crops (EFSA 2012). This European

Table Traits targeted by genome editing in potato and opportunities for improving pest and disease resistance as well as nutritional qualities of the potato

Trait	Target gene	Expected impact	References
Heat tolerance (high yield under higher temperature)	Heat-shock cognate 70 (HSc70)	Enhanced yields of potato grown under lowland tropics	Trapero-Mozos et al. (2018)
Virus resistance (PVY potyvirus resistance)	Eukaryotic translation initiation factor 4E (eIF4E)	Reduce yield loss due to PVY and enhance tuber seed quality	Arcibal et al. (2016)
Reduced accumulation of reducing sugars	Vacuolar invertase (VInv)	Improved qualities of processed potato	Clasen et al. (2016)
Reduced acrylamide in processed products	Vacuolar invertase gene VInv and the asparagine synthetase genes StAS1 and StAS2	Reduction of acrylamide formation under extreme cooking temperature	Zhu et al. (2016)
Decreased accumulation of glycoalkaloids	Sterol side chain reductase 2 (St-SSR2)	Release of advanced breeding potato lines with elevated SGA	Sawai et al. (2014)
Inbreeding tolerance	S-RNase alleles (Sp3 and Sp4)	Generation of self-compatible diploid potato for developing hybrid potato varieties	Ye et al. (2018)
LB resistance	Ethylene response factor StERF3; 6 susceptibility genes; DND1 gene	Reduction of production losses and reduced costs of production	Tian et al. (2015), Sun et al. (2016a, b)
VitA biofortification	beta carotene hydroxylase *b-ch* gene	Enrichment in beta carotene in potato (precursor of VitA)	Van Eck et al. (2007)

decision will impact agricultural biotechnology innovation negatively not only in Europe but also in developing countries.

Hybrid breeding in potato has already been tested by farmers in developing countries and has received great excitement by the potato crop improvement actors in spite of the initial skepticism (Lindhout et al. 2017). The first yield assessment of hybrid varieties was conducted in two locations; the Netherlands and the Democratic Republic of Congo (de de Vries et al. 2016). In the latter, the best hybrid variety yielded three to four times the national average in sub-Saharan Africa (SSA) countries, whereas it yielded only half of the yield of conventional varieties in the Netherlands. A confounding factor is the type of seeds and health status that will need to be factored out for more precise yield comparison between hybrid and conventional potato. Nevertheless, the possibility of combining complementary traits from the parents, obtaining heterosis from hybridization of inbred parents, avoiding pathogen load of seed tubers, and facilitating transport of true seeds leaves no doubt that hybrid varieties will attract a lot of interest in the developing world.

Concluding Remarks

Despite early optimism, and unlike in other crops, the vast insight gained from its genes and genome has not been steadily translated into substantial genetic progress in potato, through either molecular breeding or transgenic approaches. Among the issues behind the above, the genetic complexity of tetraploid potatoes, issues related with public acceptance of transgenic crops, and a critical mass smaller than in other crops stand out as the most salient ones. Regardless, we remain confident that recent scientific developments, such as an increased focus on developing hybrid varieties at the 2X level, are one of the main factors that will change the above-described trend, since a main advantage of dealing with 2X instead of 4X genetics is a much more straightforward application of molecular approaches, as demonstrated already by the routinary use of such technologies in other Solanaceous crops such as potato, and to a lesser extent, pepper. In addition, early reports on the use of genomic selection in potato have demonstrated its ability to circumvent many of the pitfalls observed when QTL were used to attempt increasing the effectiveness of potato breeding efforts. The continuous reduction of DNA sequencing will enable collecting sequencing data on a larger scale than before, further facilitating both the identification of genomic regions associated with traits of economic importance, and a better understanding of quantitative traits in potato. Regarding the use of gene editing approaches, although they provide a much more targeted ability to modify the potato's genome, the full realization of its potential to facilitate the development of varieties carrying genetic alleles not hitherto found in the germplasm available will, by and large, depend on how the public acceptance of genetic modification evolves, both in developed and developing countries.

References

Almekinders CJM, Chujoy E, Thiele G (2009) The use of true potato seed as pro-poor technology: the efforts of an International Agricultural Research Institute to innovating potato production. Potato Res 52(4):275–293. https://doi.org/10.1007/s11540-009-9142-5

Andersson M, Turesson H, Nicolia A, Fält AS, Samuelsson M, Hofvander P (2017) Efficient targeted multiallelic mutagenesis in tetraploid potato (*Solanum tuberosum*) by transient CRISPR-Cas9 expression in protoplasts. Plant Cell Rep 36(1):117–128. https://doi.org/10.1007/s00299-016-2062-3

Arcibal E, Gold KM, Flaherty S, Jiang J, Jahn M, Rakotondrafara AM (2016) A mutant eIF4E confers resistance to potato virus Y strains and is inherited in a dominant manner in the potato varieties Atlantic and Russet Norkotah. Am J Potato Res 93(1):64–71. https://doi.org/10.1007/s12230-015-9489-x

Armstrong MR, Vossen J, Lim TY, Hutten RCB, Xu J, Strachan SM, Harrower B, Champouret N, Gilroy EM, Hein I (2018) Tracking disease resistance deployment in potato breeding by enrichment sequencing. Plant Biotechnol J 17(2):540–549. https://doi.org/10.1111/pbi.12997

Aversano R, Contaldi F, Ercolano MR, Grosso V, Iorizzo M, Tatino F, Delledonne M (2015) The *Solanum commersonii* genome sequence provides insights into adaptation to stress conditions and genome evolution of wild potato relatives. Plant Cell 27:954–968. https://doi.org/10.1105/tpc.114.135954

Berdugo-Cely J, Valbuena RI, Sanchez-Betancourt E, Barrero LS, Yockteng R (2017) Genetic diversity and association mapping in the Colombian Central Collection of *Solanum tuberosum* L. *Andigenum* group using SNPs markers. Plos One 12(3):e0173039. https://doi.org/10.1371/journal.pone.0173039

Boschi F, Schvartzman C, Murchio S, Ferreira V, Siri MI, Galván GA, Dalla-Rizza M (2017) Enhanced bacterial wilt resistance in potato through expression of *Arabidopsis* EFR and introgression of quantitative resistance from *Solanum commersonii*. Front Plant Sci 8:1642. https://doi.org/10.3389/fpls.2017.01642

Bradshaw JE (2009) Potato breeding at the Scottish Plant Breeding Station and the Scottish Crop Research Institute: 1920–2008. Potato Res 52:141–172. https://doi.org/10.1007/s11540-009-9126-5

Bradshaw JE (2017) Review and analysis of limitations in ways to improve conventional potato breeding. Potato Res 60:171–193. https://doi.org/10.1007/s11540-017-9346-z

Bradshaw JE, Ramsay G (2005) Utilisation of the commonwealth potato collection in potato breeding. Euphytica 146(1–2):9–19. https://doi.org/10.1007/s10681-005-3881-4

Bradshaw JE, Stewart HE, Wastie RL, Dale MFB, Phillips MS (1995) Use of seedling progeny tests for genetical studies as part of a potato (*Solanum tuberosum* subsp tuberosum) breeding programme. Theor Appl Genet 90:899–905

Bradshaw JE, Todd D, Wilson RN (2000) Use of tuber progeny tests for genetical studies as part of a potato (*Solanum tuberosum* subsp tuberosum) breeding programme. Theor Appl Genet 100:772–781

Bravo-Almonacid FF, Segretin MF (2016) Status of transgenic crops in Argentina. In: Collinge DB (ed) Biotechnology for plant disease control. Wiley Blackwell, New York, pp 275–283

Butler NM, Atkins PA, Voytas DF, Douches DS (2015) Generation and inheritance of targeted mutations in potato (*Solanum tuberosum* L.) using the CRISPR/Cas system. PloS One 10(12):e0144591. https://doi.org/10.1371/journal.pone.0144591

Butler NM, Baltes NJ, Voytas DF, Douches DS (2016) Geminivirus-mediated genome editing in potato (*Solanum tuberosum* L.) using sequence-specific nucleases. Front Plant Sci 7:1045. https://doi.org/10.3389/fpls.2016.01045

Callaway E (2018) CRISPR plants now subject to tough GM laws in European Union. Nature 560:16. https://doi.org/10.1038/d41586-018-05814-6

Carputo D, Aversano R, Barone A, Di Matteo A, Iorizzo M, Sigillo L, Zoina A, Frusciante L (2009) Resistance to *Ralstonia solanacearum* of sexual hybrids between *Solanum commersonii* and *S. tuberosum*. Am J Potato Res 86:196–202. https://doi.org/10.1007/s12230-009-9072-4

Chen X, Lewandowska D, Armstrong MR, Baker K, Lim T-Y, Bayer M, Harrower B, McLean K, Jupe F, Witek K, Lees AK, Jones JD, Bryan GJ, Hein I (2018) Identification and rapid mapping of a gene conferring broad-spectrum late blight resistance in the diploid potato species *Solanum verrucosum* through DNA capture technologies. Theor Appl Genet 131(6):1287–1297. https://doi.org/10.1007/s00122-018-3078-6)

Clark P, Habig J, Ye J, Collinge S (2014) Petition for determination of nonregulated status for Innate™ potatoes with late blight resistance, low acrylamide potential, reduced black spot, and lowered reducing sugars: Russet Burbank event W8. JR Simplot Company Petition JRS01 (USDA Petition 14-093-01p)

Clasen BM, Stoddard TJ, Luo S, Demorest ZL, Li J, Cedrone F, Coffman A (2016) Improving cold storage and processing traits in potato through targeted gene knockout. Plant Biotechnol J 14(1):169–176. https://doi.org/10.1111/pbi.12370

de Boer JM, Datema E, Tang XM, Borm TJA, Bakker EH, van Eck HJ, van Ham RCHJ, de Jong H, Visser RGF, Bachem CWB (2015) Homologues of potato chromosome 5 show variable collinearity in the euchromatin, but dramatic absence of sequence similarity in the pericentromeric heterochromatin. BMC Genomics 16(1):374. https://doi.org/10.1186/s12864-015-1578-1

D'Hoop BB, Paulo MJ, Kowitwanich K, Sengers M, Visser RGF, van Eck HJ, van Eeuwijk FA (2010) Population structure and linkage disequilibrium unravelled in tetraploid potato. Theor Appl Genet 121(6):1151–1170. https://doi.org/10.1007/s00122-010-1379-5

Douches D, Hirsch CN, Manrique-Carpintero NC, Massa AN, Coombs J, Hardigan M, Bisognin D, De Jong W, Buell CR (2014) The contribution of the Solanaceae Coordinated Agricultural Project to potato breeding. Potato Res 57:215–224. https://doi.org/10.1007/s11540-014-9267-z

EFSA Panel on Genetically Modified Organisms (GMO) (2012) Scientific opinion addressing the safety assessment of plants developed through cisgenesis and intragenesis. EFSA J 10(2):2561[33p]. https://doi.org/10.2903/jefsa20122561

Ellis D, Chavez O, Coombs JJ, Soto JV, Gomez R, Douches DS, Anglin NL (2018) Genetic identity in genebanks: application of the SolCAP 12K SNP array in fingerprinting and diversity analysis in the global in trust potato collection. Genome 61(7):523–537. https://doi.org/10.1139/gen-2017-0201

Endelman JB, Jansky SH (2016) Genetic mapping with an inbred line-derived F2 population in potato. Theor Appl Genet 129(5):935–943. https://doi.org/10.1007/s00122-016-2673-7

Felcher KJ, Coombs JJ, Massa AN, Hansey CN, Hamilton JP, Veilleux RE, Buell CR, Douches DS (2012) Integration of two diploid potato linkage maps with the potato genome sequence. PloS One 7(4):e36347. https://doi.org/10.1371/journal.pone.0036347

Finkers-Tomczak A, Bakker E, de Boer J, van der Vossen E, Achenbach U, Golas T, Suryaningrat S, Smant G, Bakker J, Goverse A (2011) Comparative sequence analysis of the potato cyst nematode resistance locus *H1* reveals a major lack of co-linearity between three haplotypes in potato (*Solanum tuberosum* ssp.). Theor Appl Genet 122(3):595–608. https://doi.org/10.1007/s00122-010-1472-9

Gatto M, Hareau G, Pradel W, Suárez V, Qin J (2018) Release and adoption of improved potato varieties in Southeast and South Asia. International Potato Center (CIP), Lima, Peru, 45p. ISBN 978-92-9060-501-0. Social Sciences Working Paper 2018-2, 42p

Gavrilenko T, Antonova O, Shuvalova A, Krylova E, Alpatyeva N, Spooner DM, Novikova L (2013) Genetic diversity and origin of cultivated potatoes based on plastid microsatellite polymorphism. Genet Resour Crop Evol 60(7):1997–2015. https://doi.org/10.1007/s10722-013-9968-1

Gebhardt C (2013) Bridging the gap between genome analysis and precision breeding in potato. Trends Genet 29:248–256. https://doi.org/10.1016/j.tig.2012.11.006

Gebhardt C, Ballvora A, Walkemeier B, Oberhagemannand P, Schuler K (2004) Assessing genetic potential in germplasm collections of crop plants by marker-trait association: a case study for potatoes with quantitative variation of resistance to late blight and maturity type. Mol Breeding 13(1):93–102

Ghislain M, Byarugaba A, Magembe E, Njoroge A, Rivera C, Roman ML, Tovar JC, Gamboa S, Forbes G, Kreuze J, Barekye A, Kiggundu A (2018) Stacking three late blight resistance genes from wild species directly into African highland potato varieties confers complete field resistance to local blight races. Plant Biotech J 17(6):1119–1129. https://doi.org/10.1111/pbi.13042

Habig JW, Rowland A, Pence MG, Zhong CX (2018) Food safety evaluation for R-proteins introduced by biotechnology: a case study of VNT1 in late blight protected potatoes. Regul Toxicol Pharmacol 95:66–74. https://doi.org/10.1016/j.yrtph.2018.03.008

Haesaert G, Vossen JH, Custers R, De Loose M, Haverkort A, Heremans B, Hutten R, Kessel G, Landschoot S, Van Droogenbroeck B, Visser RGF, Gheysen G (2015) Transformation of the potato variety Desiree with single or multiple resistance genes increases resistance to late blight under field conditions. Crop Prot 77:163–175. https://doi.org/10.1016/j.cropro.2015.07.018

Hallerman E, Grabau E (2016) Crop biotechnology: a pivotal moment for global acceptance. Food Energy Secur 5(1):3–17. https://doi.org/10.1002/fes3.76

Halterman D, Guenthner J, Collinge S, Butler N, Douches D (2016) Biotech potatoes in the 21st century: 20 years since the first biotech potato. Am J Potato Res 93(1):1–20. https://doi.org/10.1007/s12230-015-9485-1

Hamilton JP, Hansey CN, Whitty BR, Buell CR (2011) Single nucleotide polymorphism discovery in elite North American potato germplasm. BMC Genomics 12(1):302. https://doi.org/10.1186/1471-2164-12-302

Hardigan MA, Crisovan E, Hamilton JP, Kim J, Laimbeer P, Leisner CP, Manrique-Carpintero NC, Newton L, Pham GM, Vaillancourt B, Yang X, Zeng Z, Douches DS, Jiang J, Veilleux RE, Buell CR (2016) Genome reduction uncovers a large dispensable genome and adaptive role for copy number variation in asexually propagated *Solanum tuberosum*. Plant Cell 28:388–405. https://doi.org/10.1105/tpc.15.00538

Hardigan MA, Laimbeer FPE, Newton L, Crisovan E, Hamilton JP, Vaillancourt B, Wiegert-Riningera K, Wooda JC, Douches DS, Farréa EM, Veilleux RE, Buell CR (2017) Genome diversity of tuber-bearing Solanum uncovers complex evolutionary history and targets of domestication in the cultivated potato. Proc Nat Acad Sci 114(46):E9999–E10008. https://doi.org/10.1073/pnas.1714380114

Haverkort AJ, Boonekamp PM, Hutten R, Jacobsen E, Lotz LAP, Kessel GJT, Visser RGF (2016) Durable late blight resistance in potato through dynamic varieties obtained by cisgenesis: scientific and societal advances in the DuR*Ph* project. Potato Res 59(1):35–66. https://doi.org/10.1007/s11540-015-9312-6

Haverkort AJ, Struik PC, Visser RGF, Jacobsen E (2009) Applied biotechnology to combat late blight in potato caused by *Phytophthora infestans*. Potato Res 52:249–264. https://doi.org/10.1007/s11540-009-9136-3

Haynes KG, Guedes ML (2018) Self-compatibility in a diploid hybrid population of *Solanum phureja–S. stenotomum*. Am J Potato Res 95(6):729–734. https://doi.org/10.1007/s12230-018-9680-y

Hirsch CD, Hamilton JP, Childs KL, Cepela J, Crisovan E, Vaillancourt B, Hirsch CN, Habermann M, Neal B, and Buell CR (2014) Spud DB: A resource for mining sequences genotypes and phenotypes to accelerate potato breeding. Plant Genome 7(1):1–12. https://doi.org/10.3835/plantgenome2013.12.0042

Hirsch CN, Hirsch CD, Felcher K, Coombs J, Zarka D, Van Deynze A, De Jong W, Veilleux RE, Jansky S, Bethke P, Douches DS, Buell CR (2013) Retrospective view of North American potato (*Solanum tuberosum* L.) breeding in the 20th and 21st centuries. G3: Genes, Genomes, Genetics 3(6):1003–1013. https://doi.org/10.1534/g3.113.005595

Hosaka K, Hanneman RE (1998a) Genetics of self-compatibility in a self-incompatible wild diploid potato species *Solanum chacoense*: I detection of an *S* locus inhibitor (*Sli*) gene. Euphytica 99:191–197

Hosaka K, Hanneman RE (1998b) Genetics of self-compatibility in a self-incompatible wild diploid potato species *Solanum chacoense*: 2 localization of an *S* locus inhibitor (*Sli*) gene on the potato genome using DNA markers. Euphytica 103:265–271

Jansky SH, Charkowski AO, Douches DS, Gusmini G, Richael C, Paul C, Spooner DM, Novy RG, De Jong H, De Jong WS, Bamberg JB, Thompson L, Bizimungu B, Holm DG, Brown CR, Haynes KG, Vidyasagar R, Veilleux RE, Miller JC, Bradeen JM, Jiang JM (2016) Reinventing potato as a diploid inbred line-based crop. Crop Sci 11:1–11. https://doi.org/10.2135/cropsci2015.12.0740

Jansky SH, Chungand Y, Kittipadukal P (2014) M6: a diploid potato inbred line for use in breeding and genetics research. J Plant Regist 8:195–199. https://doi.org/10.3198/jpr2013.05.0024crg

Jansky SH, Spooner DM (2017) The evolution of potato breeding. Plant Breed Rev 41:169–211

Jupe F, Pritchard L, Etherington GJ, MacKenzie K, Cock PJA, Wright F, Sharma SK, Bryan GJ, Jones JDG, Hein I (2012) Identification and localisation of the NB-LRR gene family within the potato genome. BMC Genomics 13(1):75. https://doi.org/10.1186/1471-2164-13-75

Jupe F, Witek K, Verweij W, Śliwka J, Pritchard L, Etherington GJ, Maclean D, Cock PJ, Leggett RM, Bryan GJ, Cardle L, Hein I, Jones JDG (2013) Resistance gene enrichment sequencing (RenSeq) enables reannotation of the NB-LRR gene family from sequenced plant genomes and rapid mapping of resistance loci in segregating populations. Plant J 76(3):530–544. https://doi.org/10.1111/tpj.12307

Kalaitzandonakes N, Alston JM, Bradford KJ (2006) Compliance costs for regulatory approval of new biotech crops. In: Just RE, Alston JM, Zilberman D (eds) Regulating agricultural biotechnology: economics and policy. Natural resource management and policy, vol 30. Springer, Boston, pp 37–57. https://doi.org/10.1007/978-0-387-36953-2_3

Lawson C, Kaniewski W, Haley L, Rozman R, Newell C, Sanders P, Tumer NE (1990) Engineering resistance to mixed virus infection in a commercial potato cultivar: resistance to potato virus X and potato virus Y in transgenic Russet Burbank. Nat Biotechol 8(2):127

Leisner CP, Hamilton JP, Crisovan E, Manrique-Carpintero NC, Marand AP, Newton L, Pham GM, Jiang J, Douches DS, Jansky SH, Buell CR (2018) Genome sequence of M6, a diploid inbred clone of the high-glycoalkaloid-producing tuber-bearing potato species *Solanum chacoense* reveals residual heterozygosity. Plant J 94(3):562–570. https://doi.org/10.1111/tpj.13857

Lindhout P, Meijer D, Schotte T, Hutten RCB, Visser RGF, Eck HJ (2011) Towards F1 hybrid seed potato breeding. Potato Res 54:301–312. https://doi.org/10.1007/s11540-011-9196-z

Lindhout P, De Vries M, Ter Maat M, Ying S, Viquez-Zamora M, Van Heusden S (2017) Hybrid potato breeding for improved varieties. In: Achieving sustainable cultivation of potatoes, vol. 1. Breeding, nutritional and sensory quality. Burleigh Dodds Science Publishing, Cambridge, p 24. https://doi.org/10.19103/AS.2016.0016.04

Ma J, Xiang H, Donnelly DJ, Meng FR, Xu H, Durnford D, Li XQ (2017) Genome editing in potato plants by agrobacterium-mediated transient expression of transcription activator-like effector nucleases. Plant Biotechnol Rep 11(5):249–258. https://doi.org/10.1007/s11816-017-0448-5

Maluszynski M, Ahloowalia BS, Sigurbjörnsson B (1995) Application of *in vivo* and *in vitro* mutation techniques for crop improvement. Euphytica 85(1–3):303–315

Manrique-Carpintero NC, Coombs JJ, Cui Y, Veilleux RE, Buell CR, Douches DS (2015) Genetic map and QTL analysis of agronomic traits in a diploid potato population using single nucleotide polymorphism markers. Crop Sci 55(6):2566–2579. https://doi.org/10.2135/cropsci2014.10.0745

Mihovilovich E, Aponte M, Lindqvist-Kreuze H, Bonierbale M (2014) An RGA-derived SCAR marker linked to PLRV resistance from *Solanum tuberosum* ssp. andigena. Plant Mol Biol Rep 32(1):117–128

Nicolia A, Proux-Wéra E, Åhman I, Onkokesung N, Andersson M, Andreasson E, Zhu LH (2015) Targeted gene mutation in tetraploid potato through transient TALEN expression in protoplasts. J Biotechnol 204:17–24. https://doi.org/10.1016/j.jbiotec.2015.03.021

Ning Y, Liu W, Wang GL (2017) Balancing immunity and yield in crop plants. Trends Plant Sci 22(12):1069–1079. https://doi.org/10.1016/j.tplants.2017.09.010

Ochoa CM (1990) The potatoes of South America: Bolivia. Cambridge University Press, Cambridge

Paz MM, Veilleux RE (1999) Influence of culture medium and *in vitro* conditions on shoot regeneration in *Solanum phureja* monoploids and fertility of regenerated doubled monoploids. Plant Breed 118:53–57

Perlak FJ, Stone TB, Muskopf YM, Petersen LJ, Parker GB, McPherson SA, Fischhoff DA (1993) Genetically improved potatoes: protection from damage by Colorado potato beetles. Plant Mol Biol 22(2):313–321

Peterson BA, Holt SH, Laimbeer FPE, Doulis AG, Coombs J, Douches DS, Hardigan MA, Buell CR, Veilleux RE (2016) Self-fertility in a cultivated diploid potato population examined with the Infinium 8303 potato single-nucleotide polymorphism Array. Plant Genome 9(3):1–13. https://doi.org/10.3835/plantgenome2016.01.0003

Pham GM, Newton L, Wiegert-Rininger K, Vaillancourt B, Douches DS, Buell CR (2017) Extensive genome heterogeneity leads to preferential allele expression and copy number-dependent expression in cultivated potato. Plant J 92(4):624–637. https://doi.org/10.1111/tpj.13706

Phumichai C, Mori M, Kobayashi A, Kamijimaand O, Hosaka K (2005) Toward the development of highly homozygous diploid potato lines using the self-compatibility controlling *Sli* gene. Genome 48:977–984. https://doi.org/10.1139/G05-066

Plaisted RL, Hoopes RW (1989) The past record and future prospects for the use of exotic potato germplasm. Am Potato J 66:603–627

Ramakrishnan AP, Ritland CE, Sevillano RHB, Riseman A (2015) Review of potato molecular markers to enhance trait selection. Am J Potato Res 92(4):455–472. https://doi.org/10.1007/s12230-015-9455-7

Rodríguez F, Ghislain M, Clausen AM, Jansky SH, Spooner DM (2010) Hybrid origins of cultivated potatoes. Theor Appl Genet 121(6):1187–1198. https://doi.org/10.1007/s00122-010-1422-6

del Rosario Herrera M, Vidalon LJ, Montenegro JD, Riccio C, Guzman F, Bartolini I, Ghislain M (2018) Molecular and genetic characterization of the *Ryadg* locus on chromosome XI from Andigena potatoes conferring extreme resistance to potato virus Y. Theor Appl Genet 131(9):1925–1938. doi: https://doi.org/10.1007/s00122-018-3123-5

Sawai S, Ohyama K, Yasumoto S, Seki H, Sakuma T, Yamamoto T, Takebayashi Y, Kojima M, Sakakibara H, Aoki T, Muranaka T, Kazuki Saito K, Umemotog N (2014) Sterol side chain reductase 2 is a key enzyme in the biosynthesis of cholesterol, the common precursor of toxic steroidal glycoalkaloids in potato. Plant Cell 26:3763–3774. https://doi.org/10.1105/tpc.114.130096

Scheben A, Wolter F, Batley J, Puchta H, Edwards D (2017) Towards CRISPR/Cas crops–bringing together genomics and genome editing. New Phytologist 216(3):682–698. https://doi.org/10.1111/nph.14702

Schiek B, Hareau G, Baguma Y, Medakker A, Douches DS, Shotkoski F, Ghislain M (2016) Demystification of GM crop costs: releasing late blight resistant potato varieties as public goods in developing countries. Int J Biotechnol 14:112–131

Schönhals EM, Ortega F, Barandalla L, Aragones A, De Galarreta JR, Liao JC, Sanetomo R, Walkemeier B, Tacke E, Ritter E, Gebhardt C (2016) Identification and reproducibility of diagnostic DNA markers for tuber starch and yield optimization in a novel association mapping population of potato (*Solanum tuberosum* L.). Theor Appl Genet 129(4):767–785. https://doi.org/10.1007/s00122-016-2665-7

Sharma R, Bhardwaj V, Dalamu D, Kaushik SK, Singh BP, Sharma SK, Umamaheshwari R, Baswaraj R, Kumar V, Gebhardt C (2014) Identification of elite potato genotypes possessing multiple disease resistance genes through molecular approaches. Sci Hortic 179:204–211. https://doi.org/10.1016/j.scienta.2014.09.018

Sharma SK, Bolser D, de Boer J, Sønderkær M, Amoros W et al (2013) Construction of reference chromosome-scale pseudomolecules for potato: integrating the potato genome with genetic and physical maps. G3 (Bethesda) 3(11):2031–2047. https://doi.org/10.1534/g3.113.007153

Sharma SK, MacKenzie K, McLean K, Dale F, Daniels S, Bryan GJ (2018) Linkage disequilibrium and evaluation of genome-wide association mapping models in tetraploid potato. G3 (Bethesda) 8(10):3185–3202. https://doi.org/10.1534/g3.118.200377

Slater AT, Cogan NOI, Forster JW (2013) Cost analysis of the application of marker-assisted selection in potato breeding. Mol Breeding 32(2):299–310. https://doi.org/10.1007/s11032-013-9871-7

Slater AT, Cogan NOI, Hayes BJ, Schultz L, Dale MFB, Bryan GJ, Forster JW (2014a) Improving breeding efficiency in potato using molecular and quantitative genetics. Theor Appl Genet 127:2279–2292. https://doi.org/10.1007/s00122-014-2386-8

Slater AT, Wilson GM, Cogan NOI, Forster JW, Hayes BJ (2014b) Improving the analysis of low heritability complex traits for enhanced genetic gain in potato. Theor Appl Genet 127:809–820. https://doi.org/10.1007/s00122-013-2258-7

Slater AT, Cogan NOI, Forster JW, Hayes BJ, Daetwyler HD (2016) Improving genetic gain with genomic selection in autotetraploid potato. Plant Genome 9(3):1–15. https://doi.org/10.3835/plantgenome2016.02.0021

Sood S, Bhardwaj V, Pandey SK, Chakrabarti SK (2017) History of potato breeding: improvement diversification and diversity. In: Chakrabarti SK, Xie C, Tiwari JK (eds) The potato genome. Springer, Cham, pp 31–72

Spooner DM, Núñez J, Trujillo G, del Rosario Herrera M, Guzmán F, Ghislain M (2007) Extensive simple sequence repeat genotyping of potato landraces supports a major reevaluation of their gene pool structure and classification. Proc Nat Acad Sci 104(49):19398–19403. https://doi.org/10.1073/pnas.0709796104

Spooner DM, Jansky SH, Clausen A, del Rosario Herrera M, Ghislain M (2012) The enigma of *Solanum maglia* in the origin of the Chilean cultivated potato *Solanum tuberosum* Chilotanum Group. Econ Bot 66(1):12–21

Spooner DM, Ghislain M, Simon R, Jansky SH, Gavrilenko T (2014) Systematics, diversity, genetics, and evolution of wild and cultivated potatoes. Bot Rev 80:283–383. https://doi.org/10.1007/s12229-014-9146-y

Stich B, Urbany C, Hoffmann P, Gebhardt C (2013) Population structure and linkage disequilibrium in diploid and tetraploid potato revealed by genome-wide high-density genotyping using the SolCAP SNP array. Plant Breeding 132(6):718–724. https://doi.org/10.1111/pbr.12102

Storck T, Böhme T, Schultheiss H (2012) Fortuna et al. Status and perspectives of GM approaches to fight late blight. In: Thirteenth EuroBlight workshop St. Petersburg (Russia), 9–12 Oct 2011. PPO-Special Report 15:45–48.

Sun K, Wolters AMA, Loonen AE, Huibers RP, van der Vlugt R, Goverse A, Bai Y (2016a) Down-regulation of *Arabidopsis DND1* orthologs in potato and tomato leads to broad-spectrum resistance to late blight and powdery mildew. Transgenic Res 25(2):123–138. https://doi.org/10.1007/s11248-015-9921-5

Sun K, Wolters AMA, Vossen JH, Rouwet ME, Loonen AE, Jacobsen E, Bai Y (2016b) Silencing of six susceptibility genes results in potato late blight resistance. Transgenic Res 25(5):731–742. https://doi.org/10.1007/s11248-016-9964-2

The Potato Genome Sequencing Consortium (2011) Genome sequence and analysis of the tuber crop potato. Nature 475:189–195. https://doi.org/10.1038/nature10158

The Tomato Genome Consortium (2012) The tomato genome sequence provides insights into fleshy fruit evolution. Nature 485:635–641. https://doi.org/10.1038/nature11119

Thiele G, Hareau G, Suarez V, Chujoy E, Bonierbale M, Maldonado L (2008) Varietal change in potatoes in developing countries and the contribution of the International Potato Center: 1972–2007. Social Sciences Working Paper (2008-6)

Thornton, M., 2003. The rise and fall of Newleaf potatoes. Biotechnology: Science and society at a crossroads. Nat. Agric. Biotech. Council Rep. 15:235–243.

Tian Z, He Q, Wang H, Liu Y, Zhang Y, Shao F, Xie C (2015) The potato ERF transcription factor StERF3 negatively regulates resistance to *Phytophthora infestans* and salt tolerance in potato. Plant Cell Physiol 56(5):992–1005. https://doi.org/10.1093/pcp/pcv025

Tiwari JK, Devi S, Ali N, Luthra SK, Kumar V, Bhardwaj V, Chakrabarti SK (2018) Progress in somatic hybridization research in potato during the past 40 years. Plant Cell Tiss Org 132:225–238. https://doi.org/10.1007/s11240-017-1327-z

Trapero-Mozos A, Morris WL, Ducreux LJ, McLean K, Stephens J, Torrance L, Taylor MA (2018) Engineering heat tolerance in potato by temperature-dependent expression of a specific allele of *HEAT-SHOCK COGNATE 70*. Plant Biotechnol J 16(1):197–207. https://doi.org/10.1111/pbi.12760

Uitdewilligen JG, Wolters AML, D'hoop BB, Borm TJA, Visser RGF, van Eck HJ (2013) A next generation sequencing method for genotyping-by-sequencing of highly heterozygous autotetraploid potato. PLoS One 8(5):e62355. https://doi.org/10.1371/journalpone0062355

van Berloo R, Hutten RCB, van Eck HJ, Visser RGF (2007) An online potato pedigree database resource. Potato Res 50(1):45–57. https://doi.org/10.1007/s11540-007-9028-3

Van Eck J, Conlin B, Garvin DF, Mason H, Navarre DA, Brown CR (2007) Enhancing beta-carotene content in potato by RNAi-mediated silencing of the beta-carotene hydroxylase gene. Am J Potato Res 84(4):331

Vernikos G, Medini D, Riley DR, Tettelin H (2015) Ten years of pan-genome analyses. Curr Opin Microbiol 23:148–154. https://doi.org/10.1016/j.mib.2014.11.016

Visser RGF, Bachem CWB, Borm T, de Boer J, Van Eck HJ, Finkers R, van der Linden G, Maliepaard CA, Uitdewilligen JGAML, Voorrips R, Vos P, Wolters AMA (2014) Possibilities and challenges of the potato genome sequence. Potato Res 3–4:327–330. https://doi.org/10.1007/s11540-015-9282-8

Vos PG, Uitdewilligen JGAML, Voorrips RE, Visser RGF, van Eck HJ (2015) Development and analysis of a 20K SNP array for potato (*Solanum tuberosum*): an insight into the breeding history. Theor Appl Genet 128:2387–2401. https://doi.org/10.1007/s00122-015-2593-y

Vos PG, Paulo MJ, Voorrips RE, Visser RGF, van Eck HJ, van Eeuwijk FA (2017) Evaluation of LD decay and various LD-decay estimators in simulated and SNP-array data of tetraploid potato. Theor Appl Genet 130(1):123–135. https://doi.org/10.1007/s00122-016-2798-8

de Vries M, ter Maat M, Lindhout P (2016) The potential of hybrid potato for East-Africa. Open Agric 1(1):151–156. https://doi.org/10.1515/opag-2016-0020

Walker TS (1994) Patterns and implications of varietal change in potatoes. Working paper 1994-3 Social Sciences Department, International Potato Center (CIP) Lima, Peru

Waltz E (2015) USDA approves next-generation GM potato. Nat Biotechnol 33:12–13

Wang S, Zhang S, Wang W, Xiong X, Meng F, Cui X (2015) Efficient targeted mutagenesis in potato by the CRISPR/Cas9 system. Plant Cell Rep 34(9):1473–1476. https://doi.org/10.1007/s00299-015-1816-7

Watanabe K, Arbizu C, Schmiediche PE (1992) Potato germplasm enhancement with disomic tetraploid *Solanum acaule* I. Efficiency of introgression. Genome 35(1):53–57. https://doi.org/10.1139/g92-009

Witek K, Jupe F, Witek AI, Baker D, Clark MD, Jones JD (2016) Accelerated cloning of a potato late blight–resistance gene using RenSeq and SMRT sequencing. Nat Biotechnol 34:656–660. https://doi.org/10.1038/nbt.3540

Ye M, Peng Z, Tang D, Yang Z, Li D, Xu Y, Huang S (2018) Generation of self-compatible diploid potato by knockout of *S-RNase*. Nat Plants 4:651–654. https://doi.org/10.1038/s41477-018-0218-6

Yin K, Gao C, Qiu JL (2017) Progress and prospects in plant genome editing. Nat Plants 3:17107. https://doi.org/10.1038/nplants.2017.107

Zhu S, Li Y, Vossen JH, Visser RG, Jacobsen E (2012) Functional stacking of three resistance genes against *Phytophthora infestans* in potato. Transgenic Res 21:89–99. https://doi.org/10.1007/s11248-011-9510-1

Zhu X, Gong H, He Q, Zeng Z, Busse JS, Jin W, Bethke PC, Jiang J (2016) Silencing of vacuolar invertase and asparagine synthetase genes and its impact on acrylamide formation of fried potato products. Plant Biotechnol J 14(2):709–718. https://doi.org/10.1111/pbi.12421

3

Gender Topics on Potato Research and Development

Netsayi Noris Mudege, Silvia Sarapura Escobar, and Vivian Polar

Abstract Sustainable Development Goals 5 calls for addressing gender equality and women empowerment by, among other things, eliminating all forms of discrimination against women. At CIP we interpret this to mean strengthening the use of gender approaches in research and ensuring that research products are responsive to the needs of men and women. This chapter reviews lessons learnt over the years on integrating gender into potato research and development. The chapter discusses how gender has been approached in five key themes in potato research, namely (1) conserving and accessing genetic resources, (2) genetics and crop improvement, (3) managing priority pests and disease, (4) access to seed (seed flows and networks), and (5) marketing, postharvest processing and utilization. This chapter discusses how gender relations that favor men influence women's participation in and their ability to benefit from potato production, marketing, and research for development. The review shows that potato research has been increasingly focusing on social determinants of potato farming because of the realization that purely technical solutions will not solve inefficiencies in potato production. Using a gender relations approach, the chapter attempts to draw out lessons that can contribute to the design of future potato interventions including research aimed at reducing the gender gap in agriculture in general and potato farming in particular.

N. N. Mudege (✉)
International Potato Center, Nairobi, Kenya
e-mail: n.mudege@cgiar.org

S. Sarapura Escobar
Royal Tropical Institute, Amsterdam, The Netherlands
e-mail: sarapura@uoguelph.ca

V. Polar
CGIAR Research Program on Roots, Tubers and Bananas (RTB), Lima, Peru
e-mail: v.polar@cgiar.org

Introduction

It has already been established that gender differences matter in agricultural production in various farming systems all over the world, where the ownership and management of farms and natural resources by men and women are often defined by culturally specific gender roles (Meinzen-Dick et al. 2010). Evidence indicates that agriculture and development projects should be gender responsive and take into consideration the needs, aspirations, knowledge, opportunities, constrains, and challenges faced by men and women farmers, young and old, if hunger and poverty are to be alleviated (Njenga and Gurung 2011). Additionally, it is also clear in research that gender intersects with other structures of social hierarchy such as class, race, caste, and age (German and Taye 2008). Certainly, farmers are not a uniform group. Men and women play different roles within agricultural systems occupying different socioeconomic positions linked to these roles, and may suffer from different vulnerabilities (Carr 2008). These differences and vulnerabilities should be considered when new technologies are being developed. Kingiri (2010) noted that generally in farming systems research and innovation, unequal relationships between men and women in households are taken for granted. As a result, development of new technologies may end up increasing the gender gap and benefiting men more than women because social relations of gender are not properly understood. In some cases, as noted by Quisumbing and Pandolfelli (2010), new technologies may even harm women if they are not properly thought through.

The interest of gender in research in potato-related research organizations such as CIP and partners started more than two decades ago. During this period, several studies have looked at the key technical potato production constraints. However, a limited number of studies have discussed the key technical constraints of potato production from gender perspective (see for example, Tapia and de la Torre 2000; Mera-Orcés 2001; Polar et al. 2017; Mudege et al. 2016). In a study promoting participatory technology development in potato farming and production in Ethiopia, Jibat et al. (2007) suggest that it is important to understand gender roles in agriculture production and decision-making to ensure that research address men and women's needs making the results of research more demand oriented. A key limitation of these studies is that they often look at gender roles, and pay little attention to gender relations (see for example, Tapia and de la Torre 2000; Mera-Orcés 2001).

Most of the research has focused on gender division of labor along the potato value chain. For example, a study by Muhinyuza et al. (2012) in Rwanda showed that both men and women are involved in main "potato production activities" as well as decision-making on production and marketing. "However, some activities such as weeding, cooking, and storage protection are exclusively done by women while predominantly men are totally concerned with pest management". Similar work has also been conducted in the Andes in Latin America on gender division of roles in potato production (Tapia and De La Torre 1998; Laub and Muir 2008). Such research often conducts sex disaggregated analysis of labor distribution but does not disaggregate when it comes to constraints because of the focus on technical constraints such as pests and disease (see Muhinyuza et al. 2012; Sah et al. 2007).

However, while early research focused on division of labor and social demographic variables, more recent research has started focusing more and more on gender relations, the normative environment including institutional factors along the potato value chain (Mudege and Demo 2016). Institutional factors such as access to credit and markets by men and women, and their influence on the adoption of improved potato varieties as well as productivity. As part of gender relations, gender norms and how they shape the opportunity structure for men and women in agriculture have been studied (Petesch et al. 2018). Some research also focuses on the gender gap in access to resources, (such as capital), assets (such as land and other productive assets), and knowledge (for example assess to extension services and credible information sources) which if not addressed lead to a gender gap in productivity between men and women (Quisumbing et al. 2014).

In discussing gender topics in potato research, this chapter will go beyond the usual focus on gender roles and asset gaps to look at how the normative environment creates and shape the opportunity structure for men and women. This chapter will use the gender relations approach promoted by Kabeer and Subrahmanian (1996), who defines gender relations as social interactions that embody both the material and the ideological aspects that are revealed not only in the division of labor and allocation of resource between women and men but also in how "value is given, and power is mobilized."

Thus, the gender relations approach focuses on unequal relations between men and women (Little and Panelli 2003). This is an important aspect of the approach because while understanding gender division of labor is good, it falls short in explaining the social reasons that limit women's access to resources and information, which in turn influence adoption of new technologies and accessing the benefits that they can generate. Gender inequalities which in many cases favor men often limit the resources that women have access to and what they are able or not able to do. For example, gender relations influence access to resources such as land and water sources which are essential for agriculture production and productivity.

This chapter explores how gender matters in several key topics including: (a) conservation and access to genetic resources; (b) breeding and crop improvement; (c) access to seed; (d) managing priority pests and diseases; and (e) marketing, postharvest management, processing and utilization. The chapter will look at how gender relations can influence research processes to draw lessons that can contribute in the design of future interventions that can help reduce the gender gap in agriculture and to ensure agricultural research benefits both men and women. This chapter will examine how gender issues have been considered in key potato research topics.

Conserving and Accessing Genetic Resources

The assumption that guides all conservation efforts is that genetic resources are under threat and need to be safeguarded (Brush 2004) as essential source of foods to sustain healthy diets and a source of genes to supply resistance and functional traits in breeding programs (Jones et al. 2018). Maintaining or conserving diversity in situ is an active and purposeful part of farm management (Brush et al. 1981) where men and women have differentiated roles and responsibilities. The role of women in conserving genetic biodiversity for potato has been well documented in different crops and agroecologies.

Regarding potatoes, studies in this area have been mostly conducted in the Andean region in Latin America. It has been noted that women play a key role in maintaining genetic diversity for potato particularly in selection of seed and varieties, storage of seed and utilization (GRAIN 2000; Sarapura 2013). Women in the Andes Mountains have been significantly involved in conserving genetic diversity particularly through their direct involvement in selecting and preserving different native potato varieties to meet different needs both in terms of culinary characteristics as well as to meet different social rituals and obligations. Sarapura (2013) notes that potato genetic conservation is done with care and tuber seed is stored in special places as the tubers are the nexus of relationships and commitments between the community, nature, cosmos, and deities (Sarapura 2013). Selection and conservation of native potato varieties are mostly done by women who know specific characteristics of potato varieties and have knowledge on how to select them according to different purposes the different potato varieties meet.

Tapia and de la Torre (1998), for example, note that in Peru and Bolivia women act as conservationists preserving native potatoes such as the bitter potato species (*Solanum juzepczukii* and *S. curtilobum)* which can survive temperatures as low as −3 °C and can be freeze-dried into traditional products. Female producers in the highlands and especially in peasant communities do not only play a decisive role in food security (Tapia and de la Torre 2000), but also perform a significant role in seed management and food provision (Tapia and de la Torre 2000; Aguayo and Hinrichs 2015). Women select the seed of native varieties based on the crops' in situ morphological and yield interpretation, culinary quality and crop yield, processing quality, and resistance to diseases, drought, or floods (Tapia and de la Torre 2000). Management of genetic diversity through careful management of combination of varieties enables communities to manage risks, particularly where climate stress is more frequent and intense (De Haan 2009).

Tapia and de la Torre (1998) noted that although some women adopt new improved varieties, they also keep and conserve native potatoes, because genetic diversity increases food security in the Andean highlands. In addition, potato research has been increasingly focusing on how women can be recognized and benefit from their role of conserving biodiversity. For example, Sarapura et al. (2016) illustrate that women involved in the Management Consortium of Native Potato Producers of Junin and Huancavelica in Peru (COGEPAN) took part in a project called Papa Andina, coordinated by the International Potato Center (CIP), which helped linking farmers to markets and created formal market chains for native potatoes. This project empowered women who conserved native potatoes, since they were able to use proceeds from selling native potato to buy land under their own name or be able to rent or sharecrop land to increase agricultural production. Men and women involved in the project were also able to access credit, and open bank accounts compared to counterparts who were not part of this project.

Research on potato genetic biodiversity in the Andes has illustrated that in many instances farmer knowledge on genetic biodiversity cannot be separated from the natural and cultural contexts from which it has emerged, including resources, relationships (kinship), and community relations. The traditional knowledge that women possess has been verbally transmitted from one generation to another, from person to person (mothers to daughters). It is based on the *saber campesino* (spiritual, ecological, geographical knowledge), sense and wisdom. It is entrenched in the Andean cosmovision in which the differences between elusive knowledge and physical things are frequently imprecise and vague. Research initiatives and interventions like the Papa Andina project have illustrated that the role that women potato farmers play in the Andes in safeguarding the traditional information, knowledge, traditions, and practices of producing and reproducing the Andean potatoes cannot be overestimated.

Based on our experience in potato research and review of existing literature, several opportunities and challenges related to conservation of genetic biodiversity and the role of gender relations are illustrated below.

As illustrated in Table, how men and women's work is valued and the resources they control may favor or limit the ability of women to not only participate in conserving potato genetic resources but also benefit from their efforts

Table Gender relations in conservation and access to potato genetic resources

Gender relations aspects	Aspects that favor women's engagement	Aspects that limit women's engagement
Valuing of men and women's work	• Women conserve and control genetic resources • Women's work in conservation is recognized and highly valued	• Women's role in conserving genetic resources is not highly valued by actors in the potato sector
Access to resources, assets and control of benefits	• Women have highly significant local knowledge on genetic conservation of landraces • Women are able to financially benefit from their genetic conservation efforts and be able to decide on how to use their benefits.	• Men control information means and channels • Men control access to land
Gender norms and opportunity structure	• It is women's work to save seed of different varieties and store it • Women have access to local knowledge on genetic conservation	• Men control marketing channels and land which makes it hard for women to benefit from their conservation efforts. • Men have more access to new technical information on potato production since they are targeted by extension.

Projects such as the Papa Andina Initiative integrated this traditional or "emic knowledge" with scientific and technical knowledge to give place to three different areas or spheres of innovation interconnected with the utmost essential and most original opening points for recognizing change in peasant women—(1) technology use, (2) social norm change or social innovation, and (3) economic resilience. For example, while it is acknowledged that women in peasant communities possess an intrinsic adaptive capacity to maintain, manage, and preserve the native potatoes and know how to adapt native potatoes to different climatic conditions, pathogens, and plagues (Sarapura 2013), through the Papa Andina initiative, women were able to strengthen their innovation capabilities as well as their ability to conserve genetic diversity in situ. For example, they were able to access new technical information and knowledge on potato production as well as receive market information using Information and Communication Technology (ICT) tools. Gender relations of power that restrict access to information for women may affect their ability to benefit from their genetic conservation efforts.

The analysis presented above highlights the importance of understanding gender relations in potato genetic resources conservation and use. Furthermore, it illustrates that the involvement of both men and women in the design and implementation of interventions and policies to support conservation should be prioritized.

Crop Improvement

It has been suggested that conventional breeding has not been able to benefit farmers in marginal areas such as in Rwanda because farmer traits are not considered in the breeding process, which leads to relatively low adoption (Muhinyuza et al. 2012). Hence, failing to consider the trait preferences of farmers, especially those in marginal areas, can lead to the promotion of varieties ill-suited to the needs of vulnerable groups such as women. For instance, one of the persistent gender gaps in agriculture is lower adoption of modern varieties among women producers (Ashby and Polar 2019). Overlooking traits important to women farmers and consumers may lead to women's disempowerment and aggravated household food insecurity and poverty (Tufan et al. 2018).

More broadly, the need for the involvement and participation of farmers in the development of new crop varieties for smallholder farmers was explicitly explained by DeVries and Toenniessen (2001). As noted by these authors, farmers should be involved in all aspects of variety development that include priority setting, early generation breeding, variety testing, and selection so that breeders obtain regular input from farmers that enables them to structure their selection indices accurately. Thus, farmers should not be just technology recipients and beneficiaries but actors who influence and provide key inputs to the technology development process (Machida et al. 2014).

The literature on the gender dimension of agricultural production in Africa and elsewhere points to the connection between gender and crop preferences as well as gender-related dynamics and constraints in technology adoption. Indeed, given that women and men have different roles in providing for household food security, it is not surprising that research generally portray that they have different preferences as well (Meinzen-Dick et al. 2010; Tufan et al. 2018). However, until recently the gender dimensions of trait preferences have gone largely unrecognized and unappreciated as a distinct area of research on potato breeding. To address this weakness, potato research in increasingly integrating men and women farmers' views in crop improvement initiatives by collecting sex disaggregated data to identify differences and similarities between men and women's trait preferences. A recent study in Ethiopia, for example, suggested that there were only a few significant differences between men and women's desired traits (Kolech et al. 2015, 2017). They suggested that women in one of the study sites were more concerned with long stolons than men since this was an indicator that the potato could be harvested sequentially, according to needs, and not all at once addressing women's food security concerns, while on the other hand men were more concerned with low soil fertility which lowered yields, and limited market access. In their analysis the authors concluded that men were more concerned with market demand while women concerned more with food security. Gender relations that promote engagement of men in high value market chains while relegating women to the domestic economy with concerns mainly for family food may disadvantage women in the long run. In Kabale, Uganda, one of the potatoes growing districts, it was noted that both men and women traders prefer large sized potato suitable for making French fries (Bonabana-Wabbi et al. 2013). Therefore, understanding what men, women farmers and traders prefer is important for a breeding program.

Breeding Objectives

In line with the growing evidence of the need to integrate gender considerations into crop improvement, a key objective of the potato breeding program at CIP is to characterize gender differentiated preferences for traits, in different agri-food systems, and what the consequences would be of having those traits available to help breeding strategies accelerate varietal development. Research has been conducted in Latin America and Africa to understand gender differences in trait preferences. Research in Peru for example highlighted that men often preferred improved potatoes that are high in yield, resistant to hail and frost and to diseases (particularly late blight), while women often focused on culinary quality for fresh consumption, they prefer potato varieties with shallow "eyes," and as well as yellow/cream flesh color. A study in Ethiopia (Mudege et al. forthcoming) also noted some differences in men and women's trait preferences (see Box 14.1).

Gender Relations and Adoption of Improved Potato Varieties

Due to the different roles that men and women play in families and communities, traits related to consumption exhibits some sharp gender differences. Based on our experience in potato research and review of existing literature, below we summarize examples of how gender relations may shape the opportunity space for adoption of potato varieties.

As Table illustrates, when women's preferred traits are valued and integrated into the breeding program, this may improve women's willingness to adopt new improved varieties. Thus, breeding programs need to go beyond profits to ensure that key important traits that may not have an immediate economic value but are important to women are not neglected.

In line with their gender roles, women prefer traits that lessen their burden and time in food preparation. For example, women in Ethiopia preferred potato that did not have deep eyes because it is easy to peel and prepare local dishes (Mudege et al. forthcoming). In Peru it was noted that women preferred some varieties because they could make soup or different traditional dishes or because they could be used for traditional rituals such as testing the patience of a new bride by asking her to peel a particularly difficult-to-peel potato (Tapia and De la Torre 1998). Some potato varieties are regarded as more nutritious, particularly for pregnant women (Tapia and De la Torre 1998). Preferences of certain varieties for their perceived maternal

Table How gender relations shape the likelihood for adoption and use of improved potato varieties

Gender relations aspects	Aspects that favor women's adoption and use of improved varieties	Aspects that limit women's adoption and use of improved varieties
Valuing of men and women's preferred traits	• Men and women are consulted on preferred traits and these traits are considered in a breeding program	• Gender relations that limit women's access to markets and their ability to benefit from crop marketing may limit their need and ability to adopt improved potato varieties • Traits that women value may be neglected by a breeding program if they are deemed not to have any economic value
Access to resources, assets and control of benefits		• Men control access to land and finances, thus women may not be able to adopt or benefit from improved varieties that demand high inputs such as fertilizers.
Gender norms and opportunity structure	• Women are interested in food security so are likely to adopt improved potato varieties that are high yielding	• Domestic use by women of traditional varieties for important social rituals.

health effects, i.e. more nutritious, is a fundamental revelation that can help breeding programs to better target women by developing varieties with traits that directly benefit them. Thus, to be able to breed potato that farmers can adopt more easily, research is increasingly realizing that we need to understand the socioeconomic and institutional contexts in which farmers operate.

Furthermore, farmers' trait preferences should also be understood in a holistic manner, that not only looks at gender roles but also at the sociocultural environment in which variety and trait decisions are made. When farmers consider agronomic traits such as yield, for instance, they do not do that in a discrete manner. The yield trait is important for commercial and food security purposes, as illustrated by the following example from Ethiopia (Box 14.1):

Box 14.1 : Men and Women's Differences in Potato Trait Preferences in Ethiopia

Gender mainstreamed Participatory Varietal Selection Activities (PVS) in Ethiopia showed that man and woman farmer perspectives need to be integrated to ensure that released varieties meet their needs. Men and women had different preferences in their selection of potato clones. For example, out of five important clones, men and women's preferences matched in the top two selected clones. However, in the top three clones that men and women selected, they preferred resistance to disease and pest attack, high yield as well as tuber sizes that are preferred by the market as criteria. However, while men only selected clones which they perceived as free from pests and diseases, for their second and third best clone, women selected some potato clones that had some insect and pest damage because these clones had a size which was preferred by markets, had a good shape and superficial eyes (shallow eyes) which made processing and peeling easier. For the clone which they selected as second, women stated that the clone had a disadvantage in that it had cracks which increased loss upon processing and did not make good potato stew as this type of cracked potato would disintegrate upon boiling. This study shows that although men and women are interested in marketable traits, women had additional requirements particularly related to processing that men did not have (Mudege et al. forthcoming).

In some instances, the introduction of new improved potato varieties does not require only the promotion of new potato varieties as alternatives to traditional/native or local potatoes, but also involves the construction of different social configurations, it requires new patterns of farming and the enrolment of many different social actors and different ways of interaction. For example, introduction of improved varieties in the Andes also entails training of farmers on new production technologies. New production technologies may have a gender implication. Breeding programs, on their own, cannot change social configurations to be more empowering to women. To achieve this, interventions should be linked to other programs focusing on gender transformation; for example, in terms of access to and control of resources, finance, access to markets, all of which may determine farmer's ability to adopt new improved varieties. New social configurations could mean that breeding programs are linked more to other interventions focusing more on gender transformation in communities. Therefore, gaining support for improved varieties is not a simple process, especially if the new improved varieties are being promoted in contexts where there are already established potato regimes and landscapes. As others have already argued, although end users may not have all information, they are often the experts when it comes to knowledge of the local context (in which new varieties are being introduced), hence it is imperative that their needs and aspirations are taken into considerations in all initiatives that touch on them right from idea conception (Njenga and Gurung 2011). It therefore becomes important to consider farmers' preferences, values, practices, and behaviors because these matter in understanding the complexities surrounding adoption or rejection of new improved potato, and puts farmers' agency and knowledge at the center of analysis.

Additionally, gender differentiated access and control over assets and resources can influence the crop and/or variety selected for production (Njenga and Gurung 2011). In the Andean potato-based production systems, women have less access to labor, have difficult or limited access to farm equipment for land preparation, and face restrictions to access and use inputs such as fertilizers and pest control products. These constraints shape women's preferences for lower yielding native potatoes that have lower market value but also require less inputs and labor (Polar et al. 2017). Studies on gender-differentiated crop trait preferences show evidently that varietal choice is related to access to and control of resources, rights, and responsibilities differentially shared by men and women engaged in production, processing, and marketing (Christinck et al. 2017; Bentley et al. 2018; Ashby and Polar 2019).

How to Integrate Gender Concerns into Breeding and Crop Improvement

To integrate gender concerns into the potato breeding program, CIP is using a two-pronged approach:

The first prong is to ensure that the interest of men and women are taken into account in the setting of breeding objectives. This can be done if in depth research is conducted through analysis of secondary data and/or collection of new data to identify the needs and interests of men and women who are end users and target groups. The second prong is to ensure that men and women groups are engaged in the evaluation of new potato clones either through participatory varietal selection (PVS) or participatory plant breeding (PPB) (Box 14.2). It is felt that these approaches will increase likelihood of adoption of varieties and technologies while at the same time addressing issues of gender equity since the needs of men and women will be addressed by the breeding program. In line with this, CIP has developed and tested a manual for gender mainstreamed participatory varietal selection to collect information on trait preferences as well as to allow men and women farmers to evaluate potato clones before release.

Box 14.2: PVS in the Peruvian Andes

In the mid-1980s, potato breeders in the Instituto Nacional de Innovación Agraria of Peru (INIA) and the International Potato Center (CIP) jointly evaluated advanced potato clones from a diverse late blight-resistant population. These evaluations were conducted in farmers' fields. Three hot spots for late blight were selected in the Department of Huanuco in central Peru. In return for their support, farmers received one-half of the output of the trials. Lastly, the retained seed provided farmers the opportunity to start multiplying and using any clone that fits their circumstances. In the final evaluation, six of the most promising clones from 6 years of on-station selection and 3 years of testing in farmers' fields were selected. By the time one of the selected clones, Canchan-INIAA, was released, dozens of farmers were growing the variety, and a considerable amount of seed had been distributed via the informal seed system. Both men and women were involved in the evaluation of clones. The early adoption of Canchan INIAA before its release was a result of positive evaluation by men in terms of yield, earliness, and resistance to late blight, and by women in terms of the skin color, storability, and consumption quality. Women played a significant role in conserving and managing Canchan INIAA potato variety seeds, while men had a strong role during weeding, hilling, and harvest. It was actually woman farmer- managed Canchan INIAA seed that was eventually used to release the variety (C. Fonseca, personal communication), and the inclusion of both men and women in the selection of Canchan-INIAA, not only because of resistance to late blight, but the quality attributes, may be the reason why it is still a popular variety more than 20 years after its release, and after having lost the resistance attribute.

In addition, it is important that potato gender research contributes to current work on crop ontologies. In this way, end user priorities can be integrated into breeding programs by standardizing farmer priorities while ensuring that the breeding programs are not overwhelmed. The CGIAR Research Program on Roots Tubers and Bananas (RTB) for instance has developed next-generation breeding systems based on the collection and application of genetic, metabolite, and phenotypic data together with participatory, gender-responsive research on farmers' trait preferences, aiming at establishing a connection between preferred traits and the genetics that explain them. Thus, efforts in potato research should continue to contribute to the selection of traits that can be used in genomic prediction, and use of weighted selection indices that aim to ensure that new varieties have wide and gender-equitable impact (RTB 2016). By building up and contributing to sex disaggregated crop ontology database, gender mainstreamed potato research will ensure that potato breeding objectives continue to evolve in ways that are responsive to men and women farmers in different agrifood systems. This effort is being integrated in the definition of more specific breeding product profiles that can take user and gender-differentiated preferences into account.

Managing Priority Pests and Diseases

The adoption of pest and disease management practices is an important topic to reduce production losses in potato. In some regions of the Andes, potato production is associated with heavy use of chemical inputs to manage pests and optimize profits (Mera-Orcés 2001). Similarly, in highland regions of Africa diseases and insect pests have the greatest potential for potato yield reduction, thus farmers rely heavily on pesticide use (Okonya and Kroschel 2015). However, most studies on the topic adopt a limited approach of only looking at gender roles (see Malena 1994) and how differentiated access to land, labor, finance, and education, shape women's technological needs differently (Malena 1994). In addition, writing on gender in the Andes, Paulson (2003) suggests that women are not homogenous as we have widows, married women, young, and single who have different experiences and different needs. Thus instead of focusing on a dichotomy of static gender roles, we should instead focus on the "possibilities for a more dynamic conceptualization of social roles in relation to changing social, economic, and environmental condition" (Paulson 2003). Since women differ by age, socioeconomic status, and other variables, we instead look at gender relations, which help to explain the roles of men and women and the decisions they make in relation to pests and disease management.

Gender Relations and Pest and Disease Management in Potato

Gender relations play a critical role in the management of pests and diseases in potato. Recent research in Malawi (Mudege, unpublished results) illustrates that social relations that privileged men's potato crop over women's crop for spraying meant that women's potato fields were more likely to be affected by late blight compared to men's crops. This in turn affected the availability of quality planting material for women if their crop was diseased, since farmers selected planting material from their ware potato crop for the next cropping season. Additionally, since men in men-headed households were expected to grow potato for the market, while women's potato plots were mostly for family consumption, men controlled the family budget. This had gender implications in pest and disease management. While lack of money to buy chemicals for spraying was mentioned as a key obstacle for both men and women, women were disproportionately more affected as they were only able to buy small quantities of chemicals and also because as mentioned, men's potato crop was privileged when it came to application of chemicals (see Box 14.3). While some women could afford the chemicals, many often mentioned they may not have access to knapsack sprayers that were even more expensive to purchase and difficult to rent from other farmers. Where men and women from the same household cultivated and managed different plots, the men's plot were prioritized when it came to spraying for diseases. Women often mentioned that they could not afford to buy the knapsack sprayer because in families where men and women had separate plots often there was no cooperation, leaving women with a higher burden of taking care of family consumption needs and little to invest in agriculture. Additionally, women often mentioned that because they had no money to buy quality seed, they sometimes purchase diseased seed because it is cheaper.

Box 14.3: Using Pest and Disease Control Methods in Malawi

Women focus group participants in a study conducted by CIP in Malawi debated whether it was easy for women to purchase and use chemicals if their potato crop was affected by disease:

Participant: ...some [women] don't have the money we have seen the whole field being infected by disease but people failing to get money to buy the chemical. (*People arguing, many people speaking at the same time*)

Participant: The chemical is 100 kwacha it is not expensive anyone who want can buy the chemical.

Participant: Let me speak for myself, there was a year when my potato was destroyed because I had no money for chemicals. .... My husband was working but he was not being paid so I thought that if I go and borrow money from somewhere how will I repay the money so the whole field for potatoes was destroyed.

Participant: You could have sold 2kgs of maize and got 100 kwacha.

Facilitator: Your friends are surprised.

Participant: They are surprised because they can afford but I am talking about what happened at my home I am not talking about someone else (Dedza, women farmer group members)

Even if you have money to buy the chemical the problem is we have one [knapsack sprayer] in the house and since it was bought on the husband's budget he takes it with him to the field everyday and you will have nothing to use to spray with until all your potato is destroyed (Ntcheu, women nongroup members).

Source data (Unpublished data Malawi- Integrating gender into RTB research to improve development outcomes project)

Table How gender relations shape engagement of women in pest and disease management

Gender relations aspects	Aspects that favor women's adoption and use of improved pest and disease management options	Aspects that limit women's adoption and use of improved pest and disease management options
Valuing of men and women's crops	• Cheap farmer-based pest and disease management practices • Pest control methods which may be labor-intensive but require very little use of outside inputs	• Men's potato crop is valued more than potato cultivated on women-controlled plots because men's crop is for commercial purposes and women's crop is for domestic use. Men's crops get preferential treatment when it comes to pest and disease management (this is context specific in some African countries but not the same in the Andes)
Access to knowledge and other resources	• Women have access to information about pest and disease control which often is a preserve of men	• Information on pest and disease management is packaged in ways that favor/promote men's access
Gender norms and opportunity structure	• Women are consulted and engaged in decision about crop protection • Cooperative household decision-making on agricultural investments • High outmigration of male labor which leads to feminization of most practices related to pest and disease management	• Men dominate decision-making about adopting and using pest and disease management technologies

Box 14.3 shows that power relations in the household may determine distribution of resources and women's ability to use household resources in controlling pests and diseases in women managed plots. In a different context in the Andes, there is gender differentiated specialization in terms of pest management, which is usually perceived as men's job. Therefore, men have access to information and knowledge on pesticide use and pest control in general, including IPM in the field, while women are more interested in controlling pests in the store, which is more under their control.

Sharma et al. (2017) identified the use of disease-free tubers as seed as one of the key ways of controlling bacterial wilt. Since farmers normally use saved seed from their crop, small holder-friendly seed technologies will benefit women immensely in terms of increased productivity even if the land under production does not expand because they will have lower pests and disease burden.

Based on our experience with potato research and gleaning from some of the literature we reviewed, Table shows how gender relations shape the opportunity space for women potato farmers.

If women are able to gain access to pest control information and methods are responsive to the resources available to them, women will benefit.

In many communities, such as in the Andean region, men-biased sources of information and knowledge on pest and disease management are used. For instance, information and knowledge on pest and disease management accessed through training events or the fact that men have higher literacy and can read labels of commercially available products more easily. Women on the other hand had lower literacy levels and lower command of Spanish, which limited their access to information provided in written form or through capacity building events (Polar et al. 2017). In Uganda, women and men's sources of information also differed. For example, in Eastern Uganda men obtained their information from vendors at local markets or from labels in pesticide packages, while women indicated that extension agents were their most important source of pesticide information (Erbaugh et al. 2003). However, in Uganda like in many other Africa countries extension is often underfunded. The gap in terms of access to information may need to be reduced, for example, by packaging information and presenting it in ways which makes it equally available to men and women, including use of local languages and bringing information and training closer to villages since women may not be as highly mobile as men.

While some studies in Latin America (see Polar et al. 2017; Mera-Orcés 2001) have focused on access to information by men to promote the use of pest and disease management technologies, approaches to gender and pest and disease management in potato need to go beyond just access to knowledge, information, and pest management technologies. Access to training and information in many cases intersect with control over assets and resources and power relations within households which affect use of improved pest and disease management technologies by men and women within households and communities. For example, although both men and women in Malawi had the same knowledge related to spraying of chemicals, women often mentioned that they did not have money to buy chemicals nor access to other resources that were essentially controlled by men, as illustrated in Box 14.3. This case clearly describes how access to assets and resources interplays with gender relations of power to determine the use of pest and disease management technologies. Even if economic resources are available to women, priorities are established based on whose crop is valued, who has access to and control of resources such as spraying equipment, and who makes decisions about use of available resources.

As relations of power, gender relations shape men and women's access to assets and resources. Access to and control of resources in turn may determine the type of potato pest and disease management technologies that men and women can adopt. A cross country study in Bolivia, Ecuador, and Perú on potato-based agricultural systems found that men and women have different perceptions and usage of chemical or organic inputs for pest and disease management. Women often preferred organic inputs and/or management practices because of their low cost, even if they were time-consuming; men on the other hand preferred chemical control because, they regarded chemicals as more effective against potato late blight (Polar et al. 2017). However, in adverse environments, both men and women were inclined to use chemical inputs to reduce the risk of economic losses in potatoes that were produced for markets (Polar et al. 2017; Mera-Orcés 2001). In this case the use of chemical inputs is conditioned by the women's lower access to economic resources, as well as by the purpose why the crop is cultivated: domestic versus market.

The role of women as caregivers influences the distribution of activities linked to potato production. Although pest and disease management in general is perceived as both men and women's responsibility, the actual application of pesticides is more associated with men, while food preparation for field workers is a woman's responsibility (Mera-Orcés 2001 [Ecuador]; Erbaugh et al. 2003 [Uganda]). The examples above show that it is important to analyze both division of labor and decision-making processes related to use and adoption of pest and disease management options.

Hierarchical gender relations that denote men as decision makers at community level may result in women's needs not addressed or considered in programs. Sarapura (2013) reported that in peasant communities in the Andes, decisions on applying a new technique or a new method of cultivating native potatoes had to be approved by the community council. These councils are dominated by men as leaders and number of members. Even though, women are in charge of most of the production processes related to potato, in the case of a disease outbreak, community councils decide on the way forward. For example, in the community of Racracalla in Peru, peasant producers who were concerned about a new disease, late blight, caused by *Phytophthora infestans* and did not have access to formal extension systems, tried to resolve this issue by using their community organizations. For example, they tried to deal with the disease using different methods such as high tilling, construction of canals inside the plots, use of ashes in order to repel the disease. All the solutions were agreed upon by the community councils and tried. Sarapura (2013) identified gender implications in that if extension systems are just directed to the community councils which are men dominated, women may be left behind. In this specific case, decision-making is also controlled at the collective level, where women's perceptions, needs, and limitations may not be adequately considered, since all decision makers are men.

Pest and disease management in the potato crop is a more complex issue and depends on the type of pests or disease, the alternative control methods available, the inputs to be used, the information and knowledge available and the access to external sources of information. There is the need to also differentiate the perceptions of men and women regarding alternatives, and also understand the factors that influence access to and control of resources, so that suitable control methods can be identified during participatory research and used by farmers. This aspect becomes even more critical since climate change is likely to influence the increase in pests and diseases in the potato crop in different parts of the world.

Access to Potato Seed (Seed Flows and Networks)

The lack of disease-free, quality planting material has been mentioned in several studies as one of the key reasons for low potato productivity (Hoque and Sultana 2012; Lutaladio et al. 2009; Gildemacher et al. 2009). For example, in Malawi,

Demo et al. (2008) suggests that lack of quality planting material is a key barrier to improved productivity. In many developed countries, formal systems are the source of quality planting seed for many crops including vegetatively propagated crops such as potato. Although the formal sector dominates the seed systems in developed countries, in developing countries, in spite of huge investments in the sector, "90–95% of the world's small holder farmers still obtain seed from informal sources, largely from other farmers" (Reddy et al. 2007). The situation is particularly dire for vegetatively propagated crops in many Sub-Saharan Africa countries. There has been less focus on tuber crops, legumes, and horticultural crops among the formal seed systems of SSA (Biemond et al. 2012). It had been noted that in many Southern African countries many commercial seed companies are not interested in producing seed for vegetatively propagated crops because of added complicating problems such as low multiplication rate, bulkiness, short shelf life, and difficult maintenance during the dry season.

There has been debate regarding whether formal seed systems or the integration of formal and informal elements are good for potato seed systems as and to understand which are better for women and men farmers in marginal areas. Potato projects are increasingly encouraged to collect sex disaggregated data as well as to conduct studies to understand how gender relations affect the dissemination of new seed technologies. Based on our experience in potato research and the literature we reviewed, below is a table outlining some of the ways gender relations may impact on the efficacy of seed systems for women.

Table gives examples of how gender relations permeate this sector. Approaches to seed certification clearly show how gender relations have an impact on the seed system. For example, some approaches regard formal seed certification

Table How gender relations shape engagement of women in potato seed systems

Gender relations aspects	Aspects that favor women's access and use of quality seed	Aspects that limit women's access to and use of quality seed
Which channels are valued for seed dissemination?	• Improved technologies that promote availability of affordable quality seed • Availability of quality seed in local seed networks and local markets friendly to women	• Unaffordable quality seed • Methods of availing seed that value masculine channels • Formalization of potato seed systems (certification) that dispossess women of their role as seed producers and conservers
Access to knowledge on quality seed and other resources	• Women access not only knowledge but also other resources, mainly credit, so they can run own seed businesses or gain access to quality seed for purchase	• Men dominate decision-making on seed at both household and community level • Men control the resources needed to purchase quality seed
Gender norms and opportunity structure	• Farmer-based quality seed management technics • Gender-responsive farmer-based seed producer groups at community level	

as essential to make available good quality seed to farmers. However, although certification could guarantee good quality seed, evidence from other crops have shown that in some cases women are less able to benefit from certification schemes than men. For example, women may lack the resources needed to have them certified as seed producers, thus are dispossessed from their role as seed producers and keepers. In addition, currently in many countries where CIP is intervening, new improved varieties of potato are available, but taking longer to disseminate, which means technologies that include public–private partnerships for rapid multiplication would benefit both men and women potato farmers especially if they lead to the availability of cheaper, good quality seed. Development of an innovative seed systems model called the "3G approach," which combines rapid multiplication of tuber seed (through aeroponics) with technologies for good farmer seed management in about three generations rather than in the conventional seven have the potential to increase the availability of good quality seed and may also lower the cost of seed thereby improving seed security (Demo et al. 2015).

However, a CIP study in Malawi (Mudege and Demo 2016) shows that commercialization of seed could ensure the availability of seed but may not ensure accessibility of seed. Although both men and women prefer quality seed, women often lack the income to afford clean potato seed. Both men and women expressed a willingness to pay more for good quality, but men could afford to pay much more than women. On one hand, women preferred noncash transactions—such as paying for seed with labor or asking their friends for loans of seed to be returned after harvest. On the other hand, most men said they had opportunities for odd jobs in the community and surrounding areas which they could use to raise money to purchase seed. Women said that they could rely on their friends to get seed to repay after harvest, but it was usually difficult for men to give another man seed to plant (Mudege and Demo 2016). Likewise, in the Andes it has been suggested that because women have lower access to economic resources, they prefer low cost technologies and use local seed which they can access through barter, in kind payments, rituals in festivals and farmer to farmer exchange (Thiele 1999; Tapia and de la Torre 1998; Zimmerer 2003; Sperling and McGuire 2012). As a result of limited access to cash, private sector markets may not increase women's access to quality seed, but local markets because of mechanisms where women can provide labor in exchange for seed. Thus, it is not only important to target multiplication of seed to increase availability of quality seed, but to focus on the mechanisms and pathways that can ensure women's access to seed despite the limitations they face in terms of access and control over other resources.

Furthermore, FAO (2008) acknowledges that commercialization of agriculture, including seed trade, tends to exclude women. Women are usually excluded because they lack the resources which are needed to participate in commercialized systems. Zimmerer (2003) regards women as important in seed flows but men are often engaged in seed dissemination beyond the community in the Andes. Women are key players at procuring seed within the community. Both men and women are engaged in seed flows "at the extra community level"; however, when it comes to procurement of seed from the market and development institutions women's role reduces.

Therefore, suggestions for commercial seed systems need to be evaluated from a gender perspective—in order not to do harm to women (Gibson et al. 2009). In addition, formal markets may not be able to sell fertilizers and other inputs in the smaller quantities that women can afford. Thus, vibrant local markets which are not only cash reliant but based on other forms of reciprocity and relationships within the community would be able to meet women's needs. However, when commercially produced seed is available, the effects have the potential to trickle down when those farmers who can afford to buy clean seed may have relatively clean planting material to sell to other farmers at harvest time.

To accommodate women and other poor marginalized farmers, research is also moving increasingly towards developing farmer friendly seed management technics such as negative and positive selection, which have been promoted to improve the quality of seed in the informal sector. Positive and negative selection are two methods introduced in Malawi to help farmers to access healthier planting material. Positive selection requires marking potato plants as parent stock. Plants chosen must display good growth and most importantly should show no signs of bacterial wilt and/or viruses. Negative selection is selecting plants that will not be used as parent stock. Plants marked are those infected with bacterial wilt and/or viruses (Tantowijoyo and van de Fliert 2006 see also Salazar 1996; Njukeng et al. 2007). Farmers who belong to farmers groups are taught how to identify health plants and select them as parent stock for seed. Farmers are taught how to identify the health plants which are supposed to be "big"; have many and thick stems; have dark green leaves without malformations; have many, large and well-shaped tubers; do not show obvious disease symptoms (Gildemacher et al. 2007, 2011). A study by Gildemacher et al. (2011) in Kenya shows that positive selection provided small holder farmers with better quality seed and led to high yields. However, potato research in Malawi has for example illustrated that women are often left out on training on agronomic practices and thereby lack the knowledge they need to improve their productivity and accessing clean seed (Mudege et al., 2015a, b).

Since positive and negative selection relies on visual inspection, it is not entirely reliable as it needs a farmer to be experienced in spotting the diseased plants (Chiipanthenga et al. 2012). In addition, even when farmers have knowledge, cultural practices and norms can militate against dealing with pests and diseases. For example, in the Andes it was noted that farmers were afraid to rogue diseased plants for fear that if they tampered with food crops they could be punished by the ancestors (Thiele 1999). Very few studies have also discussed what men and women know regarding seed quality and how to maintain it. Given the critical importance of quality seed it is important to know what farmers know about seed quality and how this can affect the fight against potato diseases and pests in the informal seed system. Thus, work has been conducted in different countries, including Malawi and Uganda, to understand what men and women's knowledge and perceptions are regarding seed quality.

Potato research has also looked for innovative ways to link the formal and informal seed systems to ensure accessibility of seed. For example, the Consortium of Potato Producers from Ecuador CONPAPAA initiative in Ecuador sought to produce good quality seed by providing producers (farmers) seed grown through aeroponics and training them to multiply it as quality declared seed (Kromann et al. 2016). This approach managed to make available seed which was less expensive than certified seed. Using this approach allowed women and indigenous farmers to access quality planting material through the merging of formal and informal seed systems than would not have been possible with conventional approaches. In Peru, the informal system satisfies the seed needs of 99% of the potato growers. Seed in the informal sector in Peru is locally available and cheaper than certified seed. This shows that an integrated seed system under certain conditions can outperform standalone formal and informal seed systems. Orrego and Andrade-Piedra (2016) present a case in Peru where quality seed from the formal sector was distributed through both formal and informal channels for further multiplication and dissemination. Having locally available clean and cheap seed is particularly important for women, since they have low access to monetary income. Research on potato in the Andes has already shown that women are often more engaged in sourcing seed within the Andes using family and community networks (Tapia and De la Torre 1998; Zimmerer 2003). Some women's movements in Latin America are building on traditional roles of indigenous women in seed management to positioning themselves as privileged custodians of seeds and biodiversity (Aguayo and Hinrichs 2015).

Given the current challenges to design seed systems that contribute to resilience of farming systems to threats, such as climate change, it is important to understand and take into account both women and men perceptions for seed-related alternatives, particularly, if seed businesses could become an important source for income for women and youth.

Marketing, Postharvest Processing and Utilization

This section will look at how female farmers face gender-specific challenges in relation to potato markets, and the discussion of key findings describes how gender matters in marketing. Urbanization has provided potential markets for potato and potato products due to demand from urban consumers (Bonabana-Wabbi et al. 2013). However, farmers have very limited market information and cannot take advantage of emerging opportunities to meet demand. Bonabana-Wabbi et al. (2013) suggest that because of lack of market information farmers are prone to exploitation by middleman. To address this, farmer collective marketing has often been used to improve the bargaining position of farmers in marketing (Bonabana-Wabbi et al. 2013). However, CIP research on gender and collective marketing in Malawi, illustrated that gender inequalities were often reinforced in marketing groups (Mudege et al. 2015b). For example, in groups, payment was often given to male household heads even when women had submitted the potato for sale in cases where their husbands were also group members. This was a different experience from the COGEPAN in Peru (Sarapura et al. 2016) were women directly received their money and invested it for their own benefit. While in COGEPAN, the approach

Table How gender relations shape engagement of women in potato markets

Gender relations aspects	Aspects that favor women's participation in markets and their ability to benefit	Aspects that limit women's participation in markets and their ability to benefit
Power relationship between male household heads and women	• Approaches that allow women to have direct access to markets and benefits	• Ideologies that regard male household heads as the official representatives of the family and therefore in charge of markets and marketing proceeds • Men control all decisions related to marketing.
Access to markets	• Women participate in case of low potato quantities	• In men-headed households, high volumes of potato sold are usually decided by men
Gender norms and opportunity structure	• Training on business management and market access targeting women	• Men make decision on how much to sell, where to sell and whom to sell to • Lack of mobility for women restricts access to high value markets

challenged gender relations of power, in the Malawi case, collective marketing reinforced existing gender inequalities.

In Kabale, Uganda, it was noted that men and women were equally engaged in potato trading (Bonabana-Wabbi et al. 2013). However, studies often rank types of trade in order of how many people are engaged without looking at the division of participation by sex to ensure that targeted gender responsive interventions are made. For example, Bonabana-Wabbi et al. (2013) reported that in some districts in Uganda the majority of those engaged in potato trade where in retail compared to wholesale followed by collectors, agents, and transporters. However, her study and other studies do not segment these value chain actors from a gender perspective. For example, it is not clear how many men and women are engaged at these levels and what profit margins accrue to men and women at these different nodes (Box 14.4). Research on potato marketing for example could collect this information since it is critical to know who benefits from potato technologies including improved varieties

Potato research has also looked at the gender-related constraints to marketing. For example, it was noted that in Uganda men sold more potato than women because they had contact with buyers and brokers whom they meet at local markets (Sebatta et al. 2014). In order to gain insight on the lack of participation by women in potato commercialization, the structures influencing men and women's participation need to be understood as well. For example, Jibat et al. (2007) notes that in Ethiopia women are engaged in potato markets when selling at low quantities and in local markets while men dominated bulk sales as well as sales at markets far from the village. However, the potato story is not all doom and gloom since it is possible to conduct research and implement interventions that are gender responsive as the example below shows (Box 14.4):

Box 14.4: Markets and Inclusive Value Chains

Research in potato marketing in Uganda revealed that both men and women mentioned that they had limited access to markets and to timely information on potato market prices. Women also suggested that because on unequal relationships within households, men often decided on who to sell to, how much potato to sell and where to sell as well as deciding on use of potato income without necessarily consulting women. In addition, it was clear that the market itself was structured in ways that did not favor women's participation. For example, gender norms that designated potato as a men's crop meant that women who tried to sell potato on their own without their husbands were viewed with suspicion while husbands could sell crops on their own without their wives. Lack of mobility for women, poor transportation system and infrastructure, selling potato in large bags that women could not handle, and distance to markets were all mentioned by women as barriers to participating in potato marketing. To address some of these issues within the remit of the project, both men and women farmers and traders were targeted with training of trainers on business skills including marketing and net profit and loss calculation. Sixty nine men and 34 women were trained in marketing to ensure that they understood what marketing was, demand and supply forces, customer analysis and customer feedback mechanism, strategies, segmentation, product differentiation, and marketing information in the context of potato business. Effort was made to ensure involvement of women in this training even if they were not involved in management committees of associations. Since both men and women mentioned lack of market information as limiting their ability to negotiate with buyers, the project through its partner Self Help Africa (SHA) developed mechanisms for disseminating market-related information such as potato prices in Kampala through text messaging to group leaders who would then disseminate this information to other group members to ensure that farmers negotiate from an informed position. (Summary of our experiences with the ENDURE project in Uganda).

In Uganda (Mudege et al. 2016) notes that men mentioned lack of knowledge on what the market wants and also on price intelligence. Although men participated more in markets than women, both men and women did not have enough market skills or adequate market information. For example, women in Wanale mentioned not even knowing where exactly their husbands sell the potato, whilst men mentioned that they were often told that their potato was poor quality and also that they were not sure about the prices their potato would reach. Lack of engagement by women in markets was also noted elsewhere by Mudege et al. (2015b), who showed that in Malawi wherever husband and wife made joint decisions over use of income from potato, money was used to buy seed and fertilizer and other equipment; whereas when male household heads made decisions on their own, women mentioned that they often did not benefit from potato income. Thus, potato research and interventions should not just focus on technical aspect of potato production but also include social aspects such as promoting joint decision-making to ensure that both men and women benefit. In addition, methods to improve access to markets and bargaining power, for example, the use of Information and Communication Technologies (ICT) to inform women and men about market prices or weather forecasts could increase the ability of women and men to benefit from potato markets. In reference to de facto female- headed cotton farming households, Horrell and Krishnan (2007) state that "even without additional resources, greater profitability could be achieved from their existing agricultural output through access to better selling networks and buying consortia for inputs." The same can also be true for women potato farmers, if social norms that prevent them from marketing potato or deciding on household expenditure are challenged, women may be able to invest more into potato production resulting in better quality crop and better storage. Based on our experience and literature reviewed, Table shows examples on how gender relations may influence the ability of women to engage and benefit from potato marketing.

Power relations between men and women heads of households and the position of women in the community relative to the position of men may shape the opportunity space for women to not only engage in marketing but also benefit from it.

In other geographical and social setups, such as in the Andes, research has shown that women dominate potato markets including price negotiation and control of income (Amaya and Alwang 2012), because men regard women as better negotiators. Additionally, Amaya notes that traders in the market are mostly women, and men regard it as undignified to argue with women when bargaining for a better price, thus let their wives sell potato. However, it is noted that women increasingly need information to participate in regional markets although marketing decisions are made jointly by husbands and wives. It was also noted that men monopolize the use of cellular phones to get information on markets (Amaya and Alwang 2012). Therefore, the cellular phone has not fundamentally changed gender roles: market decisions continue to be jointly made, and men continue to control access to market information.

Postharvest Utilization

Gender relations also influence women's access to postharvest technologies. In many parts of the world women are responsible for postharvest activities at household level. Knowledge and skills are passed on dynamically from generation to generation and these actions have long subsisted outside public and private sectors, R&D and agricultural extension systems (Tapia and de la Torre 2000), and oftentimes when technologies are being designed, the innate skills that men and women farmers have and the roles they play are not taken into account. However, since postharvest technologies have been developed and disseminated, there are concerns that men take over the latest technology and women can be left behind. A study in Bolivia (Polar et al. 2017) illustrates that women are the ones who are mostly engaged in grading potato by size. However, when technologies to mechanise potato selection were tested, only men were engaged in validation meetings. These technologies were later introduced but were not adopted by the women who were supposed to benefit. Women were not involved in evaluating this technology and found it difficult to use because the machine was tall and needed substantial physical strength to manually load heavy bags of potato. The machines were later adapted to meet women's needs, and their use helped reduce the amount of time women devoted to selection of tubers for the market. Ogunlana (2004) also suggests that women farmers can easily adopt innovations that can enhance their economic status if constraints pertaining to access to the technology (e.g. information and ease of utilization) are taken into consideration. Box 14.5 highlights a case where gender-related concerns were integrated during the introduction of ambient stores for storing ware potato in Uganda:

Box 14.5: Introducing Ambient Stores for Storing Ware Potato

A potato postharvest project implemented by the International Potato Center (CIP) in collaboration with the CGIAR Research Program on Roots, Tubers and Bananas (RTB) in Eastern Uganda to introduce ambient stores for storing ware potato under ambient conditions was rolled out through farmer and trader associations. Evidence from the associations demonstrated that women were underrepresented in leadership positions and almost nonexistent in storage management committees of the four associations.

Women expressed concern that if only men hold leadership positions in the management of the store, women may not be able to benefit from and to fully utilize the stores. For example, they noted that training targeted group leaders who were often men. Women insisted that for them to benefit from the stores, there should be gender balance in the people selected to manage the store. And although women had expressed the will to be active in store management, personnel from an NGO who facilitated the process of selecting store management committees revealed that most women refused to occupy these positions when elected. Some of the reasons women used to explain why they refused to occupy leadership positions included lack of time to commit for such duties, fear that their husbands would not allow them, meetings times may not be conducive for women, and some women regarded their illiteracy as a limiting factor.

While recognizing the need to have women represented in the group management committees, women also mentioned the risk of women being given token positions where they would not be involved in decision-making, stating for instance that women could be designated to deputy or committee member positions which did not have much influence in terms of decision-making. Some associations noted that since men had been the original members of the associations they dominated positions and it was difficult for women to break into top leadership.

The potato team worked with partners such as Self Help Africa (SHA) as well as farmer and traders' associations to review association management rules for gender inclusiveness. Where it was not possible, the team made a concerted effort to ensure that even if they were not in leadership positions, women representatives were also trained on store management to ensure that they are knowledgeable and actively involved. Under this project, 119 farmers and traders were trained on structure and governance (41 women and 78 men), 102 farmers (32 women and 70 men) were trained on enterprise analysis focusing on cost benefit analysis. Additionally a total of 106 farmers (69 male and 37 female) were trained on business planning in order to equip participants with skills to develop and use business plans to maximize profitability. A total of 111 (82 male and 29 female) were also trained on record keeping and store management. The aim of the module was to equip participants, in particular the store management committees, with skills to manage the ambient stores effectively and ultimately develop store management guidelines and records (Mudege and Mayanja 2016).

The case described in Box 14.5 illustrates that in some cases it is important to be gender intentional when designing potato interventions and research, otherwise researchers run the risk of not including women. It is important to identify the constraints that women face, so that these can be taken into consideration and addressed during the design of research and interventions.

Access to Extension and Training

Research has shown that women in most potato-growing areas have very limited access to training (Dersseh et al. 2016; Mudege et al. 2015a; Polar et al. 2017). For example, when women and other poor households are not targeted with training on new improved potato varieties, they did not regard lack of training as a key constraint to production (Dersseh et al. 2016). While the cause and effect relationship is unclear, it may be because oftentimes, women lack information about the importance of access to training and of improved varieties, and therefore they do not engage in training to the extent they should. Based on our experience and literature review, Table gives some examples on how gender relations shape women's engagement and access to training and information.

If training is organized where it is physically accessible to women, women are directly invited to participate, and training methodologies are taking cognizance of women and men's capabilities, women will be able to access the information they

Table How gender relations shape engagement of women in access to and benefit from potato-related extension services

Gender relations aspects	Aspects that favor women's access to and benefit from extension services	Aspects that limit women's access to and benefit from extension services
Valuing of men and women's access to extension	• Training venue and time selected according to women's physical mobility and time availability • Use of female extension workers • Using appropriate language and training methods	• Women and poor households not targeted on invited for training • Training targets men as heads of households and decision makers • Training targets group leaders. Men usually lead farmer groups
Gender norms and opportunity structure	• Access to information, technology, and resources to implement knowledge and skills acquired from training	• Men may make household decisions on who will attend training

need. However, whether women can use the knowledge and information they gain may depend on gender and decision-making power within households.

Research has shown that in East Wollega and West Shewa Zones in Ethiopia, even though the gender division of labor regards potato as a woman's crop, it was mostly men who participated in training on potato production, regardless the roles they played in its cultivation (Jibat et al. 2007). There are similar findings in the Andes where women play a key role in potato farming and management, yet they have low participation in training events compared to men (Polar et al. 2017). On the other hand, where potato cultivation is regarded as a men's activity women's contributions are overlooked (Mera-Orcés 2001) and women are denied access to the resources and training they need. In Malawi, potato research has also illustrated that women are often left out of training which further reinforces gender stereotypes that women know nothing (Mudege et al. 2015a).

The potato seed system intervention in Malawi relies on farmer participation both as group members and as lead farmers in the knowledge cascade approach. Men and women were targeted with training as part of groups. However, Mudege et al. (2015a) found that unless properly managed and monitored, delivery mechanisms that depend on participatory activities can be gender-blind. Interventions based on participation can become gender-blind if they do not recognize the differences between men and women. Gender-blind approaches "make assumptions, which lead to a bias in favor of existing gender relations … gender blind policies tend to exclude women" (March et al. 1999). For example, Mudege et al. (2015a) notes that farmer group training mostly targeted group leaders who were frequently men. If women do not have access to same extension services as men, their productivity and incomes may be limited. A study of factors influencing potential adoption of technology in potato-based systems in the Andes found that women had limited exposure to innovations through capacity building because they lacked skills in the official languages, had lower literacy levels, and the spaces where technical information was provided were dominated by men (Polar et al. 2017). Thus, to enhance women's access to and benefit from extension services, services need to be accessible to women in terms of language used, methodological tools, schedule times, and delivery spaces.

Conclusions

The analysis presented in this chapter discusses how gender relations shape the opportunity space for men and women potato farmers along the potato value chain. Power to make decisions and act on them is important and influences the ability of men and women to participate along the potato value chain and benefit from it. The analysis shows that lack of access to and control over assets and resources and decision-making can restrict women's engagement in potato production, marketing, and utilization. Institutional innovations and more gender responsive programming that consider the opportunities and constraints of men and women can contribute to equitable development of the sector. How men and women efforts are valued and the resources they control do not only affect women's ability to participate in the potato sector but also shapes their participation and ability to reap benefits from the sector at the same level as men.

Gender research related to potato production, use, and commercialization has in many cases attempted to show the role and importance of women in different potato-related activities from production to market, for example, as laborers or as custodians of genetic diversity and local knowledge. While most of this research has been concerned for example in bringing to the forefront hitherto hidden women's contributions to potato farming, it has also succeeded in developing initiatives to address both men and women's needs depending on the roles they play and the needs they have. However, research has also shown the folly of limiting gender analysis just to gender roles and access to resources and assets. This is so because even where women are the ones engaged in certain roles, they may fail to benefit from their labor because they defer to male households' heads to make decisions. Women may also not access training even for tasks they are engaged in because recruitment may favor male household heads to attend training even if they are not engaged in potato activities. As a result, it is important for potato research to also investigate gender relations and how the valuing of men and women's labor and contributions determine their ability to engage in potato production, management, and marketing. This chapter has illustrated the importance of going beyond just understanding gender roles and access to resources and to also understand gender power relations within households and communities since these determine whether women and men are able to equitably benefit from the products of research.

Non-pecuniary benefits of cultivating certain varieties of potato may influence adoption. Non-pecuniary benefits are those nonmonetary benefits related to social and ritual needs in communities where potato is part of the culture such as in the Andes. These types of benefits need to be understood as part of the value proposition to ensure that released improved varieties meet man and woman farmer's needs. Breeding programs should ensure that key important traits that may not have an immediate economic value but are important to women are not neglected. Farmers may continue to cultivate varieties that have lost some of their key attributes because these varieties may meet some needs which are not readily quantifiable. This means gender work in potato breeding should continue to contribute to knowledge on men and women user preferred traits that can be standardized and integrated into breeding programs to ensure that they continue to be responsive to farmers' needs. Failure to do so may jeopardize farmers' ability to benefit from genetic gains resulting from new improved varieties.

A supportive social and institutional environment is needed to promote adoption of improved varieties and other potato-related technologies by men and women farmers. For example, access to resources such as land, cash, and decision-making power influences whether men and women can use and adopt new technologies and methods. Having knowledge only may not result in adoption of technologies in the absence of a supportive institutional and social environment. For instance, if extension services are gender biased women will not be able to benefit from them. However, having access to information is itself not enough condition of benefitting from potato research. Unequal gender relations and other relations of inequality may prevent men and women from adopting technologies that can help them.

Commercialization and commoditization of seed runs the risk of dispossessing women of their control of seed and may also jeopardize their ability to access seed. Seed is often part of the social fabric where women may gain status or capital by distributing seed freely in their communities. Although noncash transactions such as women working for seed are often not valued by policy makers who promote commercialization and certification, they often ensure redistributive potential within communities where women and other poor and vulnerable groups are able to access seed. More research and investment may be needed to improve farmer-based seed to ensure circulation of better-quality planting material in communities. This also will be able to address the needs of women and other vulnerable groups from poor communities who prefer culturally recognized noncash transactions for accessing seed.

Certainly, more research on gender and potato still needs to be conducted because most of the research that has been conducted so far is at a relatively small scale. More large-scale research including surveys need to be conducted on a variety of topics to ensure that results are more generalizable and lead to the development of gender integration approaches suited to various regions and countries.

References

Amaya ARU, Alwang J (2012) Women rule: potato markets, cellular phones and access to information in the Bolivian highlands. Agric Econ 43:403–413

Ashby J, Polar V (2019) The implications of gender for modern approaches to crop improvement and plant breeding. In: Sachs C (ed) Gender, agriculture and agrarian transformation. Routledge, London

Aguayo EC, Hinrichs JS (2015) Curadoras de semillas: entre empoderamiento y esencialismo estratégico. Revista Estudos Feministas 23(2):347–370

Bentley JW, Andrade-Piedra JL, Demo P, Dzomeku B, Jacobsen K, Kikulwe E, Kromann P et al (2018) Understanding root, tuber, and banana seed systems and coordination breakdown: a multi-stakeholder framework. J Crop Improve. https://doi.org/10.1080/15427528.2018.1476998

Biemond CP, Stomph TJ, Kamaraa A, Abdoulaye T, Hearne S, Struik PC (2012) Are investments in an informal seed system for cowpea a worthwhile endeavour? Int J Plant Product 6(3):1735–80430

Bonabana-Wabbi J, Ayo S, Mugonola B, Taylor DB, Kirinya J, Tenywa M (2013) The performance of potato markets in South Western Uganda. J Dev Agric Econ 5(6):225–235

Brush SB (2004) Farmers' bounty: locating crop diversity in the contemporary world. Yale University Press, New Haven. https://www.jstor.org/stable/j.ctt1np9rd

Brush SB, Carney HJ, Humán Z (1981) Dynamics of Andean potato agriculture. Econ Bot 35(1):70–88. https://doi.org/10.1007/BF02859217

Carr ER (2008) Men's crops and women's crops: the importance of gender to the understanding of agricultural and development outcomes in Ghana's Central Region. World Dev 36(5):900–915

Chiipanthenga M, Maliro M, Demo P, Njoloma J (2012) Potential of aeroponics system in the production of quality potato (Solanum tuberosum l.) seed in developing countries. Afr J Biotechnol 11(17):3993–3999. http://www.academicjournals.org/AJB

Christinck A, Eva Q, Fred R, Ashby JA (2017) Gender Differentiation of Farmer Preferences for Varietal Traits in Crop Improvement: Evidence and Issues. Working Paper. https://cgspace.cgiar.org/handle/10947/4660

De Haan, S. (2009). Potato diversity at height. Multiple dimensions of farmer-driven in situ conservation in the Andes. Thesis. Wageningen University, Wageningen.

De Vries J, Toenniessen G (2001) Securing the harvest: biotechnology, breeding and seed systems for African crops. The Rockefeller Foundation, CABI Publishing, New York

Demo P, Mwenye OJ, Pankomera P, Chiipanthega M (2008) Investigation of Appropriate Fertilizer Doses for Potato Production Using Different Planting Spacing in Major Growing Areas of Malawi. http://www.cabi.org/gara/FullTextPDF/2008/20083323853.pdf

Demo P, Lemaga B, Kakuhenzire R, Schulz S, Borus D, Barker I, Woldegiorgis G, Parker M, Schulte-Geldermann E (2015) 1 Strategies to Improve Seed Potato Quality and Supply in Sub-Saharan Africa: Experience from Interventions in Five Countries

Dersseh WM, Gebresilase YT, Schulte RPO, Struik PC (2016) The analysis of potato farming systems in Chencha, Ethiopia: input, output and constraints. Am J Potato Res 93:436–447

Erbaugh JM, Donnermeyer J, Amujal M, Kyamanywa S (2003) The Role of Women in Pest Management Decision Making in Eastern Uganda 10(3) Journal of International Agricultural and Extension Education. http://citeseerx.ist.psu.edu/viewdoc/download?doi=10.1.1.577.5951&rep=rep1&type=pdf

FAO (2008) The state of food and agricultures: biofuels, prospects risks and opportunities. FAO, Rome. ftp://ftp.fao.org/docrep/fao/011/i0100e/i0100e.pdf

German L, Taye H (2008) A framework for evaluating effectiveness and inclusiveness of collective action in watershed management. J Int Dev 20(1):99–116

Gibson RW, Mwanga ROM, Namanda S, Jeremiah SC, Barker I (2009) Review of sweetpotato seed systems in East and Southern Africa. Integrated crop management working paper 2009-1. CIP, Lima, p 48. http://cipotato.org/wp-content/uploads/2014/08/004730.pdf

Gildemacher P, Demo P, Kinyae P, Nyongesa M, Mundia P (2007) Selecting the best plants to improve seed potato. LEISA Mag 23(2):10–11

Gildemacher PR, Demo P, Barker I, Kaguongo W, Woldegiorgis G, Wagoire WW, Wakahiu M, Leeuwis C, Struik PC (2009) A description of seed potato systems in Kenya, Uganda and Ethiopia. Am J Potato Res 86:373–382

Gildemacher P, Schulte-Geldermann E, Borus D, Demo P, Kinyae P, Mundia P, Struik P (2011) Seed potato quality improvement through positive selection by smallholder farmers in Kenya. Potato Res 54:253–266

GRAIN (2000) Potato: a fragile gift from the Andes GRAIN|15 September 2000|Seedling-September 2000
Horrell S, Krishnan P (2007) Poverty and productivity in female-headed households in Zimbabwe, Journal of Development Studies, Taylor & Francis Journals, 43(8):1351–1380
Hoque AM, Sultana MS (2012) Disease free seed potato production through seed plot technique at farmers' level in Bangladesh. J Plant Protect Sci 4(2):51–56
Jibat GA, Belisa M, Gudeta H (2007) Promotion of participatory technology in potato farming–Ethiopia. Chapter 2. In: Flintan F, Tedla S (eds) Natural resource management: the impact of gender and social issues. OSSEREA & IDRC, Addis Ababa, pp 19–54
Jones AD, Creed-Kanashiro H, Zimmerer KS, de Haan S, Carrasco M, Meza K, Cruz-Garcia GS et al (2018) Farm-level agricultural biodiversity in the Peruvian Andes is associated with greater odds of women achieving a minimally diverse and micronutrient adequate diet. J Nutr 148(10):1625–1637. https://doi.org/10.1093/jn/nxy166
Kabeer N, Subrahmanian R (1996) Institutions, relations and outcomes. Framework and tools for gender aware planning. IDS, Brighton. https://www.ids.ac.uk/files/Dp357.pdf
Kingiri A (2010) Gender and agricultural innovation revisiting the debate through an innovation systems perspective. Discussion Paper 06. Research Into Use (RIU), Department for International Development (DFID), UK
Kolech SA, Halseth D, Perry K, De Jong W, Tiruneh MF, Wolfe D (2015) Identification of farmer priorities in potato production through participatory variety selection. Am J Potato Res 92:648–661
Kolech SA, De Jong W, Perry K, Halseth D, Tirineh MF (2017) Participatory variety selection: a tool to understand farmers' potato variety selection criteria. Open Agric 2:453–463
Kromann P, Montesdeoca F, Andrade-Piedra J (2016) Integrating formal and informal. In: Andrade-Piedra J, Bentley J, Almekinders C, Jacobsen K, Walsh S, Thiele G (eds) Case studies of roots, tubers and bananas seed systems. CGIAR Research Program on Roots Tubers and Bananas (RTB), Lima, pp 14–32. RTB working paper no. 2016-3. ISSN 2309-6586
Laub R, Muir G (2008) Potato and gender-international year of the potato 2008. FAO factsheets_ Gender. Hidden Treasure (blog). http://www.fao.org/potato-2008/en/potato/gender.html
Little J, Panelli R (2003) Gender research in rural geography. Gend Place Cult 10(3):281–289. https://doi.org/10.1080/0966369032000114046
Lutaladio N, Ortiz O, Haverkort A, Caldiz D (2009) Sustainable potato production; guidelines for developing countries. FAO, Rome
Machida L, Derera J, Tongoona P, Langyintuo A, Mac Robert J (2014) Exploration of farmers' preferences and perceptions of maize varieties: implications on development and adoption of Quality Protein Maize (QPM) varieties in Zimbabwe. J Sustain Dev 7(2):194–207
Malena C (1994) Gender issues in integrated pest management in African agriculture. NRI Socio-economic Series 5. Chatham, United Kingdom: Natural Resources Institute. http://www.nzdl.org/gsdlmod?e=d-00000-00---off-0hdl--00-0----0-10-0---0---0direct-10---4-------0-1l--11-en-50---20-about---00-0-1-00-0--4----0-0-11-10-0utfZz-8-00&a=d&cl=CL1.7&d=HASH01c963ec65781a15133c31bf.2
March C, Smyth I, Mukhopadhyay M (1999) a guide to gender analysis framework. Oxfam, GB, Oxford
Meinzen-Dick R, Quisumbing A, Behrman J, Biermayr-Jenzano P, Wilde V, Noordeloos M, Ragasa C, Beintema N (2010) Engendering Agricultural Research, IFPRI discussion paper 00973
Mera-Orcés V (2001) Paying for survival with health: potato production practices, pesticide use and gender concerns in the Ecuadorian highlands. J Agric Educ Ext 8(1):31–40. https://doi.org/10.1080/13892240185300061
Mudege NN, Demo P (2016) Seed potato in Malawi: not enough to go around. Chapter 10. In: Andrade-Piedra J, Bentley J, Almekinders C, Jacobsen K, Walsh S, Thiele G (eds) Case studies of roots, tubers and bananas seed systems. CGIAR Research Program on Roots, Tubers and Bananas (RTB), Lima, pp 146–163. RTB working paper no. 2016-3. ISSN 2309-6586
Mudege NN, Tafadzwa C, Ted N, Eliya K, Paul D (2015a) Gender norms and access to extension services and training among potato farmers in Dedza and Ntcheu in Malawi. J Agric Educ Ext

Compet Rural Innov Transform 22(3):291–305. http://www.tandfonline.com/doi/abs/10.1080/1389224X.2015.1038282
Mudege NN, Kapalasa E, Chevo T, Nyekanyeka T, Demo P (2015b) Gender norms and the marketing of seeds and ware potatoes in Malawi. J Gender Agric Food Secur 1(2):18–41. http://www.agrigender.net/views/marketing-of-seeds-and-ware-potatoes-in-Malawi-JGAFS-122015-2.php
Mudege NN, Mayanja S, Naziri D (2016) Technical Report: Gender situational analysis of the potato value chain in Eastern Uganda and strategies for gender equity in postharvest innovations. CRP RTB. 53 p
Mudege NN, Biazin BT, Brouwer R (forthcoming) Gender situational analysis of the sweetpotato value chain in selected districts in Sidama and Gedeo Zones, Southern Ethiopia. International Potato Center, Lima, Peru
Muhinyuza JB, Shimelis H, Melis R, Sibiya J, Nzaramba MN (2012) Participatory assessment of potato production constraints and trait preferences in potato cultivar development in Rwanda. Int J Dev Sustain 1(2):358–380
Njenga M, Gurung J (2011) Enhancing gender responsiveness in putting nitrogen to work for smallholder farmers in Africa (n2africa). Women Organising for Change in Agriculture and NRM
Njukeng AP, Chewaching MG, Chofong G, Demo P, Sakwe P, Njualem D (2007) Determination of virus-free potato planting materials by positive selection and screening of tuners from seed stores in the Western Highlands of Cameroon. Afr Crop Sci Conf Proc 8:809–815
Ogunlana EA (2004) The technology adoption behavior of women farmers: The case of alley farming in Nigeria. Renewable Agriculture and Food Systems 19(1):57–65
Okonya JS, Kroschel J (2015) A Cross-Sectional Study of Pesticide Use and Knowledge of Smallholder Potato Farmers in Uganda Biomed Res Int
Orrego R, Andrade-Pedra J (2016) Aeroponic seed and native potatoes in Peru. In: Andrade-Piedra J, Bentley J, Almekinders C, Jacobsen K, Walsh S, Thiele G (eds) Case studies of roots, tubers and bananas seed systems. CGIAR research program on Roots, Tubers and Bananas (RTB), Lima, pp 33–46. RTB working paper no. 2016-3. ISSN 2309-6586
Paulson S (2003) Gendered practices and landscapes in the Andes: the shape of asymmetrical exchanges. Hum Organ 62(3):242–254
Petesch P, Bullock R, Feldman S, Badstue L, Rietveld A, Kamanzi A, Tegbaru A, Yila J (2018) Local normative climate shaping agency and agricultural livelihoods in Sub-Saharan Africa. J Gender Agric Food Secur 3(1):23
Polar V, Babini C, Velasco P, Fonseca C (2017) La technología no es neutral: Factores que influyen en la potencial adopción de technología agrícola por hombres y mujeres. International Potato Center, Lima
Quisumbing AR, Pandolfelli L (2010) Promising approaches to address the needs of poor female farmers: resources. Constrain Interv World Dev 38(4):581–592
Quisumbing AR, Meinzen-Dick R, Raney TL, Croppenstedt A, Behrman JA, Peterman A (2014) Gender in agriculture: closing the knowledge gap. Springer, New York
Reddy RC, Tonapi VA, Bezkorowajnyj PG, Navi SS, Seetharama N (2007) Seed system innovations in the semi-arid tropics of Andhra Pradesh. Research report. International Crop Research Institute for Semi Arid Tropics, International Livestock Research Institute (ILRI), ICRISAT, Patancheru, 224 pp. isbn:978-92-9066-502-1
RTB (2016) RTB proposal 2017-2022, vol 1. RTB, Volucella
Sah U, Kumar S, Pandey NK (2007) Gender analysis of potato cultivation in Meghalaya. Potato J 34(3–4):235–238
Salazar L (1996) Potato viruses and their control. International Potato Center (CIP), Lima
Sarapura S (2013). Gender and agricultural innovation in peasant production of native potatoes in the Central Andes of Peru. PhD Thesis, University of Guelph, Canada. p. 351.
Sarapura ES, Hambly-Odame H, Thiele G (2016) Gender and innovation in Peru's native potato market chains. In: Transforming gender and food systems in the Global South. IDRC, Taylor and Francis, Canada

Sebatta C, Mugisha J, Katungi E, Kashaaru A, Kyomugisha H (2014) Smallholder farmers' decision and level of participation in the potato market in Uganda. Mod Econ 5:895–906
Sharma K, Shawkat B, Miethbauer T, Schulte-Geldermann E (2017) Strategies for bacterial wilt (Ralstonia solanacearum) management in potato field: Farmers' guide. International Potato Center. Lima (Peru). 2 p
Sperling L, McGuire S (2012) Fatal gaps in seed security strategy. Food Security 4(4):569–579
Tantowijoyo W, van de Fliert E (2006) In: Widagdo H, Ketelaar JW (eds) All about potatoes. An ecological guide to potato integrated crop management. FAO Regional Office for Asia and the Pacific, Bangkok. https://research.cip.cgiar.org/typo3/web/fileadmin/icmtoolbox/ICM_Toolbox/Integrated_crop_management/All_about_potatoes_-_complete_EN_0602.pdf
Tapia ME, de la Torre A (1998) Women farmers and Andean seeds. Gender and genetic resource management, FAO, Lima. http://www.bioversityinternational.org/uploads/tx_news/Women_farmers_and_Andean_seeds_308.pdf
Tapia ME, de la Torre A (2000) La mujer campesina y las semillas andinas. FAO and IPGRI, Lima
Thiele G (1999) Informal potato seed systems in the Andes: why are they important and what should we do with them? World Dev 21(1):83–99
Tufan HA, Stefania G, Catherine M (2018) State of the Knowledge for Gender in Breeding: Case Studies for Practitioners. Working Paper, https://cgspace.cgiar.org/handle/10568/92819
Zimmerer KS (2003) Geographies of seed networks for food plants (potato, ulluco) and approaches to agrobiodiversity conservation in the Andean countries. Soc Nat Resour 16(7):583–601. https://doi.org/10.1080/08941920309185

4

Global Food Security, Contributions from Sustainable Potato Agri-Food Systems

André Devaux, Jean-Pierre Goffart, Athanasios Petsakos, Peter Kromann, Marcel Gatto, Julius Okello, Victor Suarez, and Guy Hareau

Abstract In the coming decades, feeding the expanded global population nutritiously and sustainably will require substantial improvements to the global food system worldwide. The main challenge will be to produce more food with the same or fewer resources. Food security has four dimensions: food availability, food access, food use and quality, and food stability. Among several other food sources, the potato crop is one that can help match all these requirements worldwide due to its highly diverse distribution pattern, and its current cultivation and demand, particularly in developing countries with high levels of poverty, hunger, and malnutrition. After an overview of the current situation of global hunger, food security, and agricultural growth, followed by a review of the importance of the potato in the

A. Devaux (✉)
International Potato Center, Quito, Ecuador
e-mail: a.devaux@cgiar.org

J.-P. Goffart
Walloon Agricultural Research Center, Gembloux, Belgium
e-mail: j.goffart@cra.wallonie.be

A. Petsakos
Formerly CIP, Seville, Spain

P. Kromann
International Potato Center, Nairobi, Kenya
e-mail: p.kromann@cgiar.org

M. Gatto
International Potato Center, Hanoi, Vietnam
e-mail: m.gatto@cgiar.org

J. Okello
International Potato Center, Kampala, Uganda
e-mail: j.okello@cgiar.org

V. Suarez · G. Hareau
International Potato Center, Lima, Peru
e-mail: v.suarez@cgiar.org; g.hareau@cgiar.org

current global food system and its role played as a food security crop, this chapter analyzes and discusses how potato research and innovation can contribute to sustainable agri-food systems with reference to food security indicators. It concludes with a discussion about the challenges for sustainable potato cropping considering the needs to increase productivity in developing countries while promoting better resource management and optimization.

Introduction: The Current Situation of Global Hunger, Food Security, and Agricultural Growth

A growing earth population and the increasing demand for food is placing unprecedented pressure on agriculture and natural resources. Today's food systems do not provide sufficient nutritious food in an environmentally sustainable way to the world's population (Wu et al. 2018). Around 821 million are undernourished while 1.2 billion are overweight or obese. At the same time, food production, processing, and waste are putting unsustainable pressure on environmental resources. By 2050, a global population of 9.7 billion people will demand 70% more food than is consumed today (FAO et al. 2018). Feeding this expanded population nutritiously and sustainably will require substantial improvements to the global food system—one that provides livelihoods for farmers as well as nutritious products to consumers while minimizing today's environmental footprint (Foley et al. 2011). A critical challenge is to produce more food with the same or fewer resources.

According to the Global Hunger Index (GHI), substantial progress has been made in terms of hunger reduction for the developing world (Von Grebmer et al. 2017). The GHI ranks countries on a 100-point scale with 0 being the best score (no hunger) and 100 being the worst. Whereas the 2000 GHI score for the developing world was 29.9, the 2017 GHI score is 21.8, showing a reduction of 27%. Yet, there are great disparities in hunger at the regional, national, and subnational levels, and progress has been uneven.

Sub-Saharan Africa (SSA) and South Asia (SA) have the highest 2017 GHI scores, at 29.4 and 30.9, respectively. These scores are still on the upper end of the serious category (20.0–34.9), and closer to the alarming category (35.0–49.9) than to the moderate one (10.0–19.9). These data show that persistent and widespread hunger and malnutrition remain a huge challenge in these two regions. In other parts of the developing world within the low range, are also countries with serious or alarming GHI scores, including Tajikistan in Central Asia (CA); Guatemala and Haiti in Latin America and the Caribbean (LAC); and Iraq and Yemen in the Near East and North Africa (NENA) regions. Black et al. (2013) estimate that undernutrition causes almost half of all child deaths globally.

The current rate of progress in food supply will not be enough to eradicate hunger by 2030, and not even by 2050. Despite years of progress, food security is still a serious threat. Conflicts, migration, and climate change are hitting the poorest people the hardest and effectively maintaining parts of the world in continuous crisis.

The 2017 GHI report emphasizes that hunger and inequality are inextricably linked. Most closely tied to hunger, perhaps, is poverty, the clearest manifestation of societal inequality. Both are rooted in uneven power relations that often are perpetuated and exacerbated by laws, policies, attitudes, and practices.

According to FAO (2002) "*Food security exists when all people, at all times, have physical, social and economic access to sufficient, safe and nutritious food to meet their dietary needs and food preferences for an active and healthy life.*" Food security has four key dimensions: (a) food availability, (b) food access, (c) food quality and use, and (d) food stability.

Food availability refers to the supply of food at the national or regional level which ultimately determines the price of food. Improved availability of food is necessary to reduce food insecurity and hunger but is insufficient to completely end malnutrition, particularly because access to other services such as potable water, sanitation, and health services is also required.

Food access refers to the ability to produce one's own food or buy it, which implies having the purchasing power to do so. Given that a large portion of the poor worldwide are farmers, there remains considerable attention to promoting agriculture to enhance food access. The emphasis on an agricultural pathway to increase food access is twofold, since increased agricultural production provides income to purchase food as well as direct access to food for consumption obtained from own production.

Food use and quality refers to the level of nutrition obtained through food consumption from a nutritional, sanitary, sensory, and sociocultural point of view.

Food stability incorporates the idea of having food access at all times thus incorporating issues such as price stability and securing incomes for vulnerable populations (FAO 2006a).

This widely accepted FAO definition reinforces the multidimensional nature of food security that requires multisector approaches. Such approaches should combine the promotion of broad-based agricultural growth and rural development with programs that directly target the food insecurity as well as social protection programs focused on nutrition including a gender approach (Salazar et al. 2016). Agricultural growth results in rural development and prosperity through a series of multiplier effects, that is, through backward and forward linkages, due to increased incomes. These effects typically stimulate enhanced investment in both farm and non-farm sectors (Hazell and Haggblade 1989; Pandey 2015). Growth in rural farm sector increases demand for goods and services produced by the non-farm sector, further increasing purchasing power and effective demand, thus deepening growth in non-farm sector. Further, Haggblade et al. (2007) argue that the increased income earned in rural non-farm sector can kick off a series of reverse linkages in which such income is invested in agriculture to further strengthen its growth and improve livelihoods of farm households.

During the 2014 World Economic Forum, in a debate on "Rethinking Global Food Security," Shenggen Fan, Director of the IFPRI, argued that tackling hunger and malnutrition is not only a moral issue but also one that makes economic sense. The world loses 2–3% Gross Domestic Product (GDP) per year because of hunger, while investing US$1 in tackling hunger yields a return of US$30. Ajay Vir Jakhar, Chairman of Bharat Krishak Samaj (Farmers' Forum) in India, added that farmers do not think in terms of food security at the global level, but in their own households. While policy makers tend to think in terms of global and national issues and solutions, localized solutions and help from the public and private sectors are also needed to support the bulk of farmers who are farming small plots of land and which have a critical role as engines of food productivity growth and social development. By declaring 2014 the International Year of Family Farming, the United Nations acknowledged the importance of family farming in reducing poverty and improving global food security. Localized, technical, and commercial solutions with the support of both public and private sectors are needed in combination with global food security policies.

Therefore, enhancing food security requires policies that improve households' ability to obtain food through production and better income. Growth in agricultural productivity is key to reducing rural poverty since most of the poor depend on agriculture and related activities for their livelihoods. Because the potato is one of the global crops with a most diverse distribution pattern (Haverkort et al. 2014) and is grown in areas with high levels of poverty, hunger, and malnutrition, it can be particularly effective crop for enabling smallholder families to attain food security and climb out of poverty. Hence, innovations based on potato science can be a significant vehicle for targeting the poor and hungry as part of a broader set of research and development activities.

This chapter first presents the importance of the potato in the current global food system and its value as a food security crop. It then discusses the role of agriculture and the potato for their contribution to food security in its different dimensions: analyzing opportunities and challenges on how potato research and innovation can enhance productivity and how potato agri-food systems can contribute to food security at a global scale using natural resources in a sustainable way. A list of key research and technology options that can contribute to sustainable agri-food systems intensification approaches is suggested. The chapter concludes with a discussion about the challenges for sustainable potato cropping combining the needs to increase productivity in developing countries while promoting better input management and optimization. These conclusions emphasize also the need to integrate better agriculture sustainable intensification and food security indicators.

The Potato in the Global Food System

Potato is currently grown on an estimated 19 million hectares of farmland globally, and the potato production worldwide stands at 378 million tons. The highest concentrations are found in the temperate zone of the northern hemisphere where the crop is grown in summer during the frost-free period. In these regions, potato is mainly grown as a cash crop and is therefore an important source of income. In tropical regions, the crop is significant in the highlands of the Andes, the

Table Potato production indicators

Region	2014–2016			Average annual growth rate								
	Production (000 tons)	Area (000 ha)	Yield (t/ha)	Production (%)	Area (%)	Yield (%)	Production (%)	Area (%)	Yield (%)	Production (%)	Area (%)	Yield (%)
				1961–1963 vs. 1988–1990			1988–1990 vs. 2014–2016			1961–1963 vs. 2014–2016		
Africa	25,270	1756	14.4	4.9	3.7	1.1	4.5	3.2	1.1	4.7	3.6	1.1
Asia[a]	190,617	9975	19.1	3.7	2.4	1.3	4.3	2.8	1.4	4.0	2.6	1.3
Europe	119,551	5547	21.6	−1.0	−1.8	0.8	−1.3	−2.4	1.3	−1.1	−2.2	1.1
North America	24,430	763	32.0	1.2	−0.2	1.4	0.8	0.6	0.1	1.0	0.2	0.8
Latin America and Caribbean	18,334	1023	17.9	2.2	0.0	2.2	1.5	0.1	1.4	1.8	0.0	1.8
World	378,202	19,063	19.8	0.1	−0.8	0.9	1.3	0.2	1.1	0.7	−0.3	1.0

Source: FAOSTAT (2017), accessed Oct 2018

n.a. = not available, a = U.S. Department of Agriculture, 2016

[a]Asia + Oceania

African highlands, and the Rift valley, and the volcanic mountains of West Africa and Southeast Asia, where production is both for food and cash (Muthoni et al. 2010). In the subtropics, the crop is grown as a winter crop during the heat-free period such as in the Mediterranean region, North India, and southern China. It is only in the tropical lowlands that potato is not a main staple, largely because the temperatures in these areas are too high for tuber development and growth in traditional potato cultivars (Haverkort et al. 2014). Figure illustrates the current pattern of the potato distribution worldwide (You et al. 2014; FAO 2016).

Potato is now the world's third most important food crop in terms of human consumption, after wheat and rice (FAOSTAT 2013) despite the large proportion of potato produce used for seed and as animal feed. Consumption of fresh potatoes accounts for approximately two-thirds of the harvest, and around 1.3 billion people eat potatoes as a staple food (more than 50 kg per person per year) including regions of India and China.

Potato Production and Demand Trends by Region

Across global landscapes, the versatility of the potato crop coupled with notable increases in production in many countries over the last two decades is unparalleled, although this increase has been mainly driven by area expansion and secondarily by yield improvements. Global statistics also indicate that potato production is shifting towards developing countries especially with strong increase in production in Asia and Africa, especially in East Africa. In fact, the developing world's potato production exceeded that of the developed world for the first time in 2005

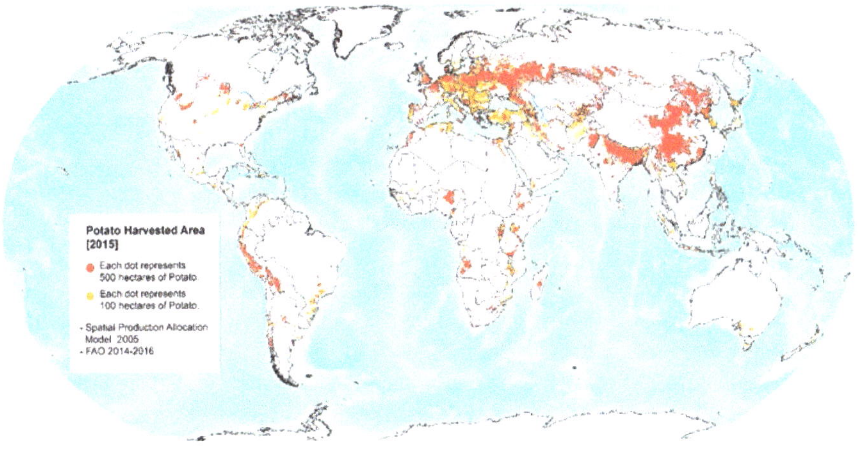

Fig. Potato distribution worldwide, harvested area (You et al. 2014; FAOSTAT 2016)

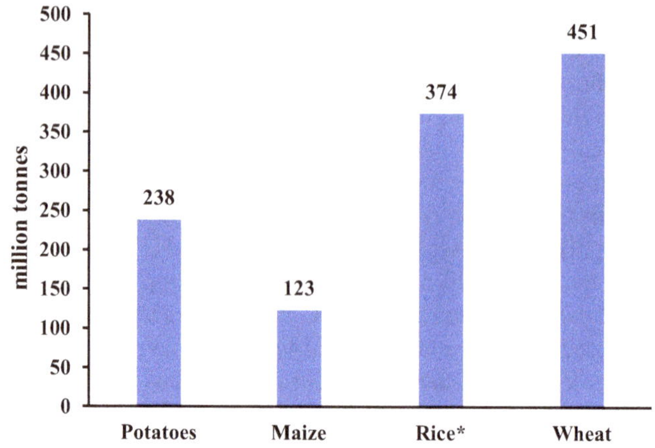

Fig. World food consumption per year for the four main nutritious crops. (Source: FAOSTAT 2013)

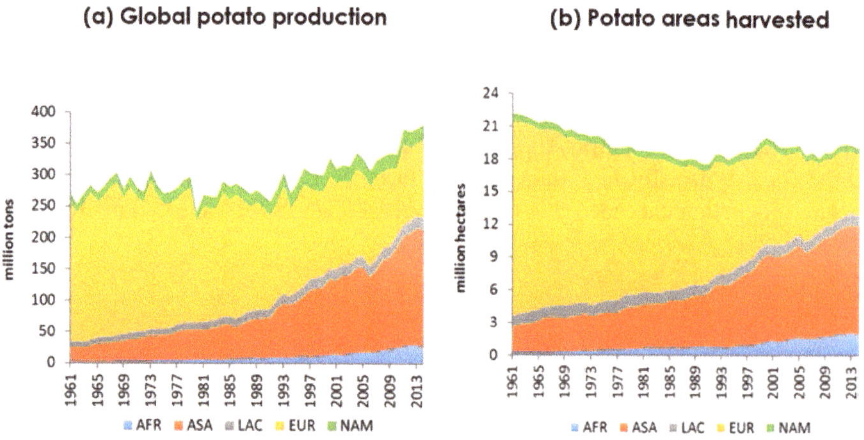

Fig. Potato supply (**a**) and (**b**) 1961–2013. *AFR* Africa, *ASA* Asia, *LAC* Latin America and Caribbean, *EUR* Europe, *NAM* North America. (Source: FAOSTAT 2017, accessed Nov 2017). (**a**) Global potato production, (**b**) potato areas harvested

(FAO 2014). It reaffirms the increasing importance of potatoes as a source of food, employment, and income in Asia, Africa, and Latin America.

As shown in Fig., Africa has registered large increases of harvested area over the last 20 years, but despite the impressive growth, total production and harvested areas are still much smaller compared to Europe and Asia. In Africa, the increase in potato production has largely been through increase of area under production, which more than doubled since 1994 and now exceeds that of the Latin

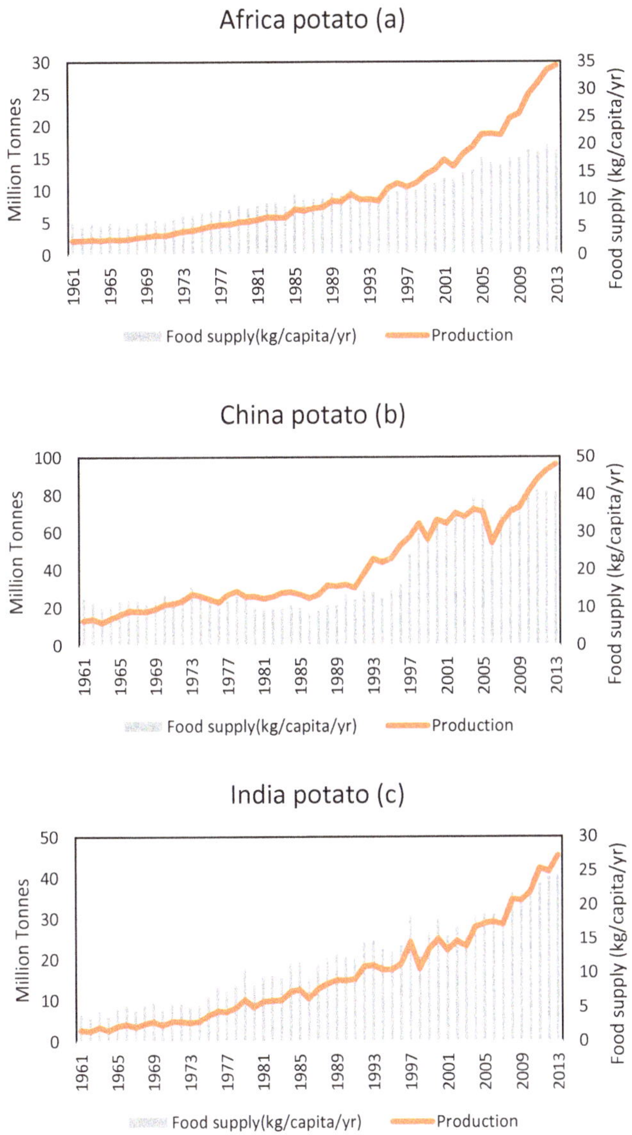

Fig. Relative development of potato production and food supply (kg capita^{-1} year^{-1}) in Africa (**a**) China (**b**), and India (**c**). FAOSTAT (2013)

America and Caribbean (LAC) region. In the tropical highlands of East Africa, farmers grow potato both for food and cash (Muthoni et al. 2010). The increase of potato production in East African countries over the last years has been impressive, suggesting a higher contribution of the crop to local food systems. In Tanzania, for instance, potato supply has almost tripled between 2000 and 2014 (FAOSTAT 2017), while in Rwanda potato is included in the national priority list of crops due to its role in national food security (approximately 125 kg per capita consummed per year; FAO 2009). As world population levels are predicted to show the greatest rise in Africa in the coming decades, increased contribution of potato to local food systems in this region is of considerable importance (Birch et al. 2012).

In Asia, China and India have experienced nearly a half century of steady growth in potato production . Both countries also have ambitious growth targets for future years. For some decades now, the Chinese state has been working to increase national potato consumption, also launching a campaign since 2014 to promote both the cultivation and the consumption of this tuber (The Wall Street Journal 2015). China became the world's largest potato producer in 1993 and currently accounts for almost one quarter of global potato production and about 28% of total cultivated areas (FAO 2015a, b). Potato in China is mainly used for food, both as a vegetable and in processed forms, while a smaller part is also consumed as animal feed (Scott and Suarez 2012a). Potato in India is mainly grown in the Indo-Gangetic plain, either as monoculture or in rotation with maize, wheat, and/or rice and it is regarded as both an important staple and a cash crop. Following the growth in production volumes, potato yields in India have also increased significantly, at an average of 2% per year, because of successful breeding programs, quality seed systems, and storage infrastructure that have reduced post-harvest losses (Scott and Suarez 2011).

Looking at other Asian countries, potato is the principal vegetable in Bangladesh and the second most important crop behind rice. Its cultivation is widely distributed across the country where it is grown mainly as a cash crop (Scott and Suarez 2012b). Potato production in Bangladesh has greatly expanded during the last decades, especially after 2000 when output surged from about 1.5 million tons to more than 8 million tons in 2013 (FAO 2015a, b). This impressive growth, besides the rising domestic demand because of population growth and the "westernization" of dietary preferences in urban areas (Pingali 2006), can also be attributed to the introduction of several improved high-yielding varieties and the development of cold storage facilities which facilitated near year-round availability of potato. At the same time, producers also gained significant price advantages (Reardon et al. 2012). In Nepal, potato is the second most important staple food crop after rice. The potato has also become a significant source of rural income in Pakistan where production is concentrated in Punjab, with spring and autumn crops accounting for 85% of the harvest. Expansion of irrigated Pakistani land has resulted in substantial increases in potato production (up 254% from 1990 to 2009) and area under cultivation.

Regarding potato production in LAC over the past 60 years, the annual average potato domestic supply has increased from 7.2 million tons in the 1961–1963 period to 19.6 million tons in 2011–2013, which represents an average annual growth rate of 2%. By way of comparison, growth rates for potato production in ASA and AFR averaged over 4% for a similar period, i.e. more than double those of LAC (Scott 2011). Most of the production is oriented towards human consumption (74%, maintaining this trend throughout the period) and it highlights a relatively low processing level of 1% (FAOSTAT 2017).

The role that potato plays in the diets in LAC vary—from basic staple, producer/consumers in the Andean highlands to complementary vegetable for urban households in most of South America, to a relatively expensive complimentary vegetable in much of Central America and the Caribbean, and to a popular fast food in the form of French fries in urban markets throughout the region (Scott 2011). Per capita consumption of potatoes in Latin America increased slightly from 22 kg/person on average between 1961 and 1963 to 25 kg/person between 2011 and 2013. But these regional trends do not reflect the important differences in trends at the subregional and country levels. Peru, is one of the countries where potato consumption has grown significantly, reaching in 2015 a figure of 85 kg/person. This is due to various public–private policies, rural infrastructure, expansion of supermarket trade focused on potatoes and a strong relationship with the gastronomy sector promoting Andean food including the native potato and its products. Brazil and Mexico have increased their consumption, although their absolute values, 18.5 and 14.8 kg/ person respectively still remain low compared to other countries in LAC. The cases of Argentina and Colombia are showing a downward trend.

The United States is the fifth largest potato producer in the world with more than 420,000 ha harvested in 2013 and a total output of almost 20 million tons (FAO 2015a, b). Although in the United States potato is no longer the traditional staple of the past, it is nevertheless gaining increased appreciation by nutritionists because of its nutrient density and its contribution to a more balanced diet (Bohl and Johnson 2010). There is also a large demand by the processing industry for producing commodities like frozen French fries and chips for both the local and foreign markets. Potato yields in the United States have more than doubled over the last 50 years, rising from 22 tons ha^{-1} in 1961 to 49 tons ha^{-1} in 2016. This increase in yields has been suggested to be primarily driven by improvements in management rather than genetic improvements, since most breeding programs have traditionally focused on quality traits such as dry matter content and storage longevity to meet the demands of the processing industry and the consumer (Douches et al. 1996).

In Europe, Germany, France, Netherlands, the United Kingdom, and Belgium are together the strongest potato producers in the European Union (EU), due to potato yields higher than 40 tons ha^{-1} in this area of northwestern Europe and to the strong links of production with the dynamic European potato processing industry. Potato is also prevalent in Eastern European countries, particularly in Russia, Ukraine, and Poland where per capita consumption has traditionally exceeded 100 kg annually. Although Eastern Europe constitutes the region with the highest use of potato as animal feed globally, feed use of potato has been steadily declining over the last 20 years and being replaced by cereals, most notably in Poland. This decline in feed use, together with the shift of diets towards low-calorie food and a trend to spend less time on cooking observed in Western European countries, has led to a significant decrease in demand for fresh potatoes, and therefore potato production in the continent is falling (European Commission 2007). Despite the aforementioned decline, some European countries like France, Denmark, and Belgium have increased production over the last decade, due to growth of the processing industry (French fries, crisps) and starch production (Eurostat 2017).

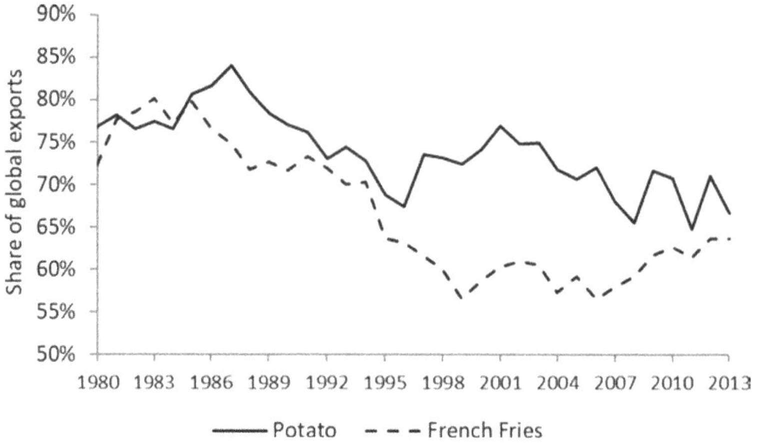

Fig. European share of global exports for potato (fresh and seed) and French fries 1980–2013. (Source: FAOSTAT 2017)

Table Potato utilization, consumption and trade indicators

	Utilization				Consumption	Trade	
					Per capita (kg year^{-1})	Import Quantity (000 tons)	Export Quantity (000 tons)
Region	Food (%)	Feed (%)	Seed (%)	Other (%)	2011–2013	2011–2013	
Africa	69	4	8	19	19.2	967	830
Asia[a]	67	12	5	11	29.1	6,042	2,683
Europe	52	19	15	11	83.4	17,890	19,477
North America	84	1	6	9	55.3	3,615	5,354
Latin America and Caribbean	73	2	8	16	23.8	2,040	439
World	64	12	9	12	34.4	30,554	28,784

Source: FAOSTAT (2017), accessed Oct 2018
n.a. = not available, a = U.S. Department of Agriculture, 2016
[a]Asia + Oceania

Moreover, the competitiveness of the potato industry has established Europe as the world's biggest net potato exporter, amounting for more than 60% of all exports of fresh potato and a similar percentage of global exports of French fries.
These statistics concern mainly intra-EU trade, and also export of seed potato to non-EU countries, primarily to Mediterranean African countries like Egypt and Algeria (FAOSTAT 2017).

Tables give a synthesis of the potato indicators by region confirming the expansion of potato in Asia, which is now the major potato producer continent. In Africa, the potato growth rate has also been strong with Egypt, Malawi, South Africa, Algeria, and Morocco producing more than two-thirds of all the potatoes in the region. In many countries of Latin America and the Caribbean, potato areas have

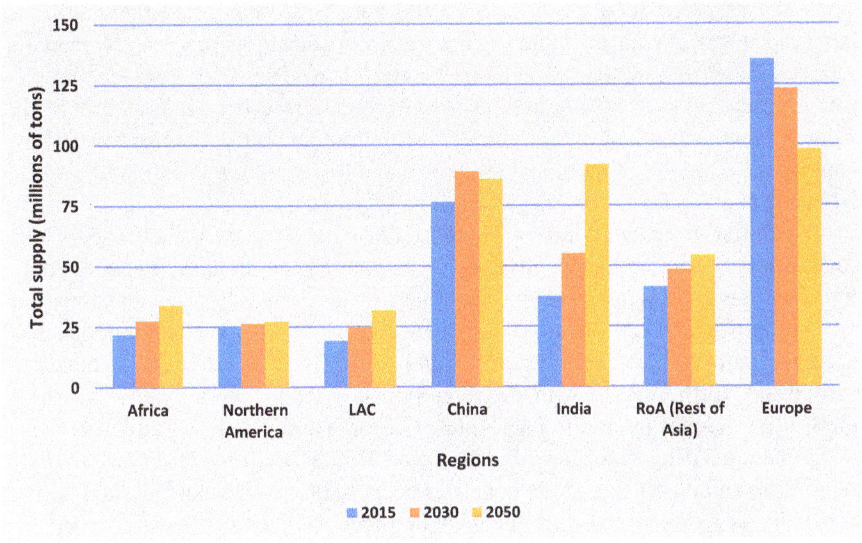

Fig. The future of potato production. Adapted from Rosegrant et al. (2017)

declined although output has risen due to improvements in productivity. Potato production growth rate in Europe and North America has declined due to a significant decrease in demand for fresh potatoes that was partially compensated by a large demand from the processing industry as described before. In North America, yields increased rapidly between 1961 and 1990 and then somehow stagnated, suggesting yields are near their potential in the region and there may exist genetic limits. In Europe, on the contrary, the relative larger yields growth rates occurred after 1990.

Future trends by region indicate a major production increase in Asia and Africa as compared to other regions. Considering some underlying assumptions such as population growth, climate change, and economic growth pathways, the UN projects a population decline in China and growth of per capita GDP which will affect diet composition. Therefore, the future supply of potato in China will not continue to grow faster than in the past. According to Rosegrant et al. (2017), it is in India where potato supply will almost triple because of the very high population growth, especially under certain socioeconomic scenarios.

The Potato Remains a Food Security Crop in the Developing World's "Nutrition Transition"

In many developing countries, and especially in urban areas, the globalization, the emergence of fast food outlets and supermarkets and the rising levels of income are driving a "nutrition transition" led by major shifts in the availability, affordability, and acceptability of different types of food, especially toward more energy-dense foods and prepared food products. It is translated into major and rapid shifts in dietary patterns. As an example of this nutrition transition process, demand for potato is increasing in many developing countries in Africa and Asia. In South Africa, potato consumption has been growing in urban areas as part of the staple food consumption of the middle class, although maize remains the primary staple in rural areas. In China, higher income and increased urbanization have led to increased demand for processed potatoes. In this context, potato plays a role in diet diversification in many countries where family agriculture and smallholders continue to supply local markets with fresh and affordable agricultural produce. Potato is still an important staple in rural food systems where it is produced, and emerging food systems which are urbanized but where consumers still rely on staples such as potatoes. In Industrial food systems, highly urbanized in Northern countries in Europe and North America with the development of the processing industry there is a lower dependence on traditional staples (Gillespie and Van den Bold 2017).

As mentioned by Haverkort and Struik (2015) potato used to be a "local for local" crop and it still is in many countries because of the bulkiness and the limited storability of the seed and ware tubers. Compared to other staples, and except for processed potato products, fresh potato is a thinly traded commodity in global markets and is absent in major international commodity exchanges. It is therefore subject to less price volatility at global scale. Thus, potato can be relied upon to smooth the disruptions in global food supply and demand that have an impact on other commodity prices, such as witnessed during the 2007–2008 and subsequent food price spikes (FAO 2009).

Potato also improves food security because it is a source of employment and income, both of which have direct links to household food access and nutrition (Pinstrup-Andersen 2014; Kanter et al. 2015). The comparatively short maturity period, nutritious characteristics, employment, and income opportunity that characterize potato make it a resilient crop that can secure vulnerable livelihoods under the effects of climate change and changing market environments. Moreover, potato yields more food more rapidly on less land than any other major crop.

Where other staple crops are available to meet energy requirements, potato should not replace them but rather supplement the diet with its vitamins and mineral content and high-quality protein. Potato can be promoted as a healthy and versatile component of a nutritious and balanced diet including other vegetables and whole grain foods. Likewise, it contributes to combat micronutrient deficiency, also referred to as hidden hunger, which is a major global public health problem, affecting an estimated two billion people globally (Bailey et al. 2015). Potato contains interesting amount of health promoting components such as vitamin C, phenolic compounds, and iron and has protein content comparable to that of cereal grains (Burlingame et al. 2009). When eaten with its skin, a single medium-sized potato of 150 g provides nearly half the daily adult requirement (100 mg) of vitamin C. It is also a good source of vitamins B_1, B_3, and B_6; minerals such as iron, potassium, phosphorus, and magnesium; and contains folate, pantothenic acid, and riboflavin. Vitamin B deficiency has both short- and long-term impacts, including poor cognitive and pregnancy outcomes and poor child development and life outcomes. Vitamin C, on the other hand, is important for body metabolism and iron absorption (FAO 2009).

As an example, potato plays an important role in the food basket of most Peruvians; it continues to make up a relatively high proportion of daily calorie availability, reflecting its importance as traditional source of energy. To enhance the nutritional contribution of the potato in rural Andean highlands, the International Potato Center (CIP) coordinated activities to promote innovation in potato-based systems with the objective of contributing to food and nutrition security of rural highland populations enhancing native potato production and promoting dietary diversity. In rural areas of Peruvian highlands, there is a high prevalence of chronic malnutrition amongst children (42% in children under 2 years according to INEI Peru 2017). Native potato varieties are commonly grown in the Andes and are a significant part of the local diets. Some of them have higher contents of micronutrients (Zn and Fe) and are rich in antioxidants compared to commercial improved varieties. Creed-Kanashiro et al. (2015) explored the relationships between agricultural production characteristics and nutritional status of young children of families in rural areas of Peru whose livelihood is based on potato production systems combining native and commercial varieties. The results showed a positive relationship between percentage of recommended dietary intake (RDI) for both Fe and Zn intakes by children and production of native potatoes for home consumption, raising small animals for consumption and sale (e.g., guinea pigs and poultry) and the area of production of commercial potatoes that allowed the families to improve their incomes and diversifying the family diet.

Finally, it should be mentioned that potatoes have been related to increased risks of obesity mainly because of their high glycemic index. Recent reviews of clinical intervention and observational studies centered on the potato concluded that these studies did not provide convincing evidence to suggest an association between intake of potato and risks of obesity, Type II diabetes (T2D), or cardiovascular disease (CVD) (Borch et al. 2016). However, as part of the trend towards urbanization and associated lifestyles, raising incomes, and greater consumption of "convenience foods", demand for fried potatoes is increasing. Overconsumption of these high-energy products, along with reduced physical activity, can lead to overweight and obesity. Therefore, the role of fried potato products in the diet must be taken into consideration in efforts to prevent overweight and diet related noncommunicable diseases, including heart disease and diabetes. For these reasons, even though the potato is a nutritious staple crop, it has often been associated to a meal component with no specific attributes in many societies, even in developing countries. This image is reinforced by some negative myths such that potato is fattening, requires intense use of fertilizers and pesticides for its cultivation and can contribute to soil erosion. Negative images can be mitigated with better information about the nutritional value of the potato and its importance in the diet, while promoting sustainable and environmentally friendly production systems, thus addressing the question of natural resource management.

Policies and Strategies for the Development of the Potato as a Food Security Crop

The role potato can play as a food security crop at national scale has been addressed in some developing countries with different policies, either sectoral and crop-specific or at the macro level. In its quest to improve food security for a rising population, the Government of China is developing a national plan to increase production and consumption of potato and promoting the crop as a staple instead of a vegetable. This status can give access to important complementary policies and resources at national and regional level and to subsidies from the central government. It also recognizes the double role of potato in current China. Potato is still a major staple for poor rural areas where local governments continue to provide subsidized inputs (e.g., clean seeds of selected varieties), while at the same time being at the forefront of an increasing private sector-led processing industry, accompanying rising incomes in urban populations and diversification of diets (Scott and Suarez 2012a, c).

In Peru, the major center of origin of the potato, a large effort began in early 2000s to develop a competitive and inclusive native potato value chain for domestic markets. Initially led by the International Potato Center, the initiative gathered several private and not-for-profit actors to add value to the native potato grown by small farmers while developing a niche market. Several new products were developed in the process, for example selected native potato varieties for fresh consumption sold as gourmet potatoes in innovative packages in large supermarket chains, snacks such as colored native potato chips, and culinary innovations in the gastronomy sector featuring native potato as a central component of sophisticated dishes. The innovations in the value chain continued and a second round of new products emerged, including for example frozen native potato fries, native potato-based liquors, and even cosmetics made from potato. Although no specific sectoral policies were behind this initiative, the development of the native potato value chain took advantage of Government policies at the macro level promoting private sector and market-led developments and the fast growth of Peru's economy and of the purchasing power of the population since the beginning of the twenty-first century. While the Government of Peru focused on public investments to promote export-oriented agricultural growth, the experience with the native potato value chain has proven successful to link small potato farmers to domestic markets and to develop a more inclusive growth strategy of the highly diverse agricultural sector of the country (World Bank 2017).

A final example of policies that have been adopted to promote the potato sector in developing countries is through seed laws and regulations. Seeds are an important input of production of the potato crop and can affect yields since they are vehicle for important diseases. Seed degeneration due to viruses is one of the most common constraints affecting potato productivity, and therefore large research and extension efforts are made to improve availability and access to quality seeds for small farmers. One particular aspect of potato seed is that it is vegetatively propagated. In most developing countries, however, seed systems were first established following developed country standards for cereals and grains, promoting the use of certified seed under a formal seed system. This has led to very low use (less than 10%) of certified potato seed in most developing countries. To increase access to quality seed by small farmers, the Food and Agriculture Organization (FAO) of the United Nations promoted the definition of a new seed category, the Quality Declared Seed (QDS), that relaxes some of the standards required for Certified Seeds and recognizes the importance of seed producers in providing seed of enough quality through the informal seed system (FAO 2006b). Ethiopia adopted the QDS definition in a new seed law passed in 2016 without distinction of crops. Peru (in 2018) and Ecuador (in 2013) have modified the seed regulation for potato to accept the use of QDS. However, differences still exist on how countries are beginning to adopt this category. While in Peru, the new regulation defines the category as QDS similar to the FAO definition, in Ecuador the regulation defines the new category as "common seed" and introduces aspects of the FAO definition for QDS.

Other countries are updating the regulations regarding seed quality assurance systems for potato to increase availability and access of quality seed by farmers. A broad range of changes are proposed, from relaxing some of the standards required for certified seed to allowing the use of private inspection services to increase the number of seed producers that can be inspected each season (e.g., Kenya). One of the motivations of these changes is the increasing recognition of the role of the potato crop for national food security. Some concerns have been raised, however, on the potential consequences of relaxing seed quality standards on the incidence of seed-borne diseases (e.g., *Ralstonia solanacearum*).

Food Security Challenges and Perspectives for Potato Research and Development

Potato in a Global Food Security Context

The potato, because of its adaptability, its yielding capacity and its nutrition contribution, and as an important component of diversified cropping systems, has a long history of helping relieve food insecurities, and contributing to improve household incomes in times of crisis and today's population expansion. Among important issues and challenges at global level, the European Association for Potato Research (EAPR) Conference in 2017 identified three broad concerns: (1) food security and food safety for a growing population considering consumer's needs; (2) sustainable and environmentally friendly production addressing the question of natural resource management taking advantage of new technologies available such as breeding techniques, biocontrol and big data management; (3) innovation in practice turning scientific results into products and processes to improve the performance of agri-food systems (Andrivon 2017).

Potato's support to food availability can be achieved through improved productivity, either by increasing yields or expanding production areas, combined with technologies that reduce post-harvest losses. Newly developed potato technologies and production concepts can potentially help create solutions in areas where there is a huge need to increase food production. Figure shows the global distribution of yields and the low yield levels in most of the developing world, where observed actual yield is usually much lower than the attainable yield. The actual yield is the expression of a potato cultivar in a specific agro-ecological environment and depends on availability of inputs, the economically optimal use of available inputs given the farmers' conditions, and externalities such as local meteorological variations and climate change as well as the technology level and the quality of crop management. Actual yields range from below 5 tons of fresh tubers ha^{-1}(median yield in Uganda: Gildemacher et al. 2009) to well above 100 tons fresh tubers ha^{-1} (in Columbian Basin, USA: Kunkel and Campbell 1987).

The yield gap, expressed as the difference between actual yield in farmers' fields and the attainable yield—using best agricultural practices—leaves a great potential for improvement considering that, in developing countries, the full expression of the crop's yielding capacity has not yet been achieved. Much improvement is needed in agronomic practices, quality seed production, and varieties tolerant/resistant to abiotic and biotic threats (Birch et al. 2012). The high nutritional value of potato mentioned above reinforces the potential of potato to respond to food security challenges.

To reach impact on food availability, access and better use in diets, proper selection of target areas for potato research and identification of the most important constraints to potato production are crucial for defining priority interventions. As a step towards achieving this and to prioritize options for potato research, CIP led an expert survey to assess priorities for potato research across the developing world (Kleinwechter et al. 2014), and an ex-ante evaluation of the economic relevance of

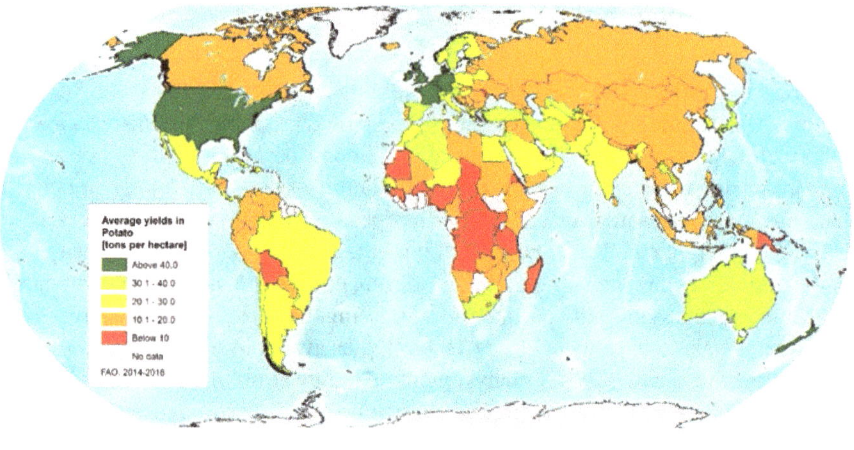

Fig. Global distribution of potato yields (tons ha). FAOSTAT, 2014–2016

these options (Hareau et al. 2014). Most of these research options for potato have been defined to be relevant in numerous countries of Asia, Africa, and LAC. They include research in potato late blight, drought tolerance/water use efficiency, seed systems and development of farmer organizations and farmers' links to markets (e.g., Harahagazwe et al. 2018 for research priorities in Africa). However, the current and upcoming contextual changes, especially considering the climate change requires to revisit some of these issues with a new perspective. Many simulation exercises based on IPCC (Inter-governmental Panel on Climate Change) scenarios and biology models are underway and suggest that future potato cropping systems could differ from those we know today with the implication that new cultivars will be required to respond these new conditions (Andrivon 2017; Quiroz et al. 2018).

Research and Innovation for Sustainable Potato Cropping

As suggested by Haverkort and Struik (2015), the future prospects of food security challenges for agricultural production can be expressed by the formula: $P = G \times E \times M \times S$ where Performance (P) is determined by Genotype or varieties (G), the Environment or agro-ecological conditions where the crop is grown (E)—which as mentioned above is evolving and will further change in the future—its Management and adaptation to local socioeconomic conditions (M) and the Societal requirements (S) driven by society's demands for food and the need to make agriculture more environmentally and consumer friendly with a focus on food safety. The societal or consumer requirements will vary according to the context. In high-income countries, consumers are looking principally for healthy and easy to prepare foods while in the developing world, the consumers' needs are driven by the food and nutrition challenges and demand concerns principally food availability in both quantity and quality. The actual performance of potato can be expressed as yield of fresh or dry matter per unit area or yield of the finished product per unit area recovered from the raw material after processing.

This Performance analysis ($P = G \times E \times M \times S$), suggested originally in the context of high-income countries, can be used for the developing world considering the need in most countries to principally respond to the hunger and malnutrition problems through a more efficient agri-food system but based on family agriculture. Family farms are the backbone of agriculture in low- and middle-income countries in Africa, Asia, and Latin America. For many years, the trend in the developed world has been towards intensification to achieve more outputs per unit of land but the sustainability of this intensification is under debate especially considering agriculture's ecological footprint. In low-income countries, sustainable intensification (SI) is a different challenge because it starts from a much lower level of inputs than in developed countries. This is especially the case in Africa where potential for increasing production through area expansion is diminishing, partly due to high population growth (Headey and Jayne 2014). For instance, Wu et al. (2018) and Jayne et al. (2014) argue that even though Africa has a high cropping intensity gap[1], closing this gap sustainably must focus on input intensification rather than area expansion. The relevant question is how to promote technology options that allow for increased output quantity and quality (especially from the nutrition point of view), while considering agriculture's environmental impact, preserving land and other resources in both developed and developing countries. In this context, sustainable intensification of potato cropping goes beyond production aspects and considers strong socioeconomic, demographic, and environmental trade-offs to optimize performance. Institutional incentives to support innovation involving diverse stakeholders, with emphasis on research partnerships, are also required to respond to hard-to-find compromises that will vary between different cropping systems (Hall et al. 2001). Multidisciplinary approaches contribute to recognize and solve practical problems at the level of the crop, the cropping system and the agri-food system to achieve sustainable food security in its four dimensions.

The issues mentioned above indicate that sustainable potato production and efficient use of resources will require future adjustments and redesigns of the cropping and processing systems. Two main options can be considered to increase food security, which remains a main goal for future potato research: (1) produce more with less through better input management and optimization; (2) to produce just as much but waste less both before and after harvest through better value chain management, better storage, processing and marketing operations, and responding to increased involvement and awareness of consumers (Andrivon 2017).

As an attempt to analyze how to combine and score different research and technology options according to their effect on sustainable agri-food system indicators in developing countries and their relation to the four dimensions of food security, Table attached suggests a list of priority research and technology options in the spheres of G and M. The assessment is based on the literature and the authors' expert opinions. We scored key options using some critical indicators of sustainable agri-food system intensification related to productivity, agriculture income, human well-being and environmental sustainability according to Smith et al. (2017). We scored the research and development options using a simple scale of high, medium, and low effects according to their relation to sustainable intensification indicators and their main contribution to one of the four dimensions of food security.

The performance of any of these technologies or development approaches will, as expressed by the Performance equation ($P = G \times E \times M \times S$), depend on external factors both socio economic and agro-ecological. Their potential contribution to food security will be strongly influenced by the environment (E) and the societal (S) requirements in the local context where they are implemented—e.g., enabling environment, policies for financial and nonfinancial services are also key components for achieving efficient food systems.

[1] Wu et al. (2018, p. 2) define cropping intensity gap as "the amount of incremental cropping intensity that is possibly available if all croplands in a given region are fully intensively used."

Table Qualitative evaluation of key research options to enhance food security dimensions through the performance of potato production systems (based on the authors appreciations and the literature related to indicators of contribution to sustainable agri-food systems intensification)

	Food security dimensions						
	Availability				Access	Use/quality	Stability
	Indicators of contribution to sustainable agri-food systems intensification						
Key research options to enhance the performance of potato production systems	Water use efficiency	Land use efficiency	Nitrogen and phosphorus use efficiency	Pesticide use efficiency	Farmer incomes	Increase in calorie and nutrient production efficiency	Reduction of environmental footprint (soil, water, air)
Breeding and variety development (Genotype)							
High yield		***	Neutral/negative. High productive crops might need additional N	Neutral/negative. High productive crops might need additional pesticides	***	**	
Disease resistance (e.g., late blight, viruses)		***		***	**	**	***
Tolerance to drought/heat/salinity	***	***	*		**	**	*
Biofortification (e.g., Fe and Zn)					*	***	
Earliness	***	***	***	**	***	***	***
Potato seed production (Management)							
High quality seed production and distribution		**	*	*	***	*	Neutral/negative high-quality seed use might motivate increased chemical input

Farmer-based seed production		*	*	*	**	*	**
Decision support tools and farming practices (Management)							
Decision support and diagnostic tools for pest and disease control		*		***	**		***
SMART agriculture approaches (sensor use for precision in NPK and water use)	***	**	***		**	*	***
Intensification of cereal-based systems with introduction of early potato varieties	*** No irrigation needed, higher productivity per unit of water	** Uses current fallow land already under agriculture	*	**	**	**	**
Sustainable resource management							
Ecosystem management and biodiversity use	*		*	**	* Policy dependent	*** Use of potato biodiversity can increase nutrient production efficiency	***
Soil and water management	***	**	**		**	***	**
Efficient and inclusive value-chains (Management and Society)							
Value-chain innovation					***	**	***
Post-harvest management (Management and Society)							
Post-harvest losses assessment and reduction					**	***	***

*** = High effect; ** = Medium effect; * = Low effect

In the sections below, some key research and technology options identified in Table are briefly described.

Potato Breeding, a Driving Force Towards More Efficient Potato Production

For genotype development (G), priority should be given to achieve a combination of traits to enhance stress tolerance and nutritional aspects to better respond to contextual changes, especially climate and local needs. The recent development of participatory breeding helps to best define the crucial trait combinations required and to facilitate acceptance of new genotypes by growers (Schulte-Geldermann et al. 2012). With the recent findings on the potato genome sequence (PGSC 2011) and the possibilities occurring with new breeding technologies (NBTs), potato breeding appears as the number one opportunity to improve potato production for global food security (Birch et al. 2012).

In many developing and in-transition countries governments have substantially invested in breeding improved varieties. A total of about 840 improved varieties[2] with various combination of traits have been released, most of which in Asian countries like China and India. CIP has considerably contributed to global crop improvement through supporting NARS and providing access to advanced breeding material. About 43% of total releases are CIP-related (i.e., NARS-bred varieties distributed/facilitated by CIP, NARS selection from CIP crosses, NARS crosses from CIP progenitors). At the regional level, CIP-related varieties are most prominent in Africa where 70% of total releases are CIP-related. This points to the importance of CIP in the regions and the support many national breeding programs require.

Table Total and CIP-related number of releases by region

Region	Year	Total	CIP-related[a]	
Asia[b]	2015	518	180	35%
Africa[c]	2010	178	124	70%
Latin America[d]	2007	141	60	43%
Total		**837**	**364**	**43%**

[a]NARS-bred varieties distributed/facilitated by CIP, NARS selection from CIP crosses, NARS crosses from CIP progenitors; calculations for Asia based on Gatto et al. (2018), for Africa on Labarta (2015), for Latin America on Thiele et al. (2008)
[b]Bhutan, Bangladesh, China, India, Indonesia, Nepal, Pakistan, Philippines, Sri Lanka, Vietnam
[c]Burundi, D.R. Congo, Ethiopia, Kenya, Madagascar, Malawi, Rwanda, Tanzania, Uganda
[d]Bolivia, Colombia, Ecuador, Peru, Venezuela

[2]The current true total number of releases is likely higher given that the most recent data available dates back more than a decade.

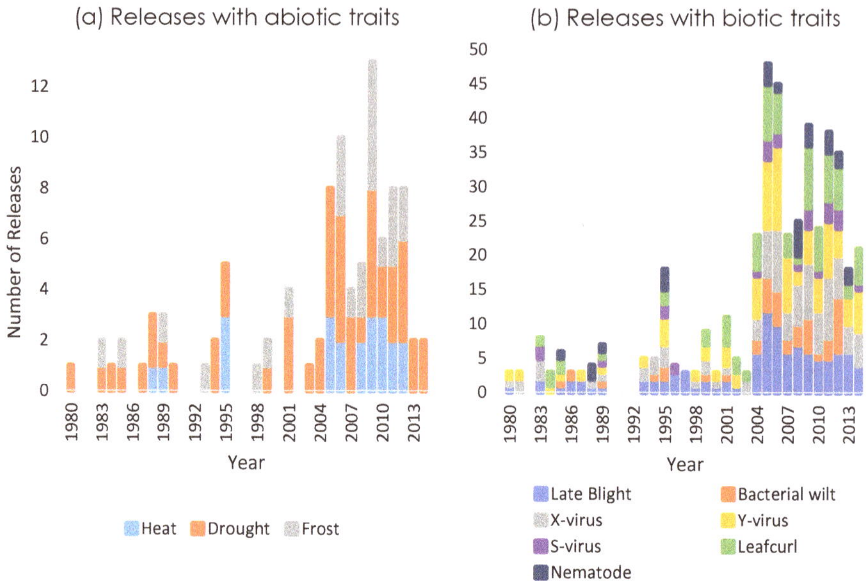

Fig. Total releases with abiotic and biotic traits in Asia between 1980 and 2014. Notes: release is for *high* resistant category only. *Medium* and *low* resistant, and *susceptible* categories are not shown. (Source: Gatto et al. 2018)

The development of early and high-yielding varieties with resistance to *P. infestans* has been a longstanding potato breeding objective. Genotypes with resistance to viruses (PVY, PLRV, PVS, PVX), nematodes (mainly *Globodera* and *Meloidogyne* species), bacterial wilt, and a broader spectrum of cultivars tolerant to abiotic stresses like heat, drought, and saline conditions, and focus on beneficial root traits, can increase productivity and expand potato production to new areas.

The importance to develop and release varieties with high tolerance to abiotic stresses and high resistances to biotic stresses has increased, as Fig. depicts for Asia. Especially starting in the early 2000, major traits have been bred into released varieties likely as a result of adjusting breeding objectives aiming increasingly at mitigating the adverse effects associated with climate change and variability.

New resilient varieties will potentially expand potato production to new areas and produce more nutritious food under current and future stress factors. Genetic biofortification through conventional and new breeding techniques can help to overcome micronutrient malnutrition and support the consumption of better-quality tubers. This crop improvement approach aims to positively influence human health, as a complement to diet supplementation and food fortification. In recent years, CIP has initiated the development of Fe and Zn biofortified potatoes, under the umbrella of the HarvestPlus Program (http://www.harvestplus.org/), a global interdisciplinary alliance for developing biofortified varieties of staple crops. Food security programs working to deploy biofortified crops will strongly benefit from nutritional education efforts and awareness programs considering gender roles in the beneficiary communities.

Seed Quality and Availability, the Key to Harvest Success

As sustainable potato production depends on a constantly renewed supply of disease-free planting material, improving quality seed production and seed distribution is another strong avenue of research opportunity related to crop management (M) in potato development. The conceptual framework underpinning the concept of seed security contemplates different types of seed insecurity: poor seed quality, lack of availability, limited access to high quality seed, lack of access to preferred and adapted varieties, inefficient seed systems (FAO 2016; CGIAR/RTB 2016). As a vegetative propagated crop, the growth, development, yield and quality of the potato is strongly influenced by the quality of the seed tubers planted affected by their physical, physiological and health status. More than 90% of seed potatoes in developing economies is produced in the farmer-based category and is considered to be of poor quality (Thomas-Sharma et al. 2015). Potato seed production systems should support the access to high quality seed potato tubers of improved varieties by combining rapid multiplication technologies (e.g., aeroponics or sand hydroponics) with decentralized seed multiplication, e.g., promotion of quality declared seed systems (FAO 2006b; Fajardo et al. 2010). It should be complemented with on-farm seed maintenance (e.g., positive selection, small seed-plot technique and improved storage) in an integrated approach (Gildemacher et al. 2011; Schulte-Geldermann et al. 2012; Thomas-Sharma et al. 2015; Obura et al. 2016). Improving technologies for farmer-based seed production and distribution of high quality planting material of existing and new varieties have the potential to reach high numbers of beneficiaries with strong impacts on poverty reduction and food availability.

Potato Crop Management and Farming Practices to Increase Productivity and Sustainability

The highly adaptable potato can fit to many types of environments (E) from sea level to high mountain conditions where small-scale farmers predominate. Beside temperature regime and solar radiation, there are many factors that affect its productivity as for example, soil characteristics, water use efficiency, nutrient availability and hazards such as night frost or heat waves that may drastically impact resource use, and thereby sustainability. The adaptation of the crop depends on the genotype but also on the crop management practices (M) that need to evolve according to the specific agro-ecological conditions, the socioeconomic context and the local production systems. Crop management is context specific and should consider local knowledge that can be improved promoting new tools and approaches.

Smart agriculture is a novel avenue for resource use optimization based on new monitoring and decision support tools. Remote sensing and global information system (GIS) tools coupled with decision support systems (DSS) and precision agriculture technologies may contribute to increased productivity while interaction among biophysical and social disciplines for sustainable food production intensification can at the same time contribute to resource use optimization. Fertilizer (N, P and K) recommendation systems are now using field scale models as well as tractor, drone and satellite embedded spectral sensors to monitor crop nutrient status to supplement fertilization according to inter and within field variability (Goffart et al. 2008, 2017). These more sophisticated technologies are still mainly used in the high-income countries, but massive and varied data management could foster new models to be developed and contribute to decision support systems under developing country conditions. CIP is adapting such system to the Andean condition developing a strategy to manage late blight that combines host resistance and a decision support system to optimize the use of fungicides. The vulnerability of the crop to many pests and diseases, which the current global climate change can worsen, remains one of the most severe threats for a wider potato diffusion and its sustainable cropping. To improve crop health, portable molecular diagnostic tools and decision support systems for early warning and control of pests and diseases (e.g., for efficient fungicide use to control late blight) will contribute to better crop production monitoring and input use efficiency. Research to develop biocontrol is very active and considered to grow substantially in the coming decade, but there are still few confirmed successes from the field, and specific management tools (Decision Support Tools) are still missing (Velivelli et al. 2014). CIP and EAPR are coordinating actions in Europe and Latin America to promote biocontrol and compare the efficiency of biocontrol agents using defined protocols (Devaux et al. 2017).

Specific intensification practices can be developed under specific cropping systems such as in the cereal-based systems in India through "Double-Transplanting (DT)" of rice and planting early maturing potato between the two rice crops as a valid alternative to the traditional potato-boro rice and kharif (monsoon) rice-boro rice. This cropping pattern contributes to enhance system productivity without sacrificing area or productivity of either of the two crops, thus creating new opportunities for potato cultivation for small-scale producers (Arya et al. 2015).

To enhance ecological sustainability, the objective is to implement management practices that increase the level of provision of ecosystem services such as natural soil fertility and biological control. Natural regulation of pests and diseases is an important element in potato agro-ecosystems. In organic farming, it has been demonstrated that the development of natural antagonist associations of the Colorado beetle, such as auxiliary insects and useful pathogens, can significantly improve the control of such a pest for the potato crop (Crowder et al. 2010). Another example of ecosystem management is the delivery of nitrogen through natural fixation and mineralization, which can be enhanced by cropping practices such as cover crops, legume-based intercropping systems and application of organic soil amendments before the potato crop. Biodiversity based agricultural approaches that rely on the design and management of on-farm agrobiodiversity to generate ecosystem services is another avenue to reduce potato's ecological footprint and increase farmer's resilience to cope with frost risks as it is the case in the Andes. The management of ecosystems at field and landscape level can provide a series of production benefits to reduce the need for off-farm inputs. The analysis of beneficial microbial communities and their impact on potato plant phenotypes expressions still needs to be developed as discussed at a EAPR-CIP workshop on biostimulant and biocontrol agents (Devaux et al. 2017).

Integrating Food Security and Value Chain Development

Although potato remains a staple food in rural areas in developing countries, it is also increasingly becoming a cash food for farmers in Asia, Africa and LAC (DeFauw Sherri et al. 2012). The majority of potato producers are smallholders who depend strongly on agriculture, including the potato crop, for income, food security and employment. Potato production reaches consumers via multilevel marketing systems, not directly from the farmer's field. Thus, the challenge to achieve food and nutrition security as well as prosperity for these smallholders will be obtained or lost by the way agricultural value chains are coordinated. Value chain development and organized markets through farmer associations, storage facilities, and better links with traders and consumers are then required to allow potato producers to access better value markets to get higher and steady incomes from their production. In the recent years, research activities to improve the efficiency of the value chain and coordination among its actors have evolved to achieve more inclusiveness in the value chain development approaches (Devaux et al. 2018). Several factors are contributing to this evolution: changes in consumer demands, new or emerging markets with strict standards, including food safety, processing technologies, and better access to market information. To respond to these changes and the need to make agriculture more environmentally and consumer friendly responding to the Society's requirements (S), research should be characterized by an interaction between natural and social sciences and should be market-driven considering the needs and challenges of all value chain stakeholders. In a compendium about perspectives on the status of innovation for Value Chain Development, Devaux et al. (2016) analyze the opportunities emerging from new markets for agricultural produce and identify challenges to smallholder participation in these markets, approaches for increasing access to markets through strengthening value chain stakeholders' relationships, enhancing innovation and improving an enabling environment. Linked to the value chain efficiency, the assessment of food losses across the value chain and the quality of marketed potatoes also require further research efforts to optimize food availability and consumer access to quality potato products.

Post-harvest Management: Reducing Food Losses

As indicated above, another way to face the food security challenge is to produce just as much, but waste less through better post-harvest management. Post-harvest management in potato, including storage, processing and value chain efficiency, is a much larger problem than cereals and deserves special attention. Reduction of food losses appears as a key opportunity. The basics of storage management have not changed, but the implementation and application of the basics are evolving worldwide, according to diversity in location, climate and market criteria, that will influence storage management structures and management decisions (Olsen 2014). In developing countries, recent studies have analyzed food loss across the potato value chain, as for example in Ecuador and Peru, by collecting qualitative and quantitative data to provide a comprehensive identification and characterization of losses. The results show that the most important losses occur in the production node, ranging from 90 to 95% of the total losses in the chain. On average farmers suffer this highest loss across the value chain ranging between 8 and 20% of their production at or before harvest before moving on to the next node of the chain. The main causes of losses are poor crop and harvest management, infested tubers by pest and diseases, high percentage of small tubers and weather conditions: frost and heavy rains (Delgado et al. 2017).

Concluding Remarks: Towards Future Potato Research for Global Food and Nutrition Security

The analysis of leveraging potato agri-food systems for global food security issues and challenges in this chapter emphasizes the need for making agricultural research programs and food system interventions more responsive to food security dimensions. The multidimensional nature of food security requires multisector innovation in approaches that allow to use the knowledge available and transform scientific results into products and processes to improve the performance of agri-food systems, considering the challenge to produce more food with the same or fewer resources.

In both developed and developing countries, innovations resulting from potato research should be incremental through a step by step improvement of an existing structure promoting technologies adapted to the local context. This is particularly true for smallholder family agriculture in developing countries where there is a great need to increase potato production in a sustainable way. While this approach has the advantage of not destabilizing an existing system, it may also suffer of a systemic lock-in or a lack of enabling environment that keeps agriculture and food systems on less efficient pathways as developed by Baret (2017). An example of lock-in is the use of pesticides and their promotion by agro-chemical companies and technical support services that influence farmers' decision making, restraining the use of more environmental friendly options such as decision support tools for efficient pest control with a more rational pesticide use. The valuable use of varieties tolerant or resistant to pests and diseases can also be limited by processing companies that promote varieties for their processing characteristics regardless of their environmental footprint. In developing countries, low infrastructure quality, weak institutions and policies create also huge limitations to the adoption of new and more sustainable technologies. To reach food security goals, a stronger emphasis must be put towards promoting evidence-based policies for communicating information and influence decision makers. It is also important to favor affordable and better-adapted technologies that can respond to the needs of small-scale farmers and, significantly limit negative impacts towards the environment. The research and technology options proposed in this chapter will require policy support, financial and nonfinancial services to have a chance to be adopted and used by local farmers. They will also necessitate a better access to discovery and creative ideas through better services from the potato research community at national and international levels.

Local calibration/validation and demonstration are two essential phases towards local end-user uptake, either involving farmers or extension services representatives. Public or private investments are also required to support such actions to enable farmers to have access to new technologies, and to be trained in their use. With the upcoming of new communication technologies such as smartphones, expansion of mobile broadband and access to local online platforms integrating large amounts of local data and links to Decision Support Systems, we have yet to fully exploit the potential of information technologies especially in developing countries. Local farmers, especially the younger ones, are expected to be able to have increasing access to such new adapted tools, i.e. for Late Blight management in the Andes. But this will only be possible globally if technological innovation is accompanied by capacity building and institutional innovation (associativity, access to credit, communication network) in rural areas.

There is a great dichotomy between research activities in developed versus developing countries that highlights the need for more exchange, knowledge sharing, and collaboration. Since 2014, the interaction between EAPR-linked research organizations and CIP has been looking at mechanisms to enhance partnership between European partners and CIP involving research partners in the Southern hemisphere to promote collaborative research activities, links between research networks such as Euro and Latin Blight (Acuña et al. 2017) as well as facilitating short and long-term training with universities in Europe (Durroux-Malpartida 2014).

To reach the strongest impact on food security, potato research and development efforts need to move towards food systems engineering, rather than focus explicitly on technology/solution development. In this paper, we are analyzing the different components that contribute to the performance of the potato using the key relation $P = G \times E \times M \times S$, enabling a list of key research and technology options to guide agriculture research and technology development toward sustainable intensification approaches responding to farmers' needs both for food security and better income. The argument is that agricultural programs need to integrate better agriculture sustainable intensification and food security indicators considering also other dimensions such as quality, diversity of products, health impacts and climate change effects. Multidisciplinary approaches and a better understanding of the evolving food systems are required to recognize and solve practical problems of the whole potato value chain to achieve sustainable food security. Policies, investments and services that support agricultural productivity, sustainability and expand risk management capacity are also required to give potato farmers the best chance to meet future needs, while increasing their adaptability and resilience to foster food security.

Acknowledgements The authors are grateful and sincerely appreciate the valuable contributions of Mr. Henry Juarez and Dr. Jorge Andrade-Piedra. Their inputs, opinions, and suggestions have greatly contributed to improve this chapter.

References

Acuña I, Restrepo S, Lucca F, Andrade-Piedra J (2017) Recent developments: late blight in Latin America. In: Schepers HTAM (ed) Proceedings of the sixteenth EuroBlight Workshop, Aarhus, Denmark, 14–17 May 2017. PAGV special report no. 18

Andrivon D (2017) Potato facing global challenges: how, how much, how well? Potato Res 60:389. https://doi.org/10.1007/s11540-018-9386-z

Arya S, Ahmed M, Bardhan Roy SK, Kadian MS, Quiroz R (2015) Sustainable intensification of potato in rice-based system for increased productivity and income of resource poor farmers in West Bengal, India. Int J Trop Agric. ISSN 0254-8755 33(2):203–208

Bailey RL, West KP Jr, Black RE (2015) The epidemiology of global micronutrient deficiencies. Ann Nutr Metab 66(Suppl 2):22–33

Baret PV (2017) Acceptance of innovation and pathways to transition towards more sustainable food systems. Potato Res 60:383. https://doi.org/10.1007/s11540-018-9384-1

Birch PRJ, Bryan GJ, Fenton B, Gilroy EM, Hein I, Jones JT, Prashar A, Taylor MA, Torrance L, Toth IK (2012) Crops that feed the world 8: potato: are the trends of increased global production sustainable? Food Secur 4:477–508. https://doi.org/10.1007/s12571-012-0220-1

Black RE, Victora CG, Walker SP, Bhutta ZA, Christian P, de Onis M, Ezzati M, Grantham-McGregor S, Katz J, Martorell R, Uauy R (2013) Maternal and child undernutrition and overweight in low-income and middle-income countries. Lancet 832(9890):427–451

Bohl WH, Johnson SB (2010) Commercial potato production in North America. The Potato Association of America Handbook. http://potatoassociation.org/wp-content/uploads/2014/04/A_ProductionHandbook_Final_000.pdf

Borch D, Juul-Hindsgaul N, Veller M, Astrup A, Jaskolowski J, Raben A (2016) Potatoes and risk of obesity, type 2 diabetes, and cardiovascular disease in apparently healthy adults: a systematic review of clinical intervention and observational studies. Am J Clin Nutr 104(2):489–498

Burlingame B, Mouillé B, Charrondiére UR (2009) Review: nutrients, bioactive non-nutrients and anti-nutrients in potatoes. J Food Compos Anal 22:494–502

CGIAR Research Program on Roots, Tubers and Bananas (2016) Multi-stakeholder framework for intervening in RTB seed systems: user's guide. Lima (Peru), 13 p. RTB Working Paper. ISSN 2309-6586, no. 2016-1

Creed-Kanashiro H, Hareau G, Devaux A, Maldonado L, Ordinola M, Fonseca C, Suarez V, Astete L, Marin M, Penny M (2015) Agriculture-nutrition linkages: analyzing nutritional outcomes of interventions in potato-based production systems of Peru. In: 2nd international conference on global food security, October 11–14, 2015. Cornell University, Ithaca

Crowder DW et al (2010) Organic agriculture promotes evenness and natural pest control. Nature 466:109–112

DeFauw Sherri L, He Z, Larkin RP, Mansour SA (2012) Sustainable potato production and global food security. In: He Z, Larkin R, Honeycutt W (eds) Sustainable potato production: global case studies. Springer, Amsterdam, 531 pp

Delgado L, Schuster M, Torero M (2017) The reality of food losses: a new measurement methodology. IFPRI discussion paper 1686. International Food Policy Research Institute (IFPRI), Washington, DC. http://ebrary.ifpri.org/cdm/ref/collection/p15738coll2/id/131530

Devaux A, Torero M, Donovan J, Horton DE (eds) (2016) Innovation for inclusive value-chain development: successes and challenges. International Food Policy Research Institute (IFPRI), Washington, DC. https://doi.org/10.2499/9780896292130

Devaux A, Goffart JP, Kromann P, Toth I, Braguard C, Declerck S (2017) Report on CIP-EAPR Workshop 2017 on biocontrol and biostimulant agents for the potato crop. Potato Res 60:291. https://doi.org/10.1007/s11540-018-9385-0

Devaux A, Torero M, Donovan J, Horton D (2018) Agricultural innovation and inclusive value-chain development: a review. J Agribus Dev Emerg Econ 8(1):99–123. https://doi.org/10.1108/JADEE-06-2017-0065

Douches DS, Maas D, Jastrzebski K, Chase RW (1996) Assessment of potato breeding progress in the USA over the last century. Crop Sci 36(6):1544–1552

Durroux-Malpartida V (2014) EAPR and CIP strengthen collaboration to tap potatoes' potential. Potato Res 57:367. https://doi.org/10.1007/s11540-014-9276-y

European Commission (2007). The potato sector in the European Union. Commission staff working document. SEC(2007) 533

Eurostat (2017) The EU potato sector—statistics on production, prices and trade. http://ec.europa.eu/eurostat/statistics-explained/index.php/The_EU_potato_sector_-_statistics_on_production,_prices_and_trade#Publications

Fajardo J, Lutaladio N, Larinde M, Rosell C, Barker I, Roca W et al (2010) Quality declared planting material: protocols and standards for vegetatively propagated crops: expert consultation, Lima, 27–29 November 2007. Food and Agriculture Organization of the United Nations, Rome

FAO (2002) The state of food insecurity in the world 2001, Rome, pp 4–7

FAO (2006a) Food security. Policy brief. FAO, Rome

FAO (2006b) Quality declared seed system. FAO plant protection and protection paper # 185, Rome

FAO (2009) International year of the potato 2008: new light on a hidden treasure. http://www.fao.org/potato-2008/en/events/book.html

FAO (2014) FAO statistical databases FAOSTAT. http://faostat3.fao.org/

FAO (2016) FAOSTAT Database. http://www.fao.org/faostat/en/#data/QC

FAO (2015a) Household seed security concepts and indicators. Discussion paper, 10 pp

FAO (2015b) FAOSTAT database. http://www.fao.org/faostat/en/#data/FBS

FAO (2016) Seed security assessment: a practitioner's guide, Rome

FAO, IFAD, UNICEF, WFP and WHO (2018) The state of food security and nutrition in the world 2018. Building climate resilience for food security and nutrition. FAO, Rome

FAOSTAT (2013) Food balance sheet. http://www.fao.org/faostat/en/#data/FBS

FAOSTAT (2017) Food and agriculture data. http://www.fao.org/faostat/en/#data/QCinfo

Foley JA et al (2011) Solution for a cultivated planet. Nature 478:337–342

Gatto M, Hareau G, Pradel W, Suarez V, Qin J (2018) Release and adoption of improved potato varieties in Southeast and South Asia. International Potato Center (CIP), Lima, Peru, 42 p. Social sciences working paper no. 2018-2. ISBN 978-92-9060-501-0

Gildemacher PR, Kaguongo W, Ortiz O, Tesfaye A, Woldegiorgis G, Wagoire WW, Kakuhenzire R, Kinyae P, Nyongesa M, Struik PC, Leeuwis C (2009) Improving potato production in Kenya, Uganda and Ethiopia: a system diagnosis. Potato Res 52(2):173–205

Gildemacher PR, Schulte-Geldermann E, Borus D, Demo P, Kinyae P, Mundia P et al (2011) Seed potato quality improvement through positive selection by smallholder farmers in Kenya. Potato Res 54:253–266

Gillespie S, Van den Bold M (2017) Agriculture, food systems, and nutrition: meeting the challenge. Global Chall 1:1600002. https://doi.org/10.1002/gch2.201600002

Goffart JP, Olivier M, Frankinet M (2008) Potato crop nitrogen status assessment to improve N fertilization management and efficiency: past–present–future. Potato Res 51(3):355–383. https://doi.org/10.1007/s11540-008-9118-x

Goffart JP, Gobin A, Delloye C, Curnel Y (2017) Crop spectral reflectance to support decision making on crop nutrition. Paper presented to the International Fertiliser Society at a conference in Cambridge, United Kingdom, on 7th December 2017. Proceedings 812, pp. 29. www.fertiliser-society.org © 2017 International Fertiliser Society—ISBN 978-0-85310-449-0

Haggblade S, Hazell PBR, Reardon T (2007) Research perspectives and prospectives on the rural nonfarm economy. In: Haggblade S, Hazell PBR, Reardon T (eds) Transforming the rural nonfarm sector: opportunities and threats in the developing world. The Johns Hopkins University Press, Baltimore

Hall A, Bockett G, Taylor S, Sivamohan MVK, Clark N (2001) Why research partnerships really matter: innovation theory, institutional arrangements and implications for developing new technology for the poor. World Dev 29(5):783–797

Harahagazwe D, Condori B, Barreda C, Bararyenya A, Byarugaba AA, Kude DA, Lung'aho C, Martinho C, Mbiri D, Nasona B, Ochieng B, Onditi J, Randrianaivoarivony JM, Tankou CM, Worku A, Schulte-Geldermann E, Mares V (CIP), de Mendiburu F, Quiroz R (CIP) (2018) How big is the potato (*Solanum tuberosum* L.) yield gap in Sub-Saharan Africa and why? A participatory approach. Open Agric 3(2):180-189. ISSN 2391-9531 .

Hareau G, Kleinwechter U, Pradel W, Suarez V, Okello J, Vikraman S (2014) Strategic assessment of research priorities for Potato. Lima (Peru). CGIAR Research Program on Roots, Tubers and Bananas (RTB). RTB working paper 2014-8. www.rtb.cgiar.org

Haverkort AJ, Struik PC (2015) Yield levels of potato crops: recent achievements and future prospects. Field Corp Res 182:76–85

Haverkort A, de Ruijter FJ, van Evert FK, Conijn JG, Rutgers B (2014) Worldwide sustainability hotspots in potato cultivation. 1. Identification and mapping. Potato Res 56:343–353

Hazell P, Haggblade S (1989) Agricultural technology and farm–nonfarm growth linkages. Agric Econ 3:345–364

Headey DD, Jayne TS (2014) Adaptation to land constraints: is Africa different? Food Policy 48:18–33

INEI Peru (2017) IV Censo Nacional Agropecuario (CENAGRO). Lima. INEI Perú (2017). Encuesta demográfica y de salud familiar 2016. Lima, Peru

Jayne TS, Chamberlin J, Headey DD (2014) Land pressures, the evolution of farming systems, and development strategies in Africa: a synthesis. Food Policy *48*:1–17

Kanter R, Walls HL, Tak M, Roberts F, Waage J (2015) A conceptual framework for understanding the impacts of agriculture and food system policies on nutrition and health. Food Secur 7(4):767–777

Kleinwechter U, Hareau G, Suarez V (2014) Prioritization of options for potato research for development—results from a global expert survey. Lima (Peru). CGIAR Research Program on Roots, Tubers and Bananas (RTB). RTB working paper 2014-7. www.rtb.cgiar.org

Kunkel R, Campbell GS (1987) Maximum potential potato yield in the Columbia Basin, USA: model and measured values. Am Potato J 64(7):355

Labarta R (2015) The effectiveness of potato and sweetpotato improvement programmes from the perspectives of varietal output and adoption in Sub-Saharan Africa. In: Walker T, Alwang J (eds) Crop improvement, adoption, and impact of improved varieties in food crops in Sub-Saharan Africa. CABI, Wallingford

Muthoni J, Mbiyu MW, Nyamongo DO (2010) A review of potato seed systems and germplasm conservation in Kenya. J Agric Food Inform 11(2):157–167

Obura B, et al (2016) Cost benefit analysis of seed potato replacement strategies among smallholder farmers in Kenya. Tropentag, September 18–21, 2016, Vienna, Austria. http://www.tropentag.de/2016/abstracts/links/Obura_eAhVYqsS.pdf. Accessed 7 Sept 2018

Olsen N (2014) Potato storage management: a global perspective. Potato Res 57:331–333. https://doi.org/10.1007/s11540-015-9283-7

Pandey A (2015) Rural farm and non-farm linkages in Uttar Pradesh. J Land Rural Stud *3*(2):203–218

PGSC (The Potato Genome Sequencing Consortium) (2011) Genome sequence and analysis of the tuber crop potato. Nature 475:189–195

Pingali P (2006) Westernization of Asian diets and the transformation of food systems: implications for research and policy. Food Policy 32(3):281–298

Pinstrup-Andersen P (2014) Food systems and human nutrition: relationships and policy interventions. In: Thompson B, Amoroso L (eds) Improving diets and nutrition: food-based approaches (Chapter 2). CABI, Wallingford, pp 8–20

Quiroz R, Ramírez DA, Kroschel J, Andrade-Piedra J, Barreda C, Condori B et al (2018) Impact of climate change on the potato crop and biodiversity in its center of origin. Open Agric 3:273–283

Reardon T, Chen K, Minten B, Adriano L (2012) The quiet revolution in staple food value chains: enter the dragon, the elephant and the tiger. Asian Development Bank (ADB) and International Food Policy Research Institute (IFPRI), Mandaluyong City

Rosegrant MW, Sulser TB, Mason-D'Croz D, Cenacchi N, Nin-Pratt A, Dunston S, Zhu T, Ringler C, Wiebe K, Robinson S, Willenbockel D, Xie H, Kwon H-Y, Thomas TS, Wimmer F, Schaldach R, Nelson GC, Willaarts B (2017) Quantitative foresight modeling to inform the CGIAR research portfolio. Project report for USAID. International Food Policy Research Institute (IFPRI), Washington, DC

Salazar L, Aramburu J, González M, Winters P (2016) Food security and productivity: impacts of technology adoption in small subsistence farmers in Bolivia. Food Policy 65:32–52

Schulte-Geldermann E, Gildemacher PR, Struik PC (2012) Improving seed health and seed performance by positive selection in three Kenyan potato varieties. Am J Potato Res 89:429–437. https://link.springer.com/content/pdf/10.1007%2Fs12230-012-9264-1.pdf. Accessed 13 Sept 2018

Scott G (2011) Growth rates for potatoes in Latin America in comparative perspective: 1961–07. Am J Potato Res 88:143–152

Scott GJ, Suarez V (2011) Growth rates for potato in India and their implications for industry. Potato J 38(2):100–112

Scott GJ, Suarez V (2012a) From Mao to McDonald's: emerging markets for potatoes and potato products in China 1961-2007. Am J Potato Res 89(3):216–231

Scott GJ, Suarez V (2012b) The rise of Asia as the center of global potato production and some implications for industry. Potato J 39(1):1–22

Scott GJ, Suarez V (2012c) Limits to growth or growth to the limits? Trends and projections for potatoes in China and their implications for industry. Potato Res 55(2):135–156

Smith A, Snapp S, Chikowoa R, Thorne P, Bekunda B, Gloverd J (2017) Measuring sustainable intensification in smallholder agroecosystems: a review. Glob Food Sec 12:127–138

The Wall Street Journal (2015) Pushing the potato: China wants people to eat more 'Earth Beans'. https://www.wsj.com/amp/articles/BL-CJB-25584?responsive=y

Thiele G, Hareau G, Suarez V, Chujoy E, Bonierbale M, Maldonado L (2008) Varietal change in potatoes in developing countries and the contribution of the International Potato Center: 1972–2007. International Potato Center (CIP), Lima, Peru. Working paper 2008-6, 46 p

Thomas-Sharma S, Abdurahman A, Ali S, Andrade-Piedra JL, Bao S, Charkowski AO et al (2015) Seed degeneration in potato: the need for an integrated seed health strategy to mitigate the problem in developing countries. Plant Pathol 65:3–16. https://www.researchgate.net/publication/280913576_Seed_degeneration_in_potato_The_need_for_an_integrated_seed_health_strategy_to_mitigate_the_problem_in_developing_countries. Accessed 13 Sept 2018

Velivelli SLS, Sessitsch A, Prestwich BD (2014) The role of microbial inoculants in integrated crop management systems. Potato Res 57(3–4):291–309

Von Grebmer K et al (2017) Global hunger index: getting to zero hunger. Welthungerhilfe/International Food Policy Research Institute/Concern Worldwide, Bonn/Washington, DC/Dublin. https://doi.org/10.2499/9780896292710

World Bank (2017) Gaining momentum in Peruvian agriculture: opportunities to increase productivity and enhance competitiveness. World Bank, Washington, DC

Wu W, Yu Q, You L, Chen K, Tang H, Liu J (2018) Global cropping intensity gaps: increasing food production without cropland expansion. Land Use Policy 76:515–525

You L, Wood-Sichra U, Fritz S, Guo Z, See L, Koo J (2014) Spatial Production Allocation Model (SPAM) 2005 v3.2. http://mapspam.info

5

Potato Seed Systems

Gregory A. Forbes, Amy Charkowski, Jorge Andrade-Piedra, Monica L. Parker, and Elmar Schulte-Geldermann

Abstract Good quality seed is almost universally considered a requirement for high productivity in all potato production systems. Much of the yield gap currently constraining productivity in low-income countries is attributed to the poor quality of seed. Potato seed sector development is thus a major concern of governments, researchers, development agencies, and civil society organizations. Potato seed systems are often characterized as formal or informal, although the informal seed system is complex and particularly in low income countries there are many linkages between the two systems. Informal seed potato systems in the Andes have existed for centuries, and for a number of reasons often produce seed of relatively high quality. In other low-income countries, informal systems produce seed of variable and frequently poor quality, contributing to very large yield gaps, characteristic of those areas. In regions of high potato productivity (e.g., the USA and Europe), formal systems, with seed of certified high quality, are dominant, although some productions subsectors (e.g., organic producers) often use seed that is not certified. Efforts to implement formal seed systems in low-income countries have been largely unsuccessful; consequently the vast majority of low-resource potato farmers source their seed via the informal system. Sectors of the development community

G. A. Forbes
Independent Consultant, Servas, Gard, France

A. Charkowski
Colorado State University, Fort Collins, CO, USA
e-mail: amy.charkowski@colostate.edu

J. Andrade-Piedra (✉)
International Potato Center, Lima, Peru
e-mail: j.andrade@cgiar.org

M. L. Parker
International Potato Center, Nairobi, Kenya
e-mail: m.parker@cgiar.org

E. Schulte-Geldermann
International Potato Center, Nairobi, Kenya

TH Bingen, University of Applied Sciences, Bingen, Germany
e-mail: e.schulte-geldermann@th-bingen.de

are pushing for alternative solutions, which generally involve some form of integrating formal and informal seed systems or semi-formal systems such as quality declared seed, and a policy structure that preserves farmers' rights to save and trade seed. Given the role that seed quality is currently playing in the low yields of potato in low-income countries, which is not the case in wealthier parts of the world, the review focuses primarily on seed sector development in resource-poor areas.

The Seed System Context

Seed system research and lexicon In keeping with the large social and economic dimensions of potato seed systems (discussed in more depth later in the article), research on seed systems is extensive and multifaceted, and terminology used to describe seed-related issues is varied and often confusing. Thiele (1999) broadly defined a seed system as "an interrelated set of components including breeding, management, replacement and distribution of seed." This definition is generally consistent with one established earlier at a workshop held in 1995 in Indonesia, "the total of physical, organizational and institutional components, their actions and interactions, that determine seed supply and use, in quantitative and qualitative terms" (Amstel et al. 1996), and with other definitions given since then (Camargo et al. 2004; Muthoni et al. 2010; Kromann et al. 2016). In recent years, patents and plant variety protection have added the additional dimension of intellectual property and germplasm ownership, and these impact plant breeding, crop management, seed replacement, and distribution of seed.

Seed systems have also been classified by type, with the major classes being formal and informal (Thiele 1999). The concept of a formal seed system is relatively clear, being characterized by components that are regulated by the public sector, usually by an inspection process known as "certification" and including controls over variety release, to ensure that available seed is of a recognized variety and with a low incidence of disease (Louwaars 1994; Amstel et al. 1996; Thiele 1999). The informal system is complex and conceptually less clear in that it basically includes all that is not formal, including self-saved seed, seed traded among farmers, and that acquired at local markets (Thiele 1999; Almekinders and Louwaars 2002; McGuire and Sperling 2016). While all seed outside the formal system is frequently referred to as "informal," it is also referred to using other terms, such as "farmers' seed" (Almekinders and Louwaars 2002), "local seed" (Almekinders et al. 1994), or "traditional seed" (McGuire and Sperling 2016). For this chapter, we will use the term "informal."

Several authors have highlighted the importance of the informal seed system in middle-low and low-income countries (hereafter referred to as low- income, Thiele 1999; Almekinders and Louwaars 2002; Louwaars and de Boef 2012; McGuire and Sperling 2016), including a particular focus on potato (Thiele 1999; Thomas-Sharma et al. 2015). The literature is also strongly supportive of the need to integrate formal and informal sectors in countries where the former provides only a small portion of seed that is needed (Louwaars 1994; Amstel et al. 1996; Thiele 1999; Louwaars and de Boef 2012; Thomas-Sharma et al. 2015; De Jonge and Munyi 2017). The strategy of integrating different seed systems has had practical implications in the development sector with reports on specific cases (e.g., Kromann et al. 2016) and through the development of programs focusing on an integrated approach to seed system development.

In spite of broad recognition of the role of the informal seed system by the academic community, seed sector development has reflected "varied and often opposing philosophies" (McGuire and Sperling 2016). Many development projects have been designed only (or primarily) to support the formal seed sector in low-income countries and have relied on outside expertise, without significantly increasing the minimal role that certified seed plays in providing planting material to low-income farmers (McGuire and Sperling 2016); this is particularly true for potato seed systems (Thomas-Sharma et al. 2015). Even in countries with strong formal seed systems, the formal seed system can fail farmers who do not grow the crop as a commodity. For example, organic farmers in the US commonly use informal seed for specialty and heirloom varieties and this is yet more evidence that formal seed systems only work for growers who specialize in large acreages of potato.

Why are potato seed systems important? Healthy seed systems have been described as providing access to quality planting material, at the time needed, at a fair price, to all who need it (Sperling 2008). Access to quality seed is widely considered one of the main requirements for bridging large yields gaps for potato still found in most low-income countries (Hidalgo et al. 2009; Schulte-Geldermann 2013). Healthy seed systems also act as to reduce risk of disease outbreaks by keeping spread of a disease in check or even as part of a pest eradication plan. Conversely, seed systems without effective quality control can be very efficient at spreading seed-borne pathogens. Seed systems are also important for the diffusion of new varieties with beneficial traits and the maintenance of crop diversity in the landscape (Pautasso et al. 2013; Arce et al. 2018). In the case of a new or emerging pathogen in a region, the seed system acts as the conduit through which locally adapted resistant varieties can be distributed (if these are available).

Arguably, the primary impetus for development of seed systems in potato is the vegetative nature of propagation of the crop, and the phenomenon of what is now referred to as degeneration. The importance and causes of seed degeneration, a process through which yield is lost in vegetatively propagated crops through pathogen accumulation in consecutive cycles of propagation, are of particular concern to informal seed systems. Globally, seed degeneration is among the leading limitations to potato yield (Thomas-Sharma et al. 2015; Bertschinger et al. 2017). In high-income countries, which have the highest potato yields, this problem has been effectively managed, at least for large commercial growers, through the utilization of seed certified to have high quality (low incidence of pathogens, varietal purity, and appropriate physiological age). This process has been highly successful for most producers in these countries by providing access to economically priced seed of high quality.

Fig. Degeneration in potato. Small plants with low yield due to the accumulation of pathogens in consecutive cycles of vegetative propagation. (Photo credits: G. Forbes)

For smallholder farmers from low- to middle-low-income countries, certified seed is often not available, or the cost is prohibitive. Instead, farmers acquire seed of unknown quality via the informal system, either from the previous year's crop, or from other informal sources such as those mentioned above. In informal systems, degeneration is often a problem because seed is not tested and may be produced under conditions of high disease pressure with little or no quality control.

Traditional Potato Seed Systems in the Andes

The Andean region is the origin of the cultivated potato and represents an interesting case for studying potato seed systems. Potato seed systems in the Andes have been informal for millennia, and even today only small amounts of formal seeds are used by Andean farmers (Hidalgo et al. 2009; Devaux et al. 2014). With even relatively low rates of disease spread, one could assume that high levels of degeneration would occur in areas where the informal system has been dominant for centuries. However, studies done in traditional Andean potato seed systems over the past 30 years often found relatively low frequencies of tubers infected with yield-limiting potato viruses (Bertschinger et al. 1990; Fankhauser 1999; Pérez et al. 2015; Navarrete et al. 2017).

Several factors have been identified that could contribute to the continued quality of seed potato in the informal systems in the Andes. Andean farmers have

Fig. Native varieties in the Andes. Seed for producing most of these varieties come from informal systems, but major yield-limiting potato viruses, which are the most common cause of seed degeneration and are often found in relatively low frequencies. (Photo credits: J.L. Gonterre, in association with the International Potato Center)

traditionally had complex farming practices that conceivably help reduce the soil-borne phases of diseases leading to degeneration, such as sectoral (Orlove and Godoy 1986) or other types of fallowing and rotation (Thurston 1990), high hilling, or reduced cultivation methods (e.g., Cartagena et al. 2004). Other factors characteristic of Andean potato systems may contribute to reduce disease spread among plants or pathogen transmission within plants. Resistance to PVY and PLRV have been found in *S. tuberosum* subsp. *andigena,* and *Ry* genes have been found in other Andean taxa making up local potato landraces (Machida-Hirano 2015), which may partly explain low incidences of PVY and PLRV found in these varieties. Bertschinger et al. (2017) also found that virus transmission from infected mother tubers to daughter tubers was greatly suppressed at high altitudes in the Andes. This is consistent with traditional practices in which farmers moved virus-infected seed to higher altitudes to reduce infection (Thiele 1999). High levels of agrobiodiversity may also help mitigate degeneration in traditional Andean potato fields. One study found that Peruvian farmers growing native "floury" cultivars between 3500 and 4250 m altitude had on average between 8 and 20 different genotypes per field (de Haan et al. 2010). This biodiversity is "uneven" in that it is highly dependent on the type of farmer, but it is also an important component of a complex seed exchange network that represents "a strong safety net through which smallholders can respond to crop failure and seed stress" (Arce et al. 2018).

Potato Seed Systems in Europe and North America

The basic outline of the seed system used in Europe and North America for potato was developed in the late nineteenth and early twentieth centuries and has its roots in how seed potatoes are grown in the Andes (Shepard and Claftin 1975; Frost et al. 2013). Considerable advances were made in the 1980s, when both pathogen testing and potato micropropagation became widespread. Despite the use and availability of technology, seed potatoes are produced primarily in the northern agricultural regions of these two continents to avoid insect virus vectors, and a wide range of bacterial and fungal diseases common in warmer climates. In Europe and North America, commercial growers who plant large acreages of potato almost exclusively use certified seed. However, in the United States, farmers who manage mixed vegetable farms, and particularly organic farmers, generally use informal seed, demonstrating that current seed systems tend to best serve growers who produce potato as a commodity.

In these seed systems, potato varieties are maintained in tissue culture as micropropagated plantlets. These initial plantlets, often called mother plants, are tested for all pathogens of concern, including the major potato viruses, potato spindle tuber viroid, and common bacterial and fungal pathogens (Frost et al. 2013). Propagation in tissue culture is relatively inexpensive, requires little space, and the plantlets grow quickly, so hundreds of thousands of plantlets can be produced annually in a relatively small facility of tens to a few hundred square meters (Naik and Buckseth 2018). The micropropagated plantlets are then planted into greenhouses or screenhouses into either pots or into hydroponic or aeroponic systems. The potatoes harvested from these greenhouse-grown plants are called nuclear seed or minitubers. The minitubers must be stored until the subsequent season to break tuber dormancy. Minitubers are planted into seed potato fields and the progeny from these plants are generally field-multiplied another 2–5 years before being sold to farmers who grow potatoes for fresh use or processing.

At each stage, specific inspections and pathogen tests are required for the certification schemes used in each country, state, or province where potatoes are grown. Generally accepted protocols are collected and verified by entities such as the European and Mediterranean Plant Protection Organization (EPPO), the North American Plant Protection Organization (NAPPO), and the United Nations Economic Commission for Europe (UNECE) to aid in trade in certified seed potato. The efforts have had some success, with increases in yield and near-elimination of diseases such as spindle tuber and bacterial ring rot (Frost et al. 2013). Policy harmonization has also engendered debate in low-income countries (De Jonge and Munyi 2017) and in Europe (Prip and Fauchald 2016). In the United States, seed potato producing states still have relatively little similarity in their certification regulations across the different states. Currently, potato virus Y is the most important potato virus on these continents, but losses due to this virus are relatively small (Zeng et al. 2018).

These certification systems primarily focus on diseases that are only spread by seed potatoes and not on important soil-borne pathogens. As a result, diseases such as powdery scab, corky ringspot, golden cyst nematode, and other difficult-to-control diseases are now widespread and increasing in importance (e.g., Beuch et al. 2014; Contina et al. 2018). A second challenge is that plant variety protection has resulted in the proliferation of similar, but protected varieties, for which little information on disease response is available. This poses a significant challenge to certification agencies, which are tasked with insuring varietal purity and disease thresholds on an ever-increasing number of new potato varieties. Finally, growers in both Europe and North America are investing in the development of inbred and hybrid diploid varieties that can be produced through true seed rather than plant micropropagation (Lindhout et al. 2011; Jansky et al. 2016). If these varieties become popular, they will alter the current seed system.

Potato Seed Systems in Low-Income Countries

Earlier we stated that low-income countries are characterized by informal seed systems with very low use of certified seed. This is generally the case, but it is worth examining in some detail efforts that have gone into establishment of certified seed programs in some of these countries, as well as a number of recent innovations aimed at improving seed systems of resource-poor farmers.

The highly conspicuous absence of certified seed in the potato seed sector in most low-income countries has recently been documented, at least in part. In specific reference to potato, Thomas-Sharma et al. (2015) list percentages of formal and informal seed in 14 low- income countries. In China and India formal seed usage is listed at 20%, but in all other countries it is below 10% and in most it is below 5%. McGuire and Sperling (2016) provide a more extensive examination of how farmers source many kinds of seed in low-income countries and note that for potato over 95% of seed comes from own stock, friends, neighbors, relatives, or local markets, i.e., the informal system. It is worth noting for context that

Fig. Informal seed potato in a local market in Bangladesh. (Photo credits: J. Andrade-Piedra)

these authors also show that in low-income countries there is a similar pattern for all vegetatively propagated crops and, somewhat surprisingly, for legumes and cereals as well.

The lack of certified seed for potato and other crops in low-income countries is not easily attributed to a lack of effort on the part of governments and development agencies. McGuire and Sperling (2016) provide an impressive list of projects funded by the World Bank and by the Alliance for a Green Revolution in Africa (AGRA) as an indication of development support to seed systems in low-income countries. In the preparation of this document we were not able to find data on investments specifically in the potato seed sector in developing countries, but there is no doubt that many millions of dollars have been spent by development agencies over the last half a century to improve the potato seed sector in low-income countries.

Seed sector actors, and specifically donors, in low-income countries take different approaches to the problem, which may be generally classified into two types: those that predominantly support development of a formal seed sector and those that support a broader approach to seed sector development (McGuire and Sperling 2011; Thomas-Sharma et al. 2015; Otieno et al. 2017). While a number of donors subscribe to broad seed sector development (Lossau et al. 2000), it is the experience of the authors that the large majority of projects, and certainly the larger projects supporting potato seed sector development, tend to focus primarily on formal seed sector development.

Perspectives on Potato Seed System Development

Policy Low-income countries are struggling with numerous policy issues related to seed. Many governments and regional organizations are developing policies and laws modeled on the guidelines of the International Union for the Protection of New Varieties of Plants (UPOV). This has led to much concern in civil society and in the research community of the impacts that such policies could have on resource-poor farmers and informal seed trade (Tripp and Louwaars 1997; De Jonge and Munyi 2017; Otieno et al. 2017; Vernooy 2017). The controversies surrounding these policies have given rise to both proponents and opponents of regional harmonization laws based on UPOV standards; opposing actors apparently rarely meet to debate options (De Jonge and Munyi 2017).

It is unclear what the eventual effects of this struggle will have on resource-poor potato farmers in these countries. Because of its vegetative nature, perishability, and bulkiness, seed potato (namely tubers whereas potato seed refers to true, botanical seed) presents particular difficulties for establishing breeding programs, implementing certified seed systems and marketing seed in a way that is commercially viable in low-income countries where the infrastructure and other elements of the business ecosystem are not favorable. This could be a major reason why there has been very little activity of major seed potato companies in low-income countries (Thomas-Sharma et al. 2015).

Funding for seed sector development represents an area where seed policy and seed sector development philosophy can affect resource-poor potato farmers. As noted, most funding in potato seed sector development over the years has been in support of the formal sector, with relatively little funding to optimize and promote on-farm seed management, which has been shown to be effective in slowing down or even reversing seed degeneration important in areas without access to certified seed (see below).

Underlying the struggle over seed regulation in low-income countries is the contrast, often seen as a dichotomy, between formal and informal seed systems. As noted, integration of seed systems (Louwaars and de Boef 2012; Kromann et al. 2016; Ferrari et al. 2017) has been proposed as a way to find common ground between those promoting commercial seed industry, plant variety protection, and harmonization of seed standards, and those promoting farmers rights to save and trade their own seed.

Although it was not intended as a mechanism to integrate seed systems, the quality declared seed (QDS) approach offers a more flexible alternative for seed quality assurance than strict certification programs. Developed by FAO (2006) and later adapted for potato and other vegetatively propagated crops (Fajardo et al. 2010), the QDS approach is being used in Ethiopia (Schulz et al. 2013) and some elements of it are applied in Ecuador (Kromann et al. 2016) and Peru (MINAGRI 2018). Seed potato produced under a QDS approach was shown to be a profitable business for seed multipliers in Kenya, but at the same time it has been ineffective in limiting the dissemination of bacterial wilt (*Ralstonia solanacearum*) and potato cyst nematode (*Globodera pallida*), which points out the need of rigorous testing and validation of the QDS approach to local conditions. However, formal seed systems also are unable to effectively limit the dissemination of pathogens such as *R. solanacearum* and *G. pallida*, so the QDS approach is not deficient compared to formal seed systems in this respect.

Technology Innovations A number of the technological innovations have been or are being evaluated and promoted to improve seed potato quality in low-income countries. Many of these are relatively old but have been recently revisited, and often adapted, for their application to certain situations, particularly where resources are scarce. Some of these technologies are reviewed here under two categories: those that relate to on-farm management of seed, and those relating to rapid multiplication of early generation seed.

On-Farm Seed Management Some relatively old technologies are receiving renewed consideration by seed specialists. Positive selection is implemented by farmers and consists of identifying and marking plants that have no visible symptoms of disease or abiotic stress. Seeds for the next planting are then taken from these plants. This sounds relatively simple and the activity itself is simple, however, the efficacy of selection can depend on many factors including, the type of virus, environmental conditions, and farmer skill. Positive selection is particularly important in areas where there is no access to seed produced under a quality control system, thus the impetus is on the farmer to manage quality control. Nonetheless, a number of studies have demonstrated significant improvements in seed quality at the farm level as a result of positive selection implemented by farmers (Gildemacher et al. 2011; Schulte-Geldermann et al. 2012; Okeyo et al. 2018; Priegnitz et al. 2018), which could be due to the fact that it is easier to identify a fully healthy plant than it is to identify symptoms potentially caused by virus or that could also be due to abiotic stresses (Gildemacher et al. 2011). When positive selection is used by farmers there is an increase in yield that could be attributed to several factors, one of these being a reduction in the incidence of virus infection (Schulte-Geldermann et al. 2012). To improve the utility of this relatively old approach, CIP and its partners developed a number of training guides for positive selection aimed at both farmers and trainers (Gildemacher et al. 2007).

Another old tactic that has received some renewed attention is the seed plot technique, which consists of producing small amounts seed of relatively high quality in a confined area that is free of or has a low incidence of soil-borne pathogens (Vashisth 1979; Bryan 1983; Kakuhenzire et al. 2005; Ali et al. 2013). There are many variations on this very simple principle that can be applied to devise flexible systems that adapt to different contexts. The initial seed may be purchased or may be derived from positive selection (Bryan 1983). The best seed coming from the seed plot, i.e., that produced with positive selection in the seed plot and further post-harvest selection, can be used for a new seed plot in the next season. The remaining seed from the seed plot is used for ware production. The seed plot technique can very easily be integrated with the purchase of small amounts of high-priced certified seed (Kinyua et al. 2015; Ochieng-Obura et al. 2016).

Rapid multiplication technologies Aeroponics is a more recent technology that has made inroads into the potato sector in the last few decades. This technique consists of a soilless culture, in which the underground part of the plant is enclosed in a dark chamber and supplied with nutrients through a misting system. Plants grown in this way produce minitubers in the dark chamber, which are harvested as they reach the desired size. Within an aeroponics system, plants may produce very high numbers of minitubers, with plants on average sometimes producing over 45 tubers (Mateus-Rodriguez et al. 2013). There is no shortage of research (and opinion) expounding the benefits of aeroponics (Muthoni et al. 2011; Chiipanthenga et al. 2012; Kakuhenzire et al. 2017; Lakhiar et al. 2018), but implementation of aeroponics in low-income countries where resources are scarce and power supplies unreliable is costly, difficult and risky (Mateus-Rodriguez et al. 2013). The introduction of aeroponics in sub-Saharan Africa has resulted in a large increase minituber production (Harahagazwe et al. 2018), although total numbers are still low and minitubers are expensive.

Unfortunately, feasibility and even economic analyses of rapid multiplication technologies used in development projects generally do not consider the role of the development community's purchase of seed, which is often a market-distorting factor in fledgling seed programs in low-income countries (Bentley and Vasques 1998; Bentley et al. 2001). Aeroponics is just one form of hydroponic plant production and other forms are also used in seed potato production (Lommen 2007; Corrêa et al. 2009), many of which are simpler than aeroponics and may be more appropriate for many low-income countries (Mateus-Rodriguez et al. 2013). For example, nutrient film hydroponic systems are widespread in North America and also present in Brazil and China. These systems are relatively easy to manage and yield and cost of production information is available (Guenthner et al. 2014). The appropriateness of a complex technology like aeroponics appears to depend on the capacity of local players and local infrastructure (Mateus-Rodriguez et al. 2013), hence sand hydroponics not relying on power and highly skilled labor is an attractive alternative.

Rapidly growing young vegetative tissue of potato can be cut and rooted in a number of ways (Bryan et al. 1981). Apical rooted cuttings have long been used in SE Asia (Vander Zaag and Escobar 1990), and particularly in Vietnam (Tran et al. 1990). This technique is being introduced into sub-Saharan Africa to provide a simple but effective technique for multiplying early generation seed (Parker et al. 2019) In the current application used in sub-Saharan Africa, two-node apical cuttings (4–5 cm long) are harvested several times at intervals of 2–3 weeks from in vitro-derived mother plants. The cuttings are then rooted in trays with a substrate of coconut sawdust, clean subsoil, and sterilized decomposed manure. Once rooted, the cuttings can be transplanted directly to the field to produce the first generation of tubers.

Cuttings are penetrating the seed system, and opportunities they present are being validated in Kenya to scale out the technology through diversified use

Strategic innovations Given the contrasting approaches of seed sector development actors (McGuire and Sperling 2016), it is not surprising that many people interested in this subject have called for greater integration, both in development and research. Thomas-Sharma et al. (2015) proposed an integrated approach to managing the problem of degeneration through a strategy called integrated seed

Fig. Cutting almost ready for transplanting (left), and soon after transplanted in nursery beds in the field (right). (Photo credits: M. Parker)

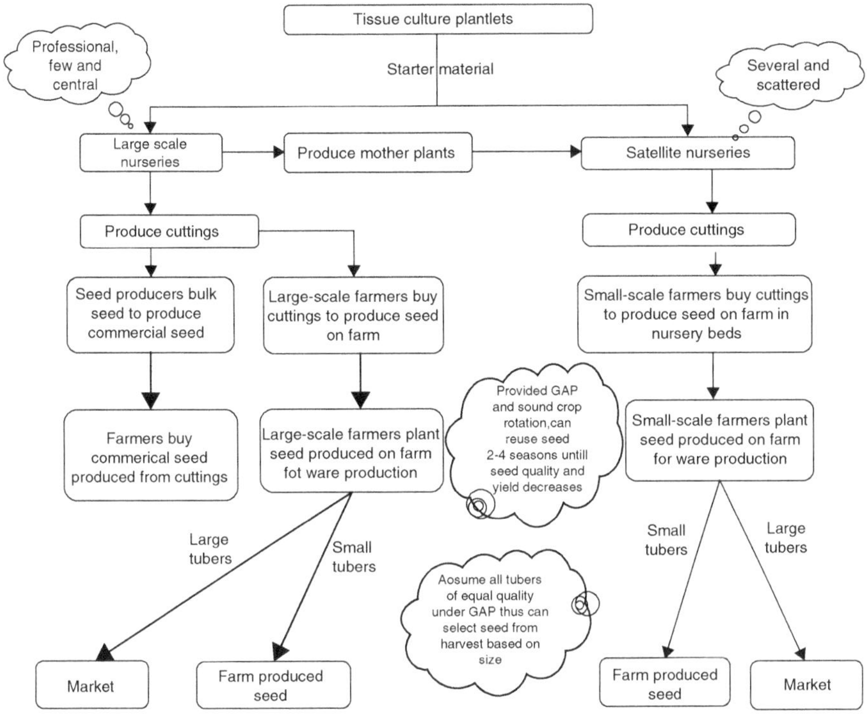

Fig. Diversified pathways to use cuttings in seed production to scale out the technology; GAP: good agricultural practices

health. This involves the integration of three different classes of tactics farmers may employ to maintain or even improve seed quality: (1) on-farm practices such as seed plot technique or positive selection; (2) the use of varieties that degenerate slowly due to their natural resistance to degeneration-causing pathogens; and (3) a more strategic use of certified seed. The latter may involve the less frequent purchase of clean replacement seed, or purchase of small quantities that may be put in a designated seed plot (Ochieng-Obura et al. 2016).

At a higher level, researchers have also called for integration throughout the seed sector, with emphasis on the interaction between formal and informal systems (Louwaars 1994; Tripp 1996; Munyi and De Jonge 2015). The most visible incarnation of this approach is the Integrated Seed Systems Development (ISSD) program.[1] This program is managed globally by the Centre for Development Innovation (CDI), Wageningen University & Research (WUR) and the Royal Tropical Institute (KIT), but has many local partners in Africa and country programs in Uganda and Ethiopia.

The CGIAR has long been another major player in seed sector development in low-income countries, with CIP leading the potato component. For many years seed

[1] See ISSD Website http://www.issdseed.org/.

programs resided within specific centers but this has been consolidated to some extent within the CGIAR Research Programs. All potato work resides within the Roots, Tubers and Banana program (RTB), which initiated a project in 2012 to address biophysical (especially seed degeneration) and socioeconomic constraints of seed production in RTB crops. This led to fruitful collaborations with advanced research institutions in the US and Europe and has produced some novel approaches to studying seed systems, including a conceptual framework for intervening in RTB seed systems (Bentley et al. 2018), a multi-crop analysis (Almekinders et al. 2019), the integrated seed heath approach explained above (Thomas-Sharma et al. 2015), epidemiological modeling of seed potato degeneration (Thomas-Sharma et al. 2017), and the geographic analysis of seed system dynamics using network analysis (Buddenhagen et al. 2017). Hence, since seed systems are complex, more research is needed to identify the right entry points and multiple angles for innovations to enhance the systems as a whole and according to local conditions.

References

Ali S, Kadian MS, Ortiz O, Singh BP, Chandla VK, Akhtar M (2013) Degeneration of potato seed in Meghalaya and Nagaland states in north-eastern hills of India. Potato J 40:122–127

Almekinders CJ, Louwaars NP (2002) The importance of the farmers' seed systems in a functional national seed sector. J New Seeds 4:15–33

Almekinders CJM, Louwaars NP, De Bruijn GH (1994) Local seed systems and their importance for an improved seed supply in developing countries. Euphytica 78:207–216

Almekinders CJM, Walsh S, Jacobsen KS, Andrade-Piedra JL, McEwan M, De Haan S, Kumar L, Staver C (eds) (2019) Why interventions in the seed systems of roots, tubers and bananas crops do not reach their full potential. Food Security. First Online 23 Jan 2019, 20 p. ISSN 1876-4525

Amstel H van, Bottema JW, Sidik M, Santen CE van (1996) Integrating seed systems for annual food crops. Proceedings of a workshop held in Malang, Indonesia October 24-27, 1995. Regional Co-ordination Center for Research and Development of Coarse Grains, Pulses, Roots and Tuber Crops in the Humid Tropics of Asia and the Pacific, Research Institute for Legume and Tuber Crops, Agronomy Society of Indonesia

Arce A, De Haan S, Burra DD, Ccanto R (2018) Unearthing unevenness of potato seed networks in the high Andes: a comparison of distinct cultivar groups and farmer types following seasons with and without acute stress. Front Sustain Food Syst 2:43

Bentley JW, Vasques D (1998) The seed potato system in Bolivia: organisational growth and missing links. ODI, London

Bentley JW, Tripp R, De La Flor RD (2001) Liberalization of Peru's formal seed sector. Agric Hum Values 18:319–331

Bentley JW, Andrade-Piedra JL, Demo P, Dzomeku B, Jacobsen K, Kikulwe E, Kromann P, Lava Kumar P, McEwan M, Mudege N, Ogero K, Okechukwu R, Orrego R, Ospina B, Sperling L, Walsh S, Thiele G (2018) Understanding root, tuber, and banana seed systems and coordination breakdown: a multi-stakeholder framework. J Crop Improv. Published online 18 Jun 2018, 23 p. ISSN 1542-7528

Bertschinger L, Scheidegger UC, Luther K, Pinillos O, Hidalgo A (1990) La incidencia de virus de papa en cultivares nativos y mejorados en la sierra peruana. Revista Latinoamericana de la Papa 3:62–79

Bertschinger L, Bühler L, Dupuis B et al (2017) Incomplete infection of secondarily infected potato plants—an environment dependent underestimated mechanism in plant virology. Front Plant Sci 8:74
Beuch U, Persson P, Edin E, Kvarnheden A (2014) Necrotic diseases caused by viruses in Swedish potato tubers. Plant Pathol 63:667–674
Bryan JE (1983) On farm seed improvement by the potato seed plot technique. Technical information bulletin 7. International Potato Center, Lima
Bryan J, Jackson M, Meléndez N (1981) Rapid multiplication techniques for potatoes. International Potato Center (CIP), Lima
Buddenhagen CE, Hernandez Nopsa JF, Andersen KF, Andrade-Piedra J, Forbes GA, Kromann P et al (2017) Epidemic network analysis for mitigation of invasive pathogens in seed systems: potato in Ecuador. Phytopathology 107:1209–1218
Camargo CP, Bragantini C, Monares A (2004) Small-scale seed production systems: a non-conventional approach. In: Seed multiplication by resource-limited farmers. Proceedings of the Latin American workshop. Goiânia, Brazil, 7–11 April 2003. FAO plant production and protection paper, vol 180, pp 63–66
Cartagena Y, Toapanta G, Valverde F (2004) Mas papas con huacho rozado. Instituto Nacional Autónomo de Investigaciones Agropecuarias (INIAP), Porteviejo
Chiipanthenga M, Maliro M, Demo P, Njoloma J (2012) Potential of aeroponics system in the production of quality potato (Solanum tuberosum l.) seed in developing countries. Afr J Biotechnol 11:3993–3999
Contina JB, Dandurand LM, Knudsen GR (2018) A spatial analysis of the potato cyst nematode Globodera pallida in Idaho. Phytopathology 108:988–1001
Corrêa RM, Pinto J, Faquin V, Pinto C, Reis ES (2009) The production of seed potatoes by hydroponic methods in Brazil. Fruit Veg Cereal Sci Biotechnol 3:133–139
de Haan S, Núñez J, Bonierbale M, Ghislain M (2010) Multilevel agrobiodiversity and conservation of Andean potatoes in central Peru: species, morphological, genetic, and spatial diversity. Mt Res Dev 30:222–231
De Jonge B, Munyi P (2017) Creating space for 'informal'seed systems in a plant variety protection system that is based on UPOV 1991. Integrated Seed Sector Development (ISSD) Africa
Devaux A, Kromann P, Ortiz O (2014) Potatoes for sustainable global food security. Potato Res 57:185–199
Fajardo J, Lutaladio N, Larinde M, Rosell C, Barker I, Roca W, Chujoy E (2010) Quality declared planting material: protocols and standards for vegetatively propagated crops. Expert consultation, Lima, 27–29 November 2007. Food and Agriculture Organization of the United Nations, Rome
Fankhauser C (1999) Main diseases affecting seed degeneration in Ecuador: new perspectives for seed production in the Andes. European Association for Potato Research, Italy, pp 50–51
FAO (2006) Quality declared seed system. FAO Plant protection and protection paper 185. FAO, Rome, 2430 p
Ferrari L, Fromm I, Jenny K, Muhirec A, Scheidegger U (2017) Formal and informal seed potato supply systems analysis in Rwanda. In: Future agriculture: social-ecological transitions and bio-cultural shifts organised by the University of Bonn and the Centre for Development Research. University of Bonn and the Centre for Development Research, Bonn
Frost KE, Groves RL, Charkowski AO (2013) Integrated control of potato pathogens through seed potato certification and provision of clean seed potatoes. Plant Dis 97:1268–1280
Gildemacher P, Demo P, Kinyae P, Wakahiu M, Nyongesa M, Zschocke T (2007) Select the best: positive selection to improve farm saved seed potatoes: trainers manual. International Potato Center, Lima
Gildemacher PR, Schulte-Geldermann E, Borus D et al (2011) Seed potato quality improvement through positive selection by smallholder farmers in Kenya. Potato Res 54:253–266
Guenthner JF, Charkowsi A, Genger R, Greenway G (2014) Varietal differences in minituber production costs. Am J Potato Res 91:376–379

Harahagazwe D, Andrade-Piedra J, Parker M, Schulte-Geldermann E (2018) Current situation of rapid multiplication techniques for early generation seed potato production in Sub-Saharan Africa. International Potato Center, CGIAR Research Program on Roots, Tubers and Bananas (RTB). RTB Working paper no. 2018-1

Hidalgo O, Manrique K, Velasco C, Devaux A, Andrade-Piedra JL (2009) Diagnostic of seed potato systems in Bolivia, Ecuador and Peru focusing on native varieties. In: Tropical roots and tubers in a changing climate: a convenient opportunity for the world. Fifteenth triennial symposium of the International Society for Tropical Root Crops, Lima, Peru, 2–6 November 2009, pp 41–46

Jansky SH, Charkowski AO, Douches DS et al (2016) Reinventing potato as a diploid inbred line-based crop. Crop Sci 56:1412–1422

Kakuhenzire R, Musoke C, Olanya M et al (2005) Validation, adaptation and uptake of potato small seed plot technology among rural, resource-limited households in Uganda. In: African crop science conference proceedings, pp 1355–1361

Kakuhenzire R, Tibanyendera D, Kashaija IN, Lemaga B, Kimoone G, Kesiime VE et al (2017) Improving minituber production from tissue-cultured potato plantlets with aeroponic technology in Uganda. Int J Agric Environ Res 3:3948–3964

Kinyua ZM, Schulte-Geldermann E, Namugga P, Ochieng-Obura B, Tindimubona S, Bararyenya A et al (2015) Adaptation and improvement of the seed-plot technique in smallholder potato production. In: Low J, Nyongesa M, Quinn S, Parker M (eds) Potato and sweetpotato in Africa: transforming the value chains for food and nutrition security. CABI, Wallingford, pp 218–225

Kromann P, Montesdeoca F, Andrade-Piedra J (2016) Integrating formal and informal potato seed systems in Ecuador. In: J Andrade Piedra, J Bentley, C Almekinders, K Jacobsen, S Walsh, G Thiele (eds) Case studies of roots, tubers and bananas seed systems, Lima (Peru), 244 p. RTB working paper. ISSN 2309–6586. no. 2016–3, pp 14–22

Lakhiar IA, Gao J, Syed TN, Chandio FA, Buttar NA (2018) Modern plant cultivation technologies in agriculture under controlled environment: a review on aeroponics. J Plant Interact 13:338–352

Lindhout P, Meijer D, Schotte T, Hutten RC, Visser RG, van Eck HJ (2011) Towards F 1 hybrid seed potato breeding. Potato Res 54:301–312

Lommen WJ (2007) The canon of potato science: 27. Hydroponics. Potato Res 50:315

Lossau A, Weiskopf B, Kasten W (2000) Support for the informal seed sector in development co-operation-conceptual issues. GTZ, Germany and Centre for Genetic Resources, Wageningen

Louwaars NP (1994) Integrated seed supply: a flexible approach. In: Seed production by small-holder farmers: proceedings of the ILCA/ICARDA research planning workshop held in ILCA, Addis Ababa, Ethiopia, pp 39–46

Louwaars NP, de Boef WS (2012) Integrated seed sector development in Africa: a conceptual framework for creating coherence between practices, programs, and policies. J Crop Improv 26:39–59

Machida-Hirano R (2015) Diversity of potato genetic resources. Breed Sci 65:26–40

Mateus-Rodriguez JR, de Haan S, Andrade-Piedra JL et al (2013) Technical and economic analysis of aeroponics and other systems for potato mini-tuber production in Latin America. Am J Potato Res 90:357–368

McGuire S, Sperling L (2011) The links between food security and seed security: facts and fiction that guide response. Dev Pract 21:493–508

McGuire S, Sperling L (2016) Seed systems smallholder farmers use. Food Security 8:179–195

Ministerio de Agricultura y Riego del Peru (MINAGRI) (2018) Reglamento específico de semillas de papa. http://minagri.gob.pe/portal/download/pdf/normas-legales/decretossupremos/2018/ds10-2018-minagri.pdf. Accessed 26 Mar 2019

Munyi P, De Jonge B (2015) Seed systems support in Kenya: consideration for an integrated seed sector development approach. J Sustain Dev 8:161

Muthoni J, Mbiyu MW, Nyamongo DO (2010) A review of potato seed systems and germplasm conservation in Kenya. J Agric Food Inform 11:157–167

Muthoni J, Mbiyu M, Kabira JN (2011) Up-scaling production of certified potato seed tubers in Kenya: potential of aeroponics technology. J Hortic Forest 3:238–243

Naik PS, Buckseth T (2018) Recent advances in virus elimination and tissue culture for quality potato seed production, Biotechnologies of crop improvement, vol 1. Springer, New York, pp 131–158

Navarrete I, Panchi N, Andrade-Piedra JL (2017) Potato crop health quality and yield losses in Ecuador. Revista Latinoamericana de la Papa 21:51–70

Ochieng-Obura B, Parker ML, Bruns C, Finckh MR, Schulte-Geldermann E (2016) Cost benefit analysis of seed potato replacement strategies among smallholder farmers in Kenya. In: Solidarity in a competing world: fair use of resources, Vienna, Austria

Okeyo GO, Sharma K, Atieno E, Narla RD, Miano DW, Schulte-Geldermann E (2018) Effectiveness of positive selection in managing seed-borne potato viruses. J Agric Sci 10:71

Orlove BS, Godoy R (1986) Sectoral fallowing systems in the Central Andes. J Ethnobiol 6:169–204

Otieno GA, Reynolds TW, Karasapan A, Lopez Noriega I (2017) Implications of seed policies for on-farm agro-biodiversity in Ethiopia and Uganda. Sustain Agric Res 6:12–30

Parker ML, Low JW, Andrade M, Schulte-Geldermann E, Andrade-Piedra J (2019) Climate change and seed systems of roots, tubers and bananas: the cases of potato in Kenya and Sweetpotato in Mozambique. In: Rosenstock TS, Nowak A, Girvetz E (eds) The climate-smart agriculture papers. Springer, Cham, pp 99–111

Pautasso M, Aistara G, Barnaud A et al (2013) Seed exchange networks for agrobiodiversity conservation. a review. Agron Sustain Dev 33:151–175

Pérez W, Valverde M, Barreto M, Andrade-Piedra J, Forbes GA (2015) Pests and diseases affecting potato landraces and bred varieties grown in Peru under indigenous farming system. Revista Latinoamericana de la Papa 19:29–41

Priegnitz U, Lommen WJ, van der Vlugt RA, Struik PC (2018) Impact of positive selection on incidence of different viruses during multiple generations of potato seed tubers in Uganda. Potato Res 62:1–30

Prip C, Fauchald OK (2016) Securing crop genetic diversity: reconciling EU seed legislation and biodiversity treaties. Rev Euro Comp Int Environ Law 25:363–377

Schulte-Geldermann E (2013) Tackling low potato yields in Eastern Africa: an overview of constraints and potential strategies. In: G Woldegiorgis, S Schulz, B Berihun (eds) Seed potato tuber production and dissemination, experiences, challenges and prospects: proceedings. National workshop on seed potato tuber production and dissemination, 12–14 Mar 2012. Ethiopian Institute of Agricultural Research (EIAR); Amhara Regional Agricultural Research Institute (ARARI); International Potato Center, Bahir Dar. pp. 72–80 . ISBN 978–99944–53-87-x

Schulte-Geldermann E, Gildemacher PR, Struik PC (2012) Improving seed health and seed performance by positive selection in three Kenyan potato varieties. Am J Potato Res 89:429–437

Schulz S, Woldegiorgis G, Hailemariam G, Aliyi A, van de Haar J, Shiferaw W (2013) Sustainable seed potato production in Ethiopia: from farm-saved to quality declared seed. Seed potato tuber production and dissemination, experiences, challenges and prospects. Proceedings. Ethiopian Institute of Agricultural Research (EIAR); Amhara Regional Agricultural Research Institute (ARARI); International Potato Center, Bahir Dar, pp 60–71

Shepard JF, Claftin LE (1975) Critical analyses of the principles of seed potato certification. Annu Rev Phytopathol 13:271–293

Sperling L (2008) When disaster strikes: a guide to assessing seed system security. CIAT, Cali

Thiele G (1999) Informal potato seed systems in the Andes: why are they important and what should we do with them? World Dev 27:83–99

Thomas-Sharma S, Abdurahman A, Ali S et al (2015) Seed degeneration in potato: the need for an integrated seed health strategy to mitigate the problem in developing countries. Plant Pathol 65:3–16

Thomas-Sharma S, Andrade-Piedra J, Carvajal Yepes M, Hernandez Nopsa JF, Jeger MJ, Jones RAC et al (2017) A risk assessment framework for seed degeneration: informing an integrated seed health strategy for vegetatively propagated crops. Phytopathology 107:1123–1135

Thurston HD (1990) Plant disease management practices of traditional farmers. Plant Dis 74:96–102

Tran VM, Nguyen VU, Vander Zaag P (1990) Rapid multiplication of potatoes: influence of environment and management on growth of juvenile apical cuttings. Am Potato J 67:789–797

Tripp R (1996) Supporting integrated seed systems: institutions, organizations and regulations. In: Amstel H v, Bottema JW, Sidik M, Santen CE v (eds) Integrating seed systems for annual food crops, Integrating seed systems for annual food crops. Proceedings of a workshop held in Malang, Indonesia October 24–27, 1995. Regional Co-ordination Center for Research and Development of Coarse Grains, Pulses, Roots and Tuber Crops in the Humid Tropics of Asia and the Pacific, Research Institute for Legume and Tuber Crops, Agronomy Society of Indonesia, Indonesia

Tripp R, Louwaars N (1997) Seed regulation: choices on the road to reform. Food Policy 22:433–446

Vander Zaag P, Escobar V (1990) Rapid multiplication of potatoes in the warm tropics: rooting and establishment of cuttings. Potato Res 33:13–21

Vashisth KS (1979) Seed plot technique: its significance in seed potato production in India. Seeds Farms (India) 5:13–14

Vernooy R (2017) Options for national governments to support smallholder farmer seed systems: the cases of Kenya, Tanzania, and Uganda

Zeng Y, Fulladolsa AC, Houser A, Charkowski AO (2018) Colorado seed potato certification data analysis shows mosaic and blackleg are major diseases of seed potato and identifies tolerant potato varieties. Plant Dis 103:192–199

6

Participatory Research (PR) at CIP with Potato Farming Systems in the Andes: Evolution and Prospects

Oscar Ortiz, Graham Thiele, Rebecca Nelson, and Jeffery W. Bentley

Abstract Participatory Research (PR) at the International Potato Center (CIP) included seven major experiences. (1) Farmer-back-to-farmer in the 1970s pioneered the idea of working with farmers to identify their needs, propose solutions, and explain the underlying scientific concepts. The ideas were of great influence at CIP and beyond. (2) With integrated pest management (IPM) pilot areas in the early 1990s, entomologists and social scientists developed technologies with farmers in Peru and other countries to control insect pests. Households that adopted just some of the techniques enjoyed high economic returns, and this showed the importance of IPM specialists, social scientists, and farmers working together. (3) Farmer field school (FFS) was adapted for participatory research in the 2000s. Farmers learned that late blight was caused by a microorganism, while testing resistant varieties and fungicides, and researchers took into account more specifically farmer knowledge for training and PR purposes. (4) CIP used participatory varietal selection (PVS) after 2004 to form consortia of farmers, local government, NGOs, and research. Farmers' preferences were disaggregated by gender. Selection criteria of other market actors were included, and new varieties were released, showing the importance of combining farmer and researcher knowledge in this process. (5) Participatory approaches to develop native potato variety value chains. After 2000, CIP used the PMCA (participatory market chain analysis) and stakeholder platforms to improve smallholders' access to markets. PMCA brought farmers and other market actors together to form stakeholder platforms which created market innovations, including

O. Ortiz (✉)
International Potato Center, Lima, Peru
e-mail: o.ortiz@cgiar.org

G. Thiele
CGIAR Research Program on Roots, Tubers and Bananas, Lima, Peru
e-mail: g.thiele@cgiar.org

R. Nelson
Cornell University, Ithaca, NY, USA
e-mail: rjn7@cornell.edu

J. W. Bentley
Independent Consultant, Cochabamba, Bolivia

new potato-based products, expanding the inclusion of diverse actors in the PR processes. (6) Advocacy for PR and policy change with the Andean Change Alliance tested PR methods including PVS and PMCA from 2007 to 2010, providing evidence to influence policies to include smallholders in research and development. (7) After 2010, nutrition-related PR documented anemia among children in the high Andes, which could be mitigated by eating native potatoes rich in zinc and iron. CIP partnered with 20 organizations to improve household incomes and nutrition. Over four decades, CIP continues evolving in using PR, showing that combining social and biological scientists' input and keeping farmers' views upfront was key for PR. The experience also showed that the participation of other actors related to the value chains was needed in order to create successful agronomic, market, and social innovations. Future participatory research at CIP may be improved by using ICT to enrich diversity and richness of information sharing among PR actors.

Introduction

Spanning nearly 40 years, PR at CIP covered most major areas of agricultural research, including new seed and storage technologies, integrated management of pests and diseases, plant breeding, in situ conservation of genetic resources, and value chains. Participatory research at CIP influenced academic and applied research at many other institutions. CIP played a pioneering role in PR in the 1970s. Since then, PR has waxed and waned (Thiele et al. 2001) and moved to higher levels of scale from farm-level management to value chains, food systems, and policy. This chapter provides an update, focusing on work in the Andes, where several PR experiences took place.

The following case studies of PR at CIP are organized roughly chronologically (although some methods overlapped in time). They start with farmer-back-to-farmer, and continue through IPM pilot units, participatory variety selection, FFS, participatory approaches for native potato variety value chains using the participatory market chain analysis (PMCA), advocacy for PR and policy change (via the Alliance for Andean Change), ending with nutrition-related PR in more recent years.

Cases of Potato-Related PR in the Andes: Learning from Experience

The Farmer-Back-to-Farmer Model

The farmer-back-to-farmer model emerged from an interdisciplinary CIP team that included both social and biophysical researchers. CIP's first director, Richard Sawyer, hired anthropologists and expected them to work in an integrated team with the center's breeders and agronomists. The combination of different views amongst the scientist around seed potato storage led to the farmer-back-to-farmer model (Rhoades and Booth 1982). This model, with its insistence at looking at actual farmer practice in a pragmatic way, was a major influence on much of the farmer participatory research that would follow at CIP and elsewhere. In many ways farmer-back-to-farmer was ahead of its time.

In the late 1970s, a team of CIP researchers was studying post-harvest losses of potato in the Mantaro Valley of Peru. CIP started its participatory research in the Mantaro Valley because it was then one of Peru's main potato producing areas and one of the nearest to the main market in Lima. CIP anthropologist Robert Werge, and CIP biological scientist Robert Booth and other researchers were developing technology that farmers could use to avoid post-harvest losses, working within the framework of Farming Systems Research (FSR, which later would be largely replaced by farmer participatory research). FSR encouraged researchers from different disciplines to work together, and in their seminal paper, Rhoades and Booth (1982) noted that interdisciplinary research could easily become merely multidisciplinary, with researchers working alone within their disciplinary boundaries, and seldom interacting. Their paper was as much about getting social scientists to work with other researchers as it was about involving farmers.

In the 1970s interaction between anthropologists, economists, and biological scientists often provoked stressful but constructive arguments (Rhoades et al. 1986). Two teams were working from different perspectives, a production team with agronomists and an economist, and a post-harvest team with agronomists and an anthropologist (Thiele et al. 2001). An early breakthrough in the Mantaro Valley potato research came when the anthropologist told the biologists that post-harvest potato losses were really of little concern to farmers, who could use the smaller tubers for seed or for animal feed. Even damaged potatoes could be salvaged for the cooking pot and shriveled tubers could be made into chuño—the Andean method for freeze-drying at high altitudes. However, seed potatoes grow long sprouts that farmers disliked having to break off before planting.

Now it was the biologist's turn to be helpful. Booth explaining to the anthropologist (Werge) that the long sprouts were induced by the darkness where the potatoes were stored, inside the farm houses, and that while seed potato needs to be sprouted, the shorter sprouts were more vigorous than longer ones. This insight helped the anthropologist to refine his questions. By working together, the biologist and the anthropologist refined their problem topic. The issue was no longer post-harvest losses, but how best to store seed potatoes on-farm.

On the Santa Ana research station in the Mantaro Valley, the biologist showed the anthropologist how potatoes stored in diffused light (not in total darkness) developed short vigorous sprouts and a greenish skin. Such tubers were ideal for planting. The anthropologist then took some of the wooden greening trays from the station to the homes of some collaborating farmers and stacked the trays under the porch roof and tried storing seed potato there, in diffused light.

Farmers liked the seed tubers with short sprouts, but observed that the fine wooden trays would be too expensive, so a CIP technologist made some simple racks from local lumber. CIP soon began teaching diffused light storage (DLS) to farmers in the Philippines (where Booth was now working) and in Peru, teaching farmers how to make and stack wooden seed trays, but also explaining the underlying scientific principle that potato sprouts are shorter and more vigorous, increasing yield if seed is stored in diffused light.

This experience with DLS became the foundation story for farmer-back-to-farmer, a four-step model. Step 1: the anthropologist and the biologist go to the field to understand the problem from the farmer's perspective based on observation and action research, and to reach a common definition of the research problem (e.g. how to improve the storage of seed potatoes). In steps 2 and 3, the researchers develop a technology through a mix of on-farm and on-station research. Finally, the researchers present a prototype technology to farmers who act as advisors on how to adapt it to suit their own conditions (Rhoades and Booth 1982).

CIP economist Douglas Horton later wrote that the greatest benefit of the Mantaro Valley Project was not improved potato production, but institutional change: within a few years CIP was conducting research in various countries with national programs, based on farmer-back-to-farmer (Horton 1986). While Rhoades and Booth never mention the word "participation" in their 1982 paper, farmer-back-to-farmer influenced much of the farmer participatory research (FPR) that would follow (Veteto and Crane 2014). Thirty-five years after it was published, the short, engaging farmer-back-to farmer paper can still be read profitably for its practical philosophy of working with farmers.

Rhoades and Booth (1982) came up with several ideas that still guide much of participatory research. Researchers must work with farmers to identify the right problem to solve, interact creatively with them, work on-farm and on-station, and present the results back to farmers for feedback. Rhoades and Booth understood the importance of telling farmers the underlying principles (why the technology works) and not just how to use it, making it easier for farmers to adapt the technique to their own circumstances. All of these ideas are as sound now as when they were written. A decade later, farmer-back-to-farmer was one of the influences on interdisciplinary participatory research at CIP, but by the 1990s, these research methods were becoming more formal.

IPM Pilot Areas

In the early 1990s CIP entomologist Fausto Cisneros led research that included biological scientists and members of the Social Sciences Department, such as Oscar Ortiz an agronomist with specialization in agricultural extension. They studied the Andean potato weevil and how to control it in 15 pilot areas, mostly in Peru, and also in Ecuador, Bolivia, and Colombia. There was also research on the potato tuber moth at six sites (in Peru, Colombia, Bolivia, and the Dominican Republic), and on the sweetpotato weevil in Cuba and the Dominican Republic. Researchers needed to adapt IPM (integrated pest management) to the local conditions (farming practices) and farmers' knowledge in pilot communities, called pilot areas. In those communities, researchers would select the most appropriate technologies, and assess farmer knowledge so that training could be provided according to knowledge gaps using practical demonstrations (Cisneros et al. 1995). The technologies were based on the results of previous research conducted by CIP in the late 1980s with Peru's national agricultural research institute, INIA. Through this pioneering experience CIP started working with nongovernmental organizations (NGOs) that were oriented to agricultural extension and development in some of the pilot units (Ortiz et al. 2009).

The researchers realized that each of the IPM technologies had to be validated through participatory research, akin to step 4 of the farmer-back-to-farmer model, presenting results back to communities for advice. The consultation with farmers was held in formal pilot areas[1] and pilot units[2] (see definitions below) in five countries. An agronomist was stationed in each pilot area to mentor the farmers on the technologies, to see which technologies worked and to get an idea of how to explain the innovations to other farmers (Cisneros et al. 1995).

This integrated pest management (IPM) research aimed to lower costs, manage pests, reduce environmental damage, and minimize health risks. The key word was "integrated," using many techniques, especially natural enemies, cultural practices, and pheromones. Insecticides were to be used as a last resort, and as little as possible. However, the researchers also realized that IPM had been vaguely described; abstractions would not convince farmers who needed concrete results (Cisneros et al. 1995). Therefore, specific training techniques were developed to support farmers' learning of complex concepts related to IPM (Ortiz et al. 1997).

The pilot area work came after some celebrated IPM research by IRRI with the rice planthopper in Indonesia. However, farmers could manage the planthopper largely by avoiding insecticide and letting natural enemies control the pest (see Heong et al. 2014). In contrast, the Andean potato weevil was becoming a more serious pest largely because of increased intensity of cropping (Ortiz 2006; Parsa 2010). The Andean potato weevil was a difficult pest that needed to be managed with several proactive and preventive options; it was not enough to abandon insecticides and let nature take its course. And, the approach was called "pilot areas" to refer to the communities where the research took place, because for the IPM to work, it needed to be applied in most of or the entire the community for it to work.

The pilot area work was led by an entomologist or an agronomist with experience in pest management who collaborated with a social scientist on training and impact assessment, testing various management options in many pilot areas, for three problems (Andean potato weevil, tuber moth, and sweetpotato weevil). The Andean potato weevil technologies were based on an understanding of local knowledge and on the pest behavior. Andean farmers recognized the adult weevil, and its grub, but did not understand that there were four life stages of the same species (from egg to adult). Entomologists had learned that the Andean potato weevil ate

[1] Communities where CIP and collaborators carried out the participatory research.

[2] Portions of farmer's fields where the management options were tested.

only potato, and that the insect could be killed in different stages (egg, larva, pupa, and adult) depending on the season of the year. Technologies included hand-picking adult weevils at night (the adult insect had nocturnal habits), uprooting the volunteer plants that harbored weevils, winter plowing to destroy pupae in the soil, piling potatoes on sheets so the larvae could not pupate in the soil, and making barriers around the field to keep the flightless weevils from walking in from neighboring plots (Cisneros et al. 1995; Ortiz et al. 2009).

Farmers rejected some technologies, including early harvesting and eliminating volunteer plants. But pilot farmers adopted barriers around the fields and piling potatoes on sheets. To teach the technologies chosen by pilot farmers, extensionists designed their own materials, including field visits, dramas, games, and manuals. The most acceptable technologies were then shared with other communities as part of a USAID-funded project, MIPANDES, implemented by the international NGO CARE with technical support from CIP (Cisneros et al. 1995; Ortiz et al. 2009).

Rhoades (1987) had claimed that only 2% of farmers surveyed had adopted DLS as it was taught, but that the other 98% had adopted the underlying principles. This may have struck the entomologists as imprecise thinking, and they wanted a more objective way of measuring adoption. Yet, the pilot areas research was designed to give farmers a hand in selecting and rejecting technologies. Farmers could also choose technologies from an integrated menu. Researchers never expected farmers to adopt all of the technologies. However, because so many techniques were taught, it was hard to decide how many components had to be adopted to count as "adoption of IPM" (Ortiz et al. 2009).

Early results showed that adopting even some of the management options led to a large reduction in pest damage. For example, in the pilot area of Huatata, Peru, weevil damage had gone from 44 to 8.5% in just 4 years. Mizque, Bolivia had a favorable environment for weevils, which damaged all of the tubers before the project, but only 10% in later years. Results from other pilot areas were comparable. Using even some of the technologies managed the pest (Cisneros et al. 1995).

Therefore, the team opted to study adoption by measuring the economic impact of the technologies. It did not matter how many technologies farmers managed, but the value of the harvest saved by using them. Farmers who adopted some of the IPM practices could achieve an average benefit of about $100 per hectare (Ortiz et al. 2009). The pilot areas ended with a large project (MIPANDES) implemented with the NGO CARE in 1995 and 1996 in Cajamarca and three other provinces of Peru to teach farmers the menu of IPM technologies for managing the Andean potato weevil and the tuber moth (Chiri et al. 1997).

The pilot areas work was an important formative experience for the FFS experience which followed it (see Sect. 13.2.3 below), particularly because several of the biological and social scientists who participated in one experience continued with the FFS, and they had realized the importance of farmer learning for IPM and the need for appropriate teaching methods (Ortiz et al. 1997). The pilot areas research proposed a systematic way of validating technologies with farmers and an objective method of measuring the results, yet this method probably did not receive all of the recognition that it deserved because there were limited journal publications. The most influential participatory methods tend to be the ones that are well documented. CIP's later research with FFS was well-documented and frequently cited, eclipsing the effort with pilot areas.

Participatory Research and Training: The FFS Experience

Following the pilot areas experience, some of the same researchers continued with IPM, but this time with a potato disease, late blight, and a new method of participatory research: FFS (farmer field school). By 1997, plant pathologist Rebecca Nelson had recently transferred to CIP from IRRI, which had been working with FFS since 1987, first with rice planthoppers and rice stemborers in Indonesia, and later with rice blast in Vietnam (Nelson et al. 2001). For the CIP team, FFS made sense because they had already experienced the importance of facilitating farmers' learning of complex concepts involved in IPM, which were even more complex when dealing with the pathogen that causes potato late blight.

IRRI entomologists had explicitly conceived of FFS as an extension method, not for research (Kevin Gallagher, pers. com.) But FFS was based on experiential learning and the learning field so it did have an informal experimental content. FFS at CIP went several steps further (perhaps too many steps, as the research design was often too complicated for farmers). In 1997, FFS became a PR method (Thiele et al. 2001) as Nelson and colleagues redesigned FFS as a research method dubbed FPR-FFS (farmer participatory research through farmer field schools).

The IPM pilot areas had worked in 16 sites from Bolivia to Cuba (see previous section) between 1990 and 1996, and the FPR-FFS started in four communities, in a single municipality, in San Miguel, Cajamarca, Peru. Two years later field schools would expand to about 20; and a similar experience took place in Cochabamba, Bolivia. CIP was still partnering with CARE, which had already worked in San Miguel during several projects, including MIPANDES and one called ANDINO, which had taught farmers about late blight (Godtland et al. 2004). The experience later expanded, between 1999 and 2004, to other communities in Peru, and also in Bolivia, Ethiopia, Uganda, China, and Bangladesh with a similar approach through an IFAD-funded project (Ortiz et al. 2011).

In the FFS, farmers observed pest ecology in the field. Farmers studied late blight lesions under small microscopes and cultured the disease in plastic bags to watch its spread (Nelson et al. 2001). Farmers had not previously known that late blight was caused by a microorganism. But the FFS provided also the opportunity to test advanced potato clones with resistance to late blight, something that had effects easily perceived by farmers (Figs. show different stages of participatory evaluation of potato clones as part of FFS implementation), so that they could select those materials that were suitable for their conditions and preferences (Ortiz et al. 2004).

Fig. FFS participant evaluating potato varieties during flowering in Cajamarca, Peru (2004) (Credits: O. Ortiz)

The Indonesian FFS focusing on insect pest management used the simplest possible design: an IPM plot next to a farmer's plot. The late blight FPR-FFS tested three potato varieties, ranging in resistance to late blight. Each variety was treated with fungicide at three different intervals, for a total of nine treatments. This research would show that late blight could be managed through varietal resistance combined with a fungicide regime that was adjusted based on the variety and the environment (Sherwood et al. 2000).

Some farmers seem to have felt that 1 year of field school was enough. The second season, 1998–1999, two of the four communities dropped out of FPR-FFS, but CIP and CARE replaced them with eight new communities. The second year these ten communities replicated their experiment with the potato varieties and fungicides from the previous year (Sherwood et al. 2000).

FPR-FFS was small-scale for its first 2 years, but after that the Dutch government funded a large IPM project with the Andean field schools, led by the FAO (Nelson et al. 2001). Between the second and third seasons of FPR-FFS, 35 extensionists

Fig. FFS farmer during variety evaluation at harvest in Cajamarca, Peru (2003) (Credits: R. Orrego)

from Peru, Bolivia, and Ecuador attended a 3-month practical ToT (training-of-trainers) FFS course in Ecuador (Sherwood et al. 2000).

After the ToT course, some of the returning extensionists taught the third year of FPR-FFS in San Miguel. It was expanded to 19 communities and three new experiments were added. Farmers continued doing trials of varieties and fungicides. Second, they also tried an IPM vs. farmer's treatment (per the original FFS design), plus a third experiment to grow potatoes from true seed (Nelson et al. 2001). In the fourth experiment, the field schools tested 50 breeding lines of potatoes from CIP. Each community tested just some of the lines, but each clone was tried by two or three communities. This was in 1999–2000, and it foreshadowed the mother-and-baby trials which would be CIP's next participatory method (see following section). Based in part on these evaluations with farmers, one of the varieties (Chata Roja) would be released in 2000 (Nelson et al. 2001).

By now the work with FPR-FFS seemed to be paying off. An economic evaluation showed that late blight management, because of its direct effect on increasing yield, was worth a net benefit of some $236 per hectare and year for the farmers of San Miguel (Ortiz et al. 2004). There was also strong implementation of CIP-affiliated field schools in Ecuador, which generated more learning material than

Fig. FFS farmer assessing culinary quality of potato varieties in Cajamarca, Peru (2005) (Credits: O. Ortiz)

could be covered in a single season; the material was compiled as a manual so that trainers could choose from it to custom-design FFS for other communities (Pumisacho and Sherwood 2000). CIP and other organizations were using FFS in other departments of Peru for bacterial wilt, insect pests, and other topics, but the method was also used in Bolivia, Ecuador, Ethiopia, and Uganda (Ortiz et al. 2008). In 2017 the IPM Association in the US gave the IPM Team Award to CIP and partners for their pioneering work with FFS research for the integrated management late blight, addressing the human and technical components of this technology (Ortiz et al. 2019).

However, CIP's research also looked at the factors that limited farmers' involvement in participatory research through FFS, and CIP observed that "the experimental designs tested sometimes were found to be overly complex for training of farmers" (Ortiz et al. 2004). There was also concern over the costs of FFS, as much as $1000 per one FFS per season (Ortiz et al. 2011). Farmers joined the field school to learn, but were discouraged by the time demands; organizations were limited by the lack of qualified trainers (Ortiz et al. 2011). CIP did not continue using FFS as a PR method after 2008, but the method has continued its evolution, and was institutionalized in several public and private agricultural organizations in Peru reaching more than 1000 up to 2008 (Orrego et al. 2010), something that has continued to evolve. More recently, the Peruvian Ministry of Education officially formalized a training and certification program for FFS facilitators (Sineace 2016).

Participatory Variety Selection (PVS)

Plant breeding may take 10 or 12 years or longer to produce a new crop variety. This enormous investment of time, money, and energy can be wasted if the new variety is not adopted. Plant breeders can anticipate farmers' preference for varieties by learning which traits farmers (and consumers) want. For example, there is no point in breeding for processing quality when farmers prefer varieties for table consumption.

In 1987–1988, a Peruvian government project was collaborating with CIP to test varieties with farmers. CIP anthropologist Gordon Prain and agronomist Urs Scheidegger had helped to set this up. Their method was heavily influenced by farmer-back-to-farmer; collaborating farmers were invited to evaluate eight or nine clones, as they saw fit, to find "friendly" varieties. Direct sales of seed to farmer collaborators were seen as an indicator of acceptance (Thiele et al. 2001).

Then Sieglinde Snapp, an agronomist at ICRISAT (International Crops Research Institute for the Semi-Arid Tropics), visited CIP/Lima to share her experiences with a new method of PVS called the mother-and-baby trial design (MBT; Snapp 1999). The MBT method, also known as the central/satellite design, involves a complete, replicated trial in a central location in a community, with satellite mini-trials in farmers' fields. Each farmer trial only includes a subset of the treatments of the central trial, with different farmers testing different combinations of lines such that all the material in the central trial is tested on multiple farms. This visit influenced the types of experiments implemented with the CIP version of FFS. Since plant breeders needed more replicability and numbers for statistical analysis, this MBT method was considered a useful advance. The PVS model thus gave way to an MBT approach, allowing the breeder to collect quantitative, replicated data from the mother trial, while gauging farmers' reactions to the lines or clones grown under farm conditions. The method was widely used at CIMMYT and elsewhere as well (Snapp 2002).

By 1999 CIP had 110 clones of a new breeding group called B1C5, which contained genes from native *Solanum tuberosum spp andigena* potatoes. The clones combined traits like high yield and resistance to late blight with tolerance for low inputs, early maturity (120 days instead of 180) and the taste qualities that Andean consumers preferred (Allauca 2011; Janampa 2012; Camacho-Henriquez et al. 2015). In 2004, CIP plant breeders sent 20–30 advanced clones to communities in Cusco in the high Andes. Over the subsequent decade, CIP facilitated the formation of a consortium of farm communities, municipalities, NGOs, universities, and INIA—Peru's national agricultural research agency so use PVS, and trained key members of staff to use the MBT method (Camacho-Henriquez et al. 2015).

Mother and baby trials were evaluated by the farmers who grew them, but also by other community members who met three times during the season (flowering, harvest, and post-harvest). Farmers' selection criteria were free-listed and then prioritized, giving villagers maximum flexibility to define the traits that they demanded. Women and men farmers worked in separate groups so that the results could be disaggregated by gender. At harvest, the yields were measured and farmers evaluated the potatoes by appearance, taste, and texture. Farmers evaluated the potatoes at post-harvest for weight loss, dormancy, and sprouting. The mother-and-baby trials were gender sensitive and appropriate for illiterate farmers (Camacho-Henriquez et al. 2015).

In 2007, as a result of the mother and baby trials, several potato varieties were released in Peru, including Pallay Poncho, Puca Lliclla, Altiplano (Arcos et al. 2015), and Kawsay (Camacho-Henriquez et al. 2015) followed by others later, e.g. INIA 325—Poderosa (CIP 2014). Some farmers in participating communities even set up businesses to produce and sell seed of the new varieties (Camacho-Henriquez et al. 2015).

In 2008–2009, as part of the Cambio Andino Program, CIP established further trials with four consortia (six in 2009–2010) with mother and baby trials in two or three places per consortium, with 10–20 clones at each site. In each region, each consortium had a leader (an NGO or another organization) that facilitated the mother-and-baby trials. By the third year (2010–2011) three clones were selected at each cite (Fonseca et al. 2011). CIP held annual workshops with the consortia members to improve the mother-and-baby method (Fonseca et al. 2011).

The mother-and-baby trial design allowed plant breeders to understand farmers' selection criteria (e.g. large tubers, resistance to frost and to late blight). There were some gender differences, with women preferring potatoes for boiling. MBT allowed farmers to receive the varieties they wanted. Participation with food manufacturers, wholesalers, and restaurants allowed selecting potatoes that met market demand (Fonseca et al. 2011); this foreshadowed methods CIP would pioneer later to engage with other market actors, besides farmers.

In order to improve the quality and standardization of the mother-and-baby method CIP published a manual to describe it in detail for trainers and facilitators (de Haan et al. 2017). MBT was originally designed to be used by plant breeders and farmers. By adding government, private sector, NGOs, vocational schools, and other actors, CIP was able to involve representative of the whole value chain in selecting varieties. This helped not just to select desired traits, but to facilitate registration of new varieties and to disseminate them with farmers, buyers, processors, and consumers. PVS at CIP went beyond the mother-and-baby approach and also combined participatory approaches with online data collection and analysis tools, which was another innovation in participatory research.

Participatory Development of Native Variety Value Chains

An analysis of different market opportunities revealed that innovation in selling native potatoes had the greatest potential to improve the incomes of smallholders in the high Andes (Devaux et al. 2009). During the 1990s, the road network expanded dramatically in Peru, allowing trucks to reach remote areas previously unreachable, and lowering transportation costs for potatoes and other produce. With long-term support from the Swiss Development Cooperation, CIP began a decade of addressing market opportunities for the native potato, developing and using two tools: the PMCA (participatory market chain analysis) and stakeholder platforms (Devaux et al. 2009). CIP researchers Thomas Bernet, André Devaux, Graham Thiele, and others created the PMCA in 2000 (Bernet et al. 2006), based on RAAKS (rapid appraisal of agricultural knowledge systems—see Engel and Salomon 2003). In the early 2000s, PMCAs conducted with support from CIP started to address the question of how to improve smallholders' access to markets, especially for farmers cultivating potato landraces. One key bottleneck was that native potatoes were unfamiliar to consumers of Lima, Peru's capital and largest city. Through its work with gastronomy schools, the PMCA helped Peruvian chefs to appreciate the culinary potential of native potatoes. This gave native potatoes a new image; and they were now acceptable to middle-class consumers (Horton and Samanamud 2013).

A single PMCA can last for 2 years and comprises three phases. Phase one includes a market survey to identify the stakeholders and includes a large meeting that brings stakeholders together in one room (Devaux et al. 2009). While previous participatory research engaged with researchers, extensionists and farmers, the PMCA links with many more actors (including ministry officials, market specialists, food technologists, transporters and wholesalers, food processors, supermarkets, and chefs—Ordinola et al. 2014). The event must be expertly facilitated to form a community from actors who all work in the same value chain, but many of whom have never met for co-innovation (Devaux et al. 2009).

At this first event, the facilitator helps processors (e.g. supermarkets and food manufacturers) to present a diagnosis and set up groups on promising possible innovations. By the end of Phase 3, innovations move to the promotion phase. In Peru, the PMCA stimulated the following innovations: colored potato chips, made from native varieties, tunta and chuño, clean and well sorted in small bags, and fresh potatoes of native varieties in net bags for supermarkets (Devaux et al. 2009).

At the start of the PMCA in Peru, and as part of the Papa Andina project, the team used what they called a "pro-poor filter," to identify market segments where resource-poor farmers would be especially likely to benefit. This led to the selection of native potatoes produced in the high Andes mostly by resource-poor farmers, as an attractive market segment for innovation. In Peru, multinational companies dominated much of the market for conventional potato chips from improved potatoes supplied by larger producers.

During the second phase of the PMCA, different actors begin to work together, researching and developing different innovations (which could be commercial products or institutional innovations, such as multi-stakeholder platforms). The third phase sees these innovations finished and ends with a large, final event where companies, farmers, and others come together to launch the new products and other innovations. It is important that the companies own the new brands. For example, Wong Supermarket created a brand, T'ikapapa, of mixed, colorful native potato varieties in an attractive net bag, as an outcome of the PMCA (Thiele et al. 2011; Devaux et al. 2009).

As the actors of the value chain begin to work together, buying and selling their products, they learn to trust each other which can lead them to take collective actions, such as Peru's National Potato Day, celebrated on 30 May every year since 2005 with displays of native varieties and gourmet food (Horton and Samanamud 2013; Ordinola et al. 2014). This achievement led the stakeholders, via the Ministry of Agriculture, to successfully petition the UN to name 2008 as the International Year of the Potato (Horton and Samanamud 2013). A new stakeholders' platform emerged from the PMCA to advocate for native potatoes and to consolidate their market position (Devaux et al. 2009; Thiele et al. 2011).

Creating products from native potatoes allowed smaller processors to open a niche market for smallholders, at a price 10–30% higher than before (Devaux et al. 2009). For example, Jalca Chips, which was directly influenced by PMCA, innovated by selling colored potato chips made from native varieties, at the Lima International Airport. This triggered other innovation processes as PMCA intended with the chips have since been copied and improved upon by other food manufacturers and are now widely sold in Peru under various brand names.

An important outcome of the PMCA in Peru was the inclusion of native varieties in the National Registry of Commercial Cultivars (Horton and Samanamud 2013). This allows native varieties to be legally sold as seed and is an important step in their continued survival.

The PMCA stimulated a broader innovation process. In Peru, the wholesale price of native potatoes is increasing as is the volume sold (Horton and Samanamud 2013). The smallholders who supply these native potatoes benefit from higher prices, so that the PMCA has many more indirect beneficiaries than direct ones (Ordinola et al. 2014).

CIP replicated the PMCA with partners in Bolivia (Proinpa) and Ecuador (INIAP). Bolivia also experienced a revived interest in native crops. The PMCA fostered the creation of bagged and clean chuño, native potato chips and net bags of fresh native potatoes sold in supermarkets (Thiele et al. 2011; Oros et al. 2011).

The PMCA stimulated the creation of stakeholder platforms as an institutional innovation. In Ecuador, the potato stakeholder platform triggered the creation of an organization, the Consortium of Small Potato Producers (CONPAPA) which linked highland farmers to others who produce quality seed and bulked potatoes for sale so smallholders can consistently supply the market for modern varieties with a high commercial demand (Thiele et al. 2011; Devaux et al. 2009; Kromann et al. 2016).

Participatory research started at CIP triggered by the interaction of social scientists and agronomists as described earlier around seed systems. Over the years, researchers learned to collaborate with farmers across a broader range of useful technologies including varieties. But PMCA and stakeholder platforms went one step further, involving the other actors of the value chain to explicitly demand new research and launch new commercial products. Farmers responded eagerly to the new market demands for native potatoes (Thiele et al. 2011).

Advocacy for PR and Policy Change (Experience of Cambio Andino)

From 2007 to 2010 a project managed by Graham Thiele at CIP and Carlos Arturo Quirós at CIAT, called Andean Change Alliance (Alianza Cambio Andino, in Spanish), partnered with a broad group of about 20 public and civil organizations, including Proinpa (Promotion and Research for Andean Products) in Bolivia, INIAP (National Agricultural and Livestock Research Institute in Ecuador), INCOPA (a CIP-led project for Innovation and Competitiveness of the Potato) in Peru, the PBA Foundation in Colombia, and many others to advocate for participatory research by showing that participatory methods enhanced research outcomes and helped give the poor a voice to ensure their demands were attended to (Thiele et al. 2012). In Colombia, Ecuador, Peru, and Bolivia, the Alliance tested various participatory methods including: Participatory Monitoring and Evaluation (SEP), Local Committees for Agricultural Research (CIALs), Participatory Variety Selection (see Sect. 13.2.4), and PMCA (Thiele et al. 2012).

To build an evidence base for advocacy for participatory research, the Andean Change Alliance developed a knowledge bank, including an online catalog of participatory methods (www.cambioandino.org) and a book of case studies (Thiele et al. 2012). The Andean Change Alliance was broad ranging including NGOs, universities, farmer associations, local and national governments of Peru, Bolivia, Ecuador, and Colombia. Advocacy was systematically and sequentially organized identifying both "pain points" around participation or a lack of it and a constituency to leverage change (Flores 2010):

- 2007—Capture demands for participatory methods within the innovation systems and organize the methods to meet that demand.
- 2008—Promote the use of participatory methods according to the demand for them and evaluate their outcomes and impacts.
- 2009—Improve the participatory methods.
- 2010—Use the evidences of the outcomes and impacts for policy advocacy to enable participatory research and improve its quality and relevance.

Advocacy was driven by evidence that the inclusion of smallholders in development research-&-development improved the interventions (Flores 2010).

From 2007 to 2009 the Andean Change Alliance facilitated the use of and evaluated participatory approaches, including eight cases of the PMCA in Peru, Bolivia, Ecuador, and Colombia (Devaux et al. this volume). Andean Change provided training, backstopping, and coaching.

Local teams consistently sought to adapt PMCA to fit local circumstances; respecting to different degrees the PMCA's fundamental principles. However, in Santa Cruz, Bolivia during a fruit PMCA and a vegetable PMCA, the facilitators tried to rush the model by skipping phase 2, and farmers rejected the contract offered to them because they had not built up enough trust with the buyer.

Table Eight cases of PMCA in the Andes, facilitated by the Andean Change Alliance

Case	New product	Completed PMCA	Analyzed[a]	Deviations/problems
1. Coffee/ Peru	Roasted coffee, packaged and labeled	Yes	Yes	
2. Cheese/ Bolivia	Mozzarella packed for retail market	Yes	Yes	NGO had low interest in creating links with other market actors
3. Potatoes/ Bolivia	Native varieties in net bags for supermarket	Yes	Yes	NGO had low interest in developing new market products and links
4. Yams/ Colombia	No	Yes	Yes	Outside knowledge had less to offer
5. Potatoes/ Ecuador	Semi-formal seed	Yes	No	Tried to work at national scale immediately. Consumers had less interest in native potatoes
6. Cheese/ Peru	No	No	No	Facilitators stressed a new production technology, not market innovation
7. Fruit/ Bolivia	No	No	No	Facilitators skipped phase 2 of PMCA
8. Veg/ Bolivia	No	No	No	Same as above

[a]I.e. discussed by Horton et al. (2013)

Horton et al. (2013) focused on four completed cases. Results were greatest where PMCA was implemented with the greatest fidelity. Most important is the engagement of market actors, not just farmers. Adapting a method (lack of fidelity) may be innovative, but if local teams are encouraged to change any and all parts of the model they may skip key components (Horton et al. 2013).

The analysis and documentation of experience with the implementation of SEP and CIALs showed positive outcomes related to the agency of farmers and farmer organizations with enhanced self-respect of farmers, mutual respect among different actors, enhanced capacity for negotiation and stronger organizations. The enhanced agency of farmers promoted structural changes that affected local traditional organizations and development institutions. Institutional innovation, changes in rules, roles, and practices involving power sharing occurred initially at the local organizational level and gradually permeated into the institutional level reaching practices and processes in the institutions that delivered the services (Polar et al. 2012). There were positive changes in the quality of agricultural advisory services as well as in coverage and inclusion. However, farmers' agency was still unable to generate institutional changes in public sector actors (Polar 2014).

Perhaps the furthest reaching, and longest lasting advocacy of the Andean Change Alliance was with the many national organizations, public and civil, which participated with farmers in research, commissioned and mentored by the Alliance (Thiele et al. 2012; Flores 2010).

Nutrition-Related Participatory Research Through Potato

In a cruel irony, people who spend much of their day producing food are often unable to properly feed their own kids. The children of many Andean farm families are chronically undernourished. The causes and contexts of poverty and a poor diet are complex. Nonetheless, in about 2010, CIP crop scientist André Devaux and colleagues, in line with government diagnosis, recognized that anemia is a major problem which research could tackle. In the Andes over 20% of children under three suffer from chronic undernourishment, mainly due to deficiencies of micro elements such has iron and zinc; 39% of children under two are anemic (Devaux et al. 2015).

Anemia can result from a diet poor in zinc and iron. Fortunately, some native potato varieties are rich in zinc and iron, and the vitamin C in potatoes helps to body to absorb these essential minerals. Potatoes are also well endowed with vitamin B, antioxidants, and have as much protein as grains, so a diet with generous portions of native potatoes can be nutritious (Devaux et al. 2012; Ordinola 2015; Creed-Kanashiro et al. 2015).

From 2011 to 2015, Devaux and colleagues led a project, IssAndes (Innovation for food security and sovereignty in the Andes) to address undernutrition with research. IssAndes started with a baseline study in rural communities in Apurímac and Huancavelica, Peru. The results confirmed that 42% of children under 2 years of age where chronically undernourished. Fifty-four percent of the children were not getting enough iron in their diets and zinc was deficient in 48% (Creed-Kanashiro et al. 2014). The baseline study also found a positive relationship between children's iron and zinc intake with native potato production, raising small animals, and area of production (Creed-Kanashiro et al. 2015). Area of production would correlate with household income, suggesting that better-off families could feed their children a more diverse diet.

Building on the work of plant breeders, Devaux and colleagues identified 200 native potato varieties in the CIP genebank with high levels of zinc and iron. The next step was to see which of those varieties would appeal to farm families. This was somewhat like the previous work with PMCA (see Sect. 13.2.5), but then the emphasis had been on finding potatoes of the right size, shape, color, and taste to appeal to the gourmet food market in Lima. Now, the goal was to find nutritious potato varieties for the poor (Devaux and Kromann 2016). Farmers in the highland communities participated in assessing CIP's improved clones from a breeding population that resembled native varieties using PVS (see Sect. 13.2.4) and selected one particular variety. This appealing variety was named "Kawsay" (Quechua for "to live"). The Peruvian government released Kawsay, as mentioned in Sect. 13.2.4, as a resistant variety with higher content of iron and zinc.

Since the baseline study had suggested that higher incomes led to healthier children (Creed-Kanashiro et al. 2015), IssAndes worked to improve livelihoods by helping farm families to create something to sell. The project helped to launch

Kiwa® nutritious: colored potato chips, made from native potatoes and sold around the world (Devaux and Kromann 2016).

Through IssAndes, over 5000 rural farm households in Peru, Bolivia, Ecuador, and Colombia, were encouraged to grow, eat, and sell native potatoes rich in zinc and iron, and to experiment with vegetable home gardens and with raising guinea pigs and other small animals.

The Prospects of PR in the Andes and Globally

The previous sections of the chapter have described different PR stages and approaches used by CIP over the last three decades, which has shown to be a valuable mechanism to bring together different stakeholders to develop technologies, methodological approaches, and business opportunities. It has been observed that at the beginning of CIP's PR experience, the types of actors involved in agricultural research for development were relatively few (farmers, extension workers, and researchers, mostly from government organizations), but the range of new stakeholders started to increase over the years, including NGOs, farmer organizations, civil society, and private sector. New PR methods such as the PMCA evolved to include the views of these diverse actors; but within a single value chain and for particular business opportunities. However, agri-food systems are becoming more complex with interconnectedness established from rural to urban areas and vice versa, and future PR approaches will need to take this into account. Therefore, methods will need to continue evolving to deal with increasingly complex agricultural innovation systems (Hall et al. 2004; Ortiz et al. 2013; Hellin 2012) to face emerging challenges, such as climate change. In addition, the ways in which farmers are connected to the world is also changing.

The penetration of information and communication technologies (ICTs) to rural areas has increased significantly in recent years, opening new opportunities for capturing farmers' views in ways not attempted before. In the developed world, farmers are already immersed in extensive and complex networks of information sharing and decision support tools facilitated by ICTs; and although in developing countries, this is not yet the case, ICTs can facilitate gathering the opinions of more farmers, and also diverse actors in the value chains, and to share those views with actors located in other regions and countries. In addition, farmers in developing countries are slowly moving towards being part of the big data movement, which will connect information in new and faster ways. For example, farmer preferences differentiated by gender could be linked with the genetics that explain those traits, making it possible to establish a connection between trait preferences and trait genetics in a way that was not possible before.

Adaptation and resilience to climate change requires farmers to adapt faster than even before, for which the flow of information from and to them should also be faster. ICTs offer the possibility of connecting a larger number of farmers with sources of information and advice that could help them to make decisions to tackle climate change—CIP already experimented with this (Sperling et al. 2008), but in the future, there will be the need to connect even more people with more information sources, and, particularly, to support making sense of the available big data for farmers' decision making.

The essence of PR approaches has been maintained over the years as a mechanism to facilitate the dialogue between scientists from different disciplines, and particularly between them and farmers or other actors of the value chain. This will continue evolving and in the future, PR methods will need to include a larger number of viewpoints, from the disciplinary view point, but also from the innovation and agri-food systems viewpoints, and new ICTs offer the possibility of improving communication, analysis, and decision making. We can envision a larger number of PVS using mother and baby trials, all interconnected in real time, so that farmer groups not only analyze their own results, but access results of many more, and also breeders can provide the genetic explanation of user preferences. The future is full of opportunities for PR.

Concluding Remarks

The early experience with participatory research at CIP, with farmer-back-to-farmer in the early 1980s, showed the importance of identifying research topics with farmers, working with communities creatively, and presenting results back to farmers. These ideas informed much of the later participatory research at CIP and elsewhere (Thiele et al. 2001).

By the early 1990s, social scientists were accepted as full team members of research teams. Research methods were more formal, unlike farmer-back-to-farmer with its "loose guidelines rather than polished and elaborate research methods" (Thiele et al. 2001). The Pilot Areas approach now measured results with economic impact (Ortiz et al. 2009), which allowed evaluating several technologies at once.

The IPM research (Pilot Areas and FFS) showed the importance of understanding gaps in farmers' knowledge and sharing ideas with them to further collaboration. By the late 1990s, FFS became a formal research method at CIP. FFS and PVS (mother-and-baby) started at about the same time and were influenced by mother-and-baby trials at CIMMYT and work by CIP with "friendly" varieties (Prain et al. 1992).

The Pilot Areas had started working with NGOs. FFS and PVS continued to collaborate with NGOs, but with PVS the NGOs were incorporated into larger consortia, with more forming training and workshops for staff. Food manufacturers, wholesalers, restaurants, and the rest of the value chain were brought into research.

In the early 2000s, the PMCA involved other market chain actors, besides farmers. With PMCA it was no longer enough to select a new variety or a technology: PMCA stimulates innovations by actors. PMCA contributed to a progressive shift in consumers' perceptions of native potato varieties to reposition them as prestigious, even gourmet food, and improving livelihoods.

In the 2010s, the Alliance for Andean Change explicitly intended to promote broader uptake of PMCA, PVS, Participatory Monitoring and Evaluation, and other participatory methods, validating them in four Andean countries and using the evidence of success to advocate for the mainstreaming of participatory methods with national research systems. It was not possible to document the influence of Andean Change on national systems, because the project ended prematurely, but there may have been a lasting influence on many partner organizations in Peru, Bolivia, Ecuador, and Colombia.

By 2012, multidisciplinary research linked with CIP plant breeding which provided native varieties with high zinc and iron content. These varieties were then evaluated with farmers, using techniques refined by previous work with PVS. The nutrition research followed on the successful work with PMCA, but now the emphases was not on finding a more marketable potato, but on choosing a more nutritious one. Like the Andean Change Alliance, the nutrition work with the IssAndes project collaborated with 20 organizations in four Andean countries (actually, many of the same organizations that had been part of Andean Change). INIAP, the government agricultural research institute in Ecuador, built a large aeroponics unit devoted in part to producing seed of native potato varieties. This can be seen as an example of successful advocacy from Andean Change and previous CIP work (e.g. with PVS).

While participatory research at CIP may seem fragmented and at times uncertain (Thiele et al. 2001), CIP never stopped using participatory research. CIP social scientists, breeders, IPM specialists and agronomists have continued working together, innovating, creating influential participatory methods that integrate disciplines, bring farmers on board, and widen participation to include all the actors of the value chain, which in the future will continue evolving to face more complex challenges and link more diverse stakeholders taking advantage of emerging technologies (ITCs) and the possibility of using the big data revolution.

References

Allauca Saguano SM (2011) Análisis de los criterios de preferencias en la selección de nuevas variedades de papa, con enfoque en la influencia del mercado. Thesis, Universidad Nacional Agraria, La Molina, Lima, Peru, 239 p

Arcos J, Gastelo M, Holguín V (2015) INIA 317 – Altiplano, variedad de papa con buena adaptación en la región altiplánica del Perú. Revista Latinoamericana de la Papa 19(2):68–75

Bernet T, Thiele G, Zschocke T (2006) Participatory market chain approach (PMCA)—user guide. International Potato Center (CIP) – Papa Andina, Lima

Camacho-Henriquez A, Kraemer F, Galluzzi G, de Haan S, Jäger M, Christinck A (2015) Decentralized collaborative plant breeding for utilization and conservation of neglected and underutilized crop genetic resources. In: Al-Khayri JM, Jain SM, Johnson DV (eds) Advances in plant breeding strategies: breeding, biotechnology and molecular tools, vol 1. Springer, Heidelberg, pp 25–61

Chiri A, Ccama F, Fano H, Dale W (1997) Final evaluation of the integrated pest management for Andean Communities (MIPANDES) project (no. 527-0372). Report submitted to CARE/Peru

CIP (2014) Liberan nueva variedad de papa 'INIA 325 – PODEROSA' en Perú. https://cipotato.org/es/press-room/blog/liberan-nueva-variedad-papa-inia-325-poderosa-peru/. Accessed 9 Aug 2017

Cisneros F, Alcázar J, Palacios M, Ortiz O (1995) A strategy for developing and implementing integrated pest management. CIP Circular 21(3):2–7

Creed-Kanashiro HM, Astete-Robilliard L, Abad Arrue M, Marin M, Bartolini R (2014) Línea de Base Nutricional Perú. Proyecto IssAndes. CIP, Lima, p 65

Creed-Kanashiro H, Hareau G, Devaux A, Maldonado L, Ordinola M, Fonseca C, Suarez V, Astete L, Marin M, Penny M (2015) Agriculture-nutrition linkages: nutritional outcomes in potato based production systems of Peru. Poster presented at the conference on global food security

De Haan S, Salas E, Fonseca C, Gastelo M, Amaya N, Bastos C, Hualla V, Bonierbale M (2017) Selección participativa de variedades de papa (SPV) usando el diseño mamá y bebé: una guía para capacitadores con perspectiva de género. CIP, Lima, 82 p

Devaux A, Kromann P (2016) Strengthening innovation in agri-food systems for food security in the Andes. CIP, Lima, 2 p

Devaux A, Horton D, Velasco C, Thiele G, López G, Bernet T, Reinoso I, Ordinola M (2009) Collective action for market chain innovation in the Andes. Food Policy 34(1):31–38

Devaux A, Andrade-Piedra J, Ordinola M, Velasco C, Hareau G, López G, Rojas A, Flores P, Fonseca C, Kromann P (2012) Agricultura, seguridad alimentaria y nutrición en los Andes: potenciales aportes de la innovación en papa. Paper read at the Congreso de la Asociación Latinoamericana de la Papa, 17–20 Sep 2012, Uberlândia, Brazil

Devaux A, Flores P, Velasco C, Babini C, Ordinola M (2015) Innovation to enhance agriculture, nutrition, and health linkages. IssAndes project brief. International Potato Center, Lima, 6 p

Engel P, Salomon M (2003) Facilitating innovation for development: a RAAKS resource box, Amsterdam

Flores R (2010) Avances, oportunidades y temas estratégicos de incidencia política en la región. Paper read at the Taller de Evidencias y Argumentos, Metodologías de la Alianza Cambio Andino y Metodologías Invitadas. Cambio Andino, Lima, 8–10 Nov 2010

Fonseca C, De Haan S, Miethbauer T, Maldonado L, Ruiz R (2011) Selección participativa de variedades de papa en Perú. In: Thiele G, Quirós CA, Ashby J, Hareau G, Rotondo E, López G, Paz Ybarnegaray R, Oros R, Arévalo D, Bentley J (eds) Métodos participativos para la inclusión de los pequeños productores rurales en la innovación agropecuaria: experiencias y alcances en la región andina 2007–2010. Programa Alianza Cambio Andino, Lima, pp 169–184

Godtland E, Sadoulet E, de Janvry A, Murgai R, Ortiz O (2004) The impact of farmer-field-schools on knowledge and productivity: a study of potato farmers in the Peruvian Andes. Econ Dev Cult Change 53:63–92

Hall A, Mytelka L, Oyeyinka B (2004) Innovation systems: what's involved for agricultural research policy and practice? ILAC brief 2

Hellin J (2012) Agricultural extension, collective action and innovation systems: lessons on network brokering from Peru and Mexico. J Agric Educ Exten 18(2):141–159

Heong KL, Escalada MM, Chien HV, Cuong LQ (2014) Restoration of rice landscape biodiversity by farmers in Vietnam through education and motivation using media. Surv Perspect Integr Environ Soc 7(2)

Horton D (1986) Farming systems research: twelve lessons from the Mantaro Valley Project. Agric Admin 23:93–107

Horton D, Samanamud K (2013) Peru's native potato revolution. Papa Andina innovation brief 2. International Potato Center, Lima, 6 p

Horton D, Rotondo E, Paz Ybarnegaray R, Hareau G, Devaux A, Thiele T (2013) Lapses, infidelities, and creative adaptations: lessons from evaluation of a participatory market development approach in the Andes. Eval Program Plann 39:28–41

Janampa Martínez AM (2012) Selección participativa bajo el diseño mamá &bebé de 20 clones de papa *Solanum tuberosum* spp. *andígena* (población B1C5), con resistencia horizontal a la rancha (*Phytophthora infestans*). Thesis, Universidad para el Desarrollo Andino, Lircay, Huancavelica, Peru, 127 p

Kromann P, Montesdeoca F, Andrade-Piedra J (2016) Integrating formal and informal potato seed systems in Ecuador. In: Andrade-Piedra J, Bentley J, Almekinders C, Jacobsen K, Walsh S, Thiele G (eds) Case studies of root, tuber and banana seed systems. RTB working paper no. 2016-3. RTB, Lima, pp 13–32

Nelson R, Orrego R, Ortiz O, Tenorio J, Mundt C, Fredrix M, Vien NV (2001) Working with resource-poor farmers to manage plant diseases. Plant Dis 85(7):684–695

Ordinola M (2015) Agricultura, nutrición y seguridad alimentaria: un enfoque diferente. Agro Enfoque 199:65–67

Ordinola M, Devaux A, Bernet T, Manrique K, Lopez G, Fonseca C, Horton D (2014) The PMCA and potato market chain innovation in Peru. Papa Andina innovation brief 3. International Potato Center, Lima, 8 p

Oros R, Rodríguez F, Gonzales F, Thiele G (2011) Caso de implementación EPCP 2: la papa nativa en Norte Potosí, Bolivia. In: Thiele G, Quirós CA, Ashby J, Hareau G, Rotondo E, López G, Paz Ybarnegaray R, Oros R, Arévalo D, Bentley J (eds) Métodos participativos para la inclusión de los pequeños productores rurales en la innovación agropecuaria: experiencias y alcances en la región andina 2007–2010. Programa Alianza Cambio Andino, Lima, pp 99–109

Orrego R, Ortiz I, Tenorio J (2010) Interactuando para aprender: el caso de las escuelas de campo de agricultores (ECAs) en el Perú. LEISA 26(4):33–35

Ortiz O (2006) Evolution of agricultural extension and information dissemination in Peru: an historical perspective focusing on potato-related pest control. Agric Human Values 23(4):477–489

Ortiz O, Alcazar J, Palacios M (1997) La enseñanza del manejo integrado de plagas en el cultivo de papa: la experiencia del CIP en la Zona Andina del Perú. 9/10:1–22

Ortiz O, Garret KA, Heath JJ, Orrego R, Nelson RJ (2004) Management of potato late blight in the Peruvian highlands: evaluating the benefits of farmer field schools and farmer participatory research. Plant Dis 88(5):565–571

Ortiz O, Frias G, Ho R, Cisneros H, Nelson R, Castillo R, Orrego R, Pradel W, Alcázar J, Bazán M (2008) Organizational learning through participatory research: CIP and CARE in Peru. Agric Human Values 25(3):419–431

Ortiz O, Kroschel J, Alcázar J, Orrego R, Pradel W (2009) Evaluating dissemination and impact of IPM: lessons from case studies of potato and sweet potato IPM in Peru and other Latin American countries. In: Peshin R, Dhawan AK (eds) Integrated pest management: dissemination and impact. Springer, New York, pp 419–434

Ortiz O, Orrego R, Pradel W, Gildemacher P, Castillo R, Otiniano R, Gabriel J, Vallejo J, Torres O, Woldegiorgis G, Damene B, Kakuhenzire R (2011) Incentives and disincentives for stakeholder involvement in participatory research (PR): lessons from potato-related PR from Bolivia, Ethiopia, Peru and Uganda. Int J Agric Sustain 9(4):522–536

Ortiz O, Orrego R, Pradel W, Gildemacher P, Castillo R, Otiniano R, Gabriel J, Vallejo J, Torres O, Woldegiorgis G, Damene B, Kakuhenzire R, Kasahija I, Kahiu I (2013) Insights into potato innovation systems in Bolivia, Ethiopia, Peru and Uganda. Agric Syst 114(2013):73–83

Ortiz O, Nelson R, Olanya M, Thiele G, Orrego, R, Pradel W, Kakuhenzire R, Woldegiorgis G, Gabriel J, Vallejo J, Xie, K (2019) Human and Technical Dimensions of Potato Integrated Pest Management Using Farmer Field Schools: International Potato Center and Partners' Experience With Potato Late Blight Management. Journal of Integrated Pest Management 10(1):1–8

Parsa S (2010) Native herbivore becomes key pest after dismantlement of a traditional farming system. Am Entomol 56(4):242–251

Polar V (2014) Participation for empowerment: an analysis of agricultural innovation in two contrasting settings of Bolivia. PhD thesis. School of Oriental and African Studies, University of London, 331 p, La Paz – SOAS. http://eprints.soas.ac.uk/20311/1/Polar_Funez_3625.pdf

Polar V, Fernández J, Ashby J, Quiros CA, Roa JI (2012) Participatory methods and the co-production of agricultural advisory services. Results from four case studies in Bolivia and Colombia. International Potato Center (CIP), Lima. Working paper 2012-1, 107 p. https://cgspace.cgiar.org/bitstream/handle/10568/90952/CIP-Participatory-methods-and-the-co-production-of-agricultural-advisory.pdf?sequence=1

Prain G, Uribe F, Scheidegger U (1992) "The friendly potato": farmer selection of potato varieties for multiple uses. In: Moock JL, Rhoades RE (eds) Diversity, farmer knowledge and sustainability. Cornell University Press, Ithaca, pp 52–68

Pumisacho M, Sherwood S (2000) Herramientas de aprendizaje para facilitadores: manejo integrado del cultivo de papa. INIAP, CIP, FAO, Quito

Rhoades R (1987) Farmers and experimentation. ODI AgREN discussion paper 21

Rhoades RE, Booth RH (1982) Farmer-back-to-farmer: a model for generating acceptable agricultural technology. Agric Admin 11(2):127–137

Rhoades RE, Horton DE, Booth RH (1986) Anthropologist, biological scientist and economist: the three musketeers or three stooges of farming systems research. In: Jones JR, Wallace BJ (eds) Social sciences and farming systems research. Methodological perspectives on agricultural development. Westview, Boulder, pp 21–40

Sherwood S, Nelson R, Thiele G, Ortiz O (2000) Farmer field schools for ecological potato production in the Andes. LEISA-Leusden 16:24–25

Sineace (National System for Evaluation, Accreditation and Certification of Educational Quality—Ministry of Education, Peru) (2016) Facilitadores de Escuelas de Campo de Agricultores podrán certificar sus competencias. https://www.sineace.gob.pe/facilitadores-de-escuelas-de-campo-de-agricultores-podran-certificar-sus-competencias/. Accessed 11 Jan 2018

Snapp S (1999) Mother and baby trials: a novel trial design being tried out in Malawi. TARGET. The newsletter of the Soil Fertility Research Network for Maize-Based Cropping Systems in Malawi and Zimbabwe. January 1999 issue. CIMMYT, Harare

Snapp S (2002) Quantifying farmer evaluation of technologies: the mother and baby trial design. In: Bellon MR, Reeves J (eds) Quantitative analysis of data from participatory methods in plant breeding. CIMMYT, Mexico City, pp 9–17

Sperling F, Validivia C, Quiroz R, Valdivia R, Angulo L, Seimon A, Noble I (2008) Transitioning to climate resilient development: perspectives from communities in Peru, Environment department papers, no.115. Climate change series. Washington, DC, World Bank, 103p

Thiele G, van de Fliert E, Campilan D (2001) What happened to participatory research at the International Potato Center? Agric Human Values 18:429–446

Thiele G, Devaux A, Reinoso I, Pico H, Montesdeoca F, Pumisacho M, Andrade-Piedra J, Velasco C, Flores F, Esprella R, Thomann A, Manrique K, Horton D (2011) Multi-stakeholder platforms for linking small farmers to value chains: evidence from the Andes. Int J Agric Sustain 9(3):423–433

Thiele G, Quirós CA, Ashby J, Hareau G, Rotondo E, López G, Paz Ybarnegaray R, Oros R, Arévalo D, Bentley J (eds) (2012) Métodos participativos para la inclusión de los pequeños productores rurales en la innovación agropecuaria: experiencias y alcances en la región andina 2007-2010. Programa Alianza Cambio Andino, Lima. https://cgspace.cgiar.org/handle/10568/65712

Veteto JR, Crane TA (2014) Tending the field: special issue on agricultural anthropology and Robert E. Rhoades. Cult Agric Food Environ 36(1). https://doi.org/10.1111/cuag.12023

7

The Potato and its Contribution to the Human Diet and Health

Gabriela Burgos, Thomas Zum Felde, Christelle Andre, and Stan Kubow

Abstract Potato has contributed to human diet for thousands of years, first in the Andes of South America and then in the rest of the world. Its contribution to the human diet is affected by cooking, potato intake levels, and the bioavailability of potato nutrients. Generally, the key nutrients found in potatoes including minerals, proteins, and dietary fiber are well retained after cooking. Vitamins C and B_6 are significantly reduced after cooking while carotenoids and anthocyanins show high recoveries after cooking due to an improved release of these antioxidants.

In many developed countries potatoes are consumed as a vegetable with intakes that vary from 50 to 150 g per day for adults. On the other hand, in some rural areas of Africa and in the highlands of Latin American countries, potato is considered a staple crop and is consumed in large quantities with intakes that vary from 300 to 800 g per day for adults. These marked differences in the potato intake affect significantly the contribution of potato nutrients to the human dietary requirements.

In recent years, information about nutrient bioaccessibility and bioavailability from potatoes has become available indicating higher bioaccessibility of minerals and vitamins in potato as compared with other staple crops such as beans or wheat. Bioavailability refers to the fraction of an ingested nutrient that is available for utilization in normal physiological functions and/or for body storage while bioaccessibility refers to the amount that is potentially absorbable from the gut lumen.

In addition, potatoes have shown promising health-promoting properties in human cell culture, experimental animal and human clinical studies, including anticancer, hypocholesterolemic, anti-inflammatory, anti-obesity, and antidiabetic

G. Burgos (✉) · T. Zum Felde
International Potato Center, Lima, Peru
e-mail: g.burgos@cgiar.org; t.zumfelde@cgiar.org

C. Andre
The New Zealand Institute for Plant and Food Research Limited/Luxembourg Institute of Science and Technology, Auckland, New Zealand
e-mail: christelle.andre@plantandfood.co.nz

S. Kubow
McGill University, Quebec, Canada
e-mail: stan.kubow@mcgill.ca

properties with phenolics, anthocyanins, fiber, resistant starch, carotenoids as well as glycoalkaloids contributing to the health benefits of potatoes.

Introduction

Diverse studies have demonstrated that potato is an important source of carbohydrates, resistant starch, quality proteins, vitamins C and B_6 as well as potassium (Camire et al. 2009). Potato is also a source of antioxidants that can contribute to prevent both degenerative and age-related diseases with lutein and zeaxanthin being present in high levels in yellow-fleshed potatoes (Burgos et al. 2009) and anthocyanins being present in purple and red-fleshed potato landraces (Burgos et al. 2013b) commonly grown and eaten in the Andean highlands of Peru, Bolivia, Ecuador, and Colombia. Potatoes also contain glycoalkaloids, which in high concentrations can be toxic to humans but in low concentrations can have beneficial effects such as inhibition of the growth of cancer cells (Friedman 2015). The nutritional composition of potatoes is summarized in Fig. The concentration of energy, starch, protein, lipids, dietary fiber, potassium, phosphorus, magnesium, iron, zinc, vitamin C, vitamin B_6, chlorogenic acid, and glycoalkaloids has a range of variation independent from the flesh color. Yellow-fleshed potatoes have a carotenoid concentration higher than white-fleshed potatoes while purple potatoes have a higher anthocyanin concentration than red- or white-fleshed potatoes.

Like other plant foods, the nutritional composition of potatoes is affected by different pre-harvest (environment, cultural practices, maturity at harvest, biotic and abiotic stresses, etc.) and post-harvest (processing, storage, transport, etc.) conditions.

Potato has contributed to the human diet for thousands of years, first in the Andean region and then in the rest of the world. Its contribution is affected by cooking, the amount of potato intake, and the bioavailability of the nutrients. Generally, the key phytonutrients found in potatoes including minerals, proteins, and dietary fibers are well retained after cooking. Vitamins C and B_6 are significantly reduced after cooking while carotenoids and anthocyanins show high recovery after cooking due to an improved release of these antioxidants from the food matrix after cooking (Tian et al. 2016). In this chapter, the range of nutrient concentrations is expressed on a fresh weight (FW) basis and ranges refer to both raw and cooked potatoes. However, for calculating their contribution to the diet, only values of cooked potatoes are considered.

The worldwide mean potato intake is equivalent to 93 g per day (FAO 2013). However, this value has a large range of variation. In many developed countries potatoes are consumed as a vegetable and served as a part of a larger meal with intakes that vary from 50 to 150 g per day for adults. On the other hand, in some rural areas of Africa and in the highlands of Latin American countries, potato is considered a staple crop and consumed alone in large quantities as a complete meal with intakes that vary from 300 to 800 g per day for adults (De Haan et al. 2019).

Energy:	96 to 123 Kcal	Magnesium:	16 to 40 mg
Starch:	16 to 20 g	Iron:	0.29 to 0.69 mg
Protein:	1.76 to 2.95 g	Zinc:	0.29 to 0.48 mg
Lipids:	0.1 to 0.5 g	Vitamin C:	7.8 to 20.6 mg
Dietary fiber:	1.8 g to 2.1 g	Vitamin B6:	0.299 mg
Potassium:	150 to 1386 mg	Chlorogenic acid:	19 to 399 mg
Phosphorus:	42 to 120 mg	Glycoalkaloids:	0.7 to 18.7 mg

Fig. Nutritional composition of potatoes per 100 g FW

Implications of the contribution of potatoes to the human diet as related to the magnitude of potato intake are also described in this chapter.

Bioavailability refers to the fraction of an ingested nutrient that is available for utilization in normal physiological functions and/or for its contribution towards body stores (La Frano et al. 2014). Many factors affect the bioavailability of a compound; these may be divided into exogenous factors such as the complexity of the food matrix, the chemical form of the compound of interest, structure and amount of co-ingested compounds as well as endogenous factors including mucosal mass, intestinal transit time, rate of gastric emptying, intestinal and hepatic metabolism, and the extent of conjugation and protein-binding in blood and tissues (Holst and

Williamson 2008). A prerequisite for bioavailability of any compound is its bioaccessibility in the gut, defined as the amount that is potentially absorbable from the lumen (Fernández-Garcia et al. 2009). Bioavailability can be limited by a low bioaccessibility, which can be affected by the nature of the food matrix, location within the plant, food processing, gastric and luminal digestion, in addition to the physicochemical properties of the compound itself. In this chapter, the bioaccessibility and bioavailability of phytonutrients in potato will be reported and discussed.

The contribution of potato to human health will be described in terms of the evidence concerning the anticancer, hypocholesterolemic, anti-inflammatory, anti-obesity, and anti-diabetic role of potatoes.

Contribution to Diet

Energy

The energy provided by 100 g of boiled tubers of potatoes varies from 96.33 to 123.17 kcal (De Haan et al. 2019), which is similar to the energy provided by 100 g of cooked rice (130 kcal) but lower than the energy provided by 100 g of wheat (361 kcal), 100 g of cooked cassava (160 kcal) and soybeans (173 kcal) (King and Slavin 2013). Potato has a low energy density with 100 g of boiled potatoes contributing between 4 and 6% of the requirement of energy of an adult of between 50 and 90 kg of weight (considering 1.90 as basal metabolic rate factor, FAO/OMS/UNU 2004). However, preparing and serving potatoes with ingredients with a high fat content raises greatly the caloric value of the dish. One hundred grams of potato chips and French fries provide 529 and 564 kcal, respectively.

In areas where potato is considered as a staple food, the amount of potato intake is high and consequently the contribution of potatoes towards meeting dietary requirements is much higher. In Huancavelica, a location in the Peruvian central highlands, women have an average daily consumption of 840 and 645 g during abundance and scarcity period of potato, respectively. In those regions, potatoes are mainly eaten as boiled and provide between 28 and 38% of the recommended total energy requirements for women (De Haan et al. 2019).

Carbohydrates

Starch

Starch is the predominating carbohydrate in potato ranging from 16.5 to 20.0 g/100 g FW (Liu et al. 2007). Biochemically, potato starch is composed of amylose and amylopectin with the latter molecule typically making up 70–80% of the available starch in the tuber and the remaining portion being composed of amylose (Zeeman et al. 2010). Starch can also be classified by levels of digestibility within the human intestinal tract, i.e. rapidly digested (RDS), slowly digested (SDS), or resistant (RS) starch (Englyst et al. 1992). RDS and SDS represent the portion of starch digested within the first 20 and 21–120 min post-ingestion, respectively. The remaining resistant starch (RS) is undigested and fermented when it reaches the large intestine with the production of short-chain fatty acids (Raigond et al. 2014). Because of the resistance of the amylose structure to digestion, more of the RS component is expected to be composed of amylose rather than amylopectin. The rapid breakdown of amylopectin to digestion is the reason that it is more prevalent in RDS and SDS fractions (Bach et al. 2013). Potential health benefits attributed to SDS include satiety, improved physical performance, glucose tolerance enhancement and blood lipid level reduction in healthy individuals and in those with hyperlipidemia (Miao et al. 2015). Possible health benefits of RS include prevention of colon cancer, hypoglycemic effects, substrate provision for growth of gut probiotic microorganisms, reduction of gall stone formation, hypocholesterolemic effects, inhibition of fat accumulation, and increased absorption of minerals (Sajilata et al. 2006).

Monro et al. (2009) determined the RDS, SDS, and RS concentration of freshly cooked potatoes from nine potato varieties and found concentrations ranging from 9 to 15 g/100 g FW, from 0 to 1.72 g/100 g FW, and from 0.58 to 1.05 g/100 g FW, respectively. These authors also found that cooking and then cooling potatoes significantly increased SDS and RS (up to 7.7 g and 1.96 g/100 g FW, respectively), while RDS was significantly reduced to 7.3 g/100 g FW. This latter phenomenon is referred to as starch retrogradation, which is based upon rearrangement of the molecules of amylose and amylopectin causing decreased starch digestion (Leeman et al. 2005).

Glycemic index (GI) is a measure of the extent of the change in blood glucose content (glycemic response) following consumption of digestible carbohydrate, relative to a standard such as glucose (Venn and Green 2007). A higher GI value represents a more rapid entry of a larger quantity of glucose from a test food into the bloodstream. Based on in vivo postprandial GI, high RDS content in foods has been significantly correlated with a high glycemic index response (Champ 2004), whereas low RDS levels are associated with low to medium GI values (Lynch et al. 2007). A wide variability in GI values of potatoes has been noted ranging from high to medium to low values based on cultivar differences (Ek et al. 2012). Such variations could partly be related to differences in the amylopectin to amylose ratio as amylose-rich starches are digested more slowly due to their difficulty to gelatinize and swell as opposed to starches with a high amylopectin content (Brennan 2005). Bach et al. (2013) defined low RDS and high SDS as the optimal profile for potatoes that leads to low GI values, and identified two genotypes with this profile. Tuber cooking followed by cooling (forming retrograded starch) is also another way for the consumer to obtain lower postprandial glucose levels, and thereby benefit from reduced GI following potato intake (Fernandes et al. 2005). Lowering the dietary GI load has been associated with body weight loss, improved blood pressure, and decreased risk of cardiovascular diseases, whereas habitual intake of high GI foods has been linked to type 2 diabetes and other chronic heart issues (McGill et al. 2013). As the GI does not take into account the typical portion size, the GI value and the quantity of carbohydrates are combined to generate the glycemic load (GL) value, which can better quantify the glycemic impact of a food (Salmeron et al. 1997). Initial studies involving potatoes were limited by the sole use of GI for their glycemic evaluation (Crapo et al. 1977; Soh and Brand-Miller 1999), which categorized them with a high GI. In contrast, potatoes have been generally noted to have a medium to low glycemic impact based on the GL estimation (Lynch et al. 2007).

Sugars

Potato tubers also contain significant quantities of free sugars with glucose and fructose as the principal monosaccharides and sucrose as the major disaccharide. Glucose, fructose, and sucrose concentrations in raw tubers of tetraploid potatoes range from 3.25 to 255 mg/100 g FW, from 2.5 to 153.7 mg/100 g FW, and from 43 to 159.7 mg/100 g FW, respectively (Amrein et al. 2003; Rodríguez et al. 2010). Higher levels of glucose, fructose, and sucrose have been recently reported for diploid potato group *Phureja* with concentrations ranging from 11.5 to 701 mg/100 g FW for glucose, from 7.25 to 605 mg/100 g FW for fructose and from 159 to 737 mg/100 g FW for sucrose (Duarte-Delgado et al. 2016). The reducing sugars glucose and fructose as well as free asparagine are acrylamide precursors. Acrylamide is formed through the Maillard reaction during high temperature cooking such as frying, roasting, or baking (Muttucumaru et al. 2008). Acrylamide has been classified as "probably carcinogenic to humans" by the WHO and the International Agency for Research on Cancer. Therefore, the reducing sugar content in potatoes has been recommended not to be greater than 100 mg/100 g FW in order to keep acrylamide formation on a low level (Kumar et al. 2004). Importantly, cold storage (2–4 °C) may induce an accumulation of reducing sugars in tuber tissue leading to undesirable browning, production of bitter flavors, and increased levels of acrylamide with cooking (Neilson et al. 2017).

Protein

According to Camire et al. (2009), the protein content of potatoes generally ranges from 1 to 1.5 g/100 g FW depending on the cultivar. De Haan et al. (2019) reported higher levels of protein in cooked tubers of Peruvian floury landraces (1.76–2.95 g/100 g FW). Potato protein content is generally low compared with other major staples like maize and beans although potato yields more protein per unit growing area than do cereals (Bamberg and Del Rio 2005). Also, the quality of the potato protein, which reflects its digestibility and indispensable amino acid content, is very good. The biological value of potato protein—the proportion retained for growth or maintenance divided by the amount absorbed—is high. Depending on the cultivar, the biological value of potato protein is between 90 and 100 and is very similar to the biological value of whole egg protein (100) and is higher than that of soybeans (84) and legumes (73) (Camire et al. 2009).

The levels of lysine, methionine, threonine, and tryptophan are likely to limit the protein quality of mixed diets consumed by humans. Potatoes exceed the recommended levels of these indispensable amino acids, demonstrating that potato protein is of high quality. Compared with pasta, white rice, and whole grain cornmeal, potatoes are the only staple food meeting the recommended lysine level. However, sulfur-containing amino acids (methionine + cysteine) are lower in potatoes than in the other common plant staple foods (King and Slavin 2013).

Lipids

Total lipids of potatoes are low and range from 0.1 to 0.5 g/100 g FW and consist mainly of phospholipids (47%), glycol and galactolipids (22%), which are structural elements of biological membranes as well as neutral lipids (21%) such as acylglycerols and free fatty acids (Ramadan and Oraby 2016). More than 94% of the tuber lipids contain esterified fatty acids. The essential polyunsaturated fatty acids with one to three double bounds consist of mainly linoleic acid (C18:2 cis-9,12, an *n*-6 fatty acid) and linolenic acid (C18:3 cis-9,12,15, an *n*-3 fatty acid) (70–75%), precursors of a wide range of bioactive compounds generated endogenously (Galliard 1973). The composition of the fatty acids of the potato lipids is nutritionally advantageous. For example, potato consumption in the United Kingdom was estimated to provide 10 and 13% of the dietary *n*-6 and *n*-3 polyunsaturated fatty acid intake, respectively (Gibson and Kurilich 2013). In contrast, potato intake provided only 4% of saturated fatty acid and 6% trans fatty acid intake, which was largely attributed to the addition of fats and oils such as butter and margarine to potato dishes.

Fiber

Dietary fiber represents the undigested and unabsorbed carbohydrate part in the diet. These resistant carbohydrates may be fermented in the large intestine. Soluble fibers lower serum lipids, whereas insoluble fibers increase stool weight (Slavin 2008). Potatoes contain dietary fiber in their cell walls, especially in the thickened cell walls of the peel (Camire et al. 2009). Cooked potatoes without the skin provide 1.8 g fiber/100 g, FW, whereas cooked potatoes with the skin provide 2.1 g fiber/100 g FW. Potatoes contain less fiber than whole-grain cornmeal (7.3 g/100 g), but more fiber than white rice (0.3 g/100 g) or whole-wheat cereal (1.6 g/100 g). Although potatoes cannot therefore be considered a high-fiber food, they can be a significant source of fiber for individuals regularly eating potatoes, particularly in developed countries where fiber intake is generally far below recommended levels (Auestad et al. 2015). In that regard, potatoes have been indicated to contribute 14.4–26.2% of daily fiber intake in men and women living in the USA based on the National Health and Nutrition Examination Survey (NHANES) data (2009–2010).

Minerals

Potassium is the most abundant mineral in potato with concentrations varying from 150 to 1386 mg/100 g FW (Nassar et al. 2012). Potassium functions as an important electrolyte in the nervous system. High intake levels of potassium can help control high blood pressure and may decrease the risk of stroke (Bethke and Jansky 2018). One hundred grams of boiled potatoes can contribute up to 16% of the Adequate Intake (AI) of potassium recommended for adults (4700 mg per day).

Phosphorus and magnesium are also present in potato in moderate quantities ranging from 42 to 120 mg/100 g FW and from 16 to 40 mg/100 g FW, respectively (Bonierbale et al. 2010). One hundred grams of boiled potatoes can contribute up to 11% of the Estimated Average Requirement (EAR) of phosphorus and magnesium for adults (42–120 and 265–340 mg per day, respectively). Calcium is present in minor quantities in potato ranging from 2 to 20 mg/100 g FW; contributing no more of 2% of the EAR of calcium for adults (800–1100 mg per day).

Iron and zinc concentrations from raw potatoes range from 0.25 to 0.83 mg/100 g FW and from 0.23 to 0.39 mg/100 g FW, respectively (Burgos et al. 2007). Iron and zinc concentrations are significantly affected by the growing environment. Interestingly, Lombardo et al. (2013) reported that soil composition affects the mineral concentration of crops, with a sandy texture of the soil favoring the iron oxidation processes to insoluble polymers and consequently reducing iron availability to the plant.

Burgos et al. (2007) reported iron and zinc concentration in cooked potatoes ranging from 0.29 to 0.69 mg/100 g FW and from 0.29 to 0.48 mg/100 g FW, respectively. These values are lower than iron and zinc concentrations reported for cereals and legumes but bioavailability of iron and zinc from potatoes may be higher due to the presence of high levels of ascorbic acid—which facilitates iron absorption in the human body—and low levels of phytic acid, an inhibitor of iron and zinc absorption. It has been recently demonstrated that the bioaccessibility of iron in potato is higher than that reported in crops such as wheat and beans. Approximately 63–79% of the potato iron is released from the food matrix after in vitro gastrointestinal digestion, and therefore available for intestinal absorption (Andre et al. 2015).

In the Andean highlands, where there is little access to meat and the levels of anemia and malnutrition are high, potatoes are an important dietary source of iron due to their high consumption. For example, in Huancavelica, in the Peruvian highlands, women and children consume on average 840 and 200 g of potato per day, respectively (De Haan et al. 2019). Similarly, in parts of Rwanda and other African countries, women consume an average of 400 g of potatoes per day. Therefore, improving the iron and zinc concentrations of potato and their bioavailability would have a real impact to contribute to reduce malnutrition and improve life quality in these and other areas where anemia and/or stunted growth are still pervasive.

The International Potato Center (CIP) has been working for the past 15 years on potato mineral biofortification to increase the concentration of iron and zinc in this crop. The CIP Biofortification Potato Program started from a baseline of 0.48 mg/100 g FW for iron and 0.35 mg/100 g FW for zinc. After three cycles of breeding and recurrent selection, concentrations of the first biofortified potatoes reach 0.73 mg iron and 0.63 mg zinc/100 g FW. Considering 400 g of potato consumption for women of the Andes, the consumption of biofortified potatoes would cover 41 and 37% of the EAR of iron and zinc in women.

Presently, CIP is combining the first products of its biofortification breeding program with advanced breeding lines to release new varieties that will be able to withstand major potato pests and diseases, tolerate heat and drought, providing high yields, and respond to preferences of farmers and consumers.

Vitamins

Potatoes are a good source of ascorbic acid (vitamin C) and pyridoxine (vitamin B_6). Vitamin C as an antioxidant plays an important role in protection against oxidative stress. Vitamin C is an important free radical scavenger of reactive oxygen species such as hydroxyl radicals, superoxide anions, singlet oxygen, and hydrogen peroxide that can cause tissue damage resulting from lipid peroxidation, DNA breakage or base alterations, which may contribute to degenerative diseases such as heart disease or cancer (Bates 1997). In addition, due to its participation in the oxidation of transition metal ions, vitamin C also plays an important role in enhancing the bioavailability of non-haem iron (Teucher et al. 2004) and serves as a cofactor in the synthesis of collagen needed to support cardiovascular function, maintenance of cartilage, bones, and teeth, as well as wound healing (Naidu 2003).

Fresh potatoes have varying concentrations of vitamin C, which can reach 50 mg/100 g FW (Han et al. 2004) when they are freshly harvested. Significant variation in vitamin C concentrations of potatoes occur due to genotype and environment and genotype by environment interactions (Andre et al. 2007; Burgos et al. 2009).

Cooking and storage reduce the concentration of vitamin C in potato tubers. In addition, there are differences in the degree of reduction of vitamin C content depending on the cooking types. Retention levels of vitamin C after boiling in 20 native landraces varied between 50 and 90%. The losses may be caused by: (1) leaching into cooking water, (2) destruction by heat treatment, and (3) oxidation. It is interesting to note that the peel forms a barrier preventing loss of nutrients during cooking. As a consequence, boiling potato when it is peeled results in 10% more loss of vitamin C or phenolic compounds than if it is cooked with the peel (Woolfe and Poats 1987). Retention levels after storing under farming conditions has been shown to vary between 22 and 62%, depending on the variety (Burgos et al. 2009). Retention levels of vitamin C in 12 genotypes grown in Colorado state in the USA after 7 months of cold storage was less than 50% (Külen et al. 2012).

One hundred grams of cooked potatoes with vitamin C levels around 20 mg/100 g FW can provide between 27 and 33% of the EAR of vitamin C for an adult (75 for males and 60 for females, according to FAO/WHO 2001). One hundred grams of cooked potatoes contains lower concentrations of vitamin C than 100 g of cooked broccoli (68–108 mg/100 g FW; depending on the way of cooking; Yuan et al. 2009), 100 g of cooked spinach (44–79 mg/100 g FW; depending on the way of cooking; Zeng 2013) and 100 g of raw red pepper (up to 200 mg/100 g; Wahyuni et al. 2011). However, it is noteworthy that the final contribution of a particular food to the total intake of vitamin C depends on the total amount consumed in the diet and so potatoes may therefore contribute to a significant extent to the total dietary intake of vitamin C (Love and Pavek 2008). In that respect, potatoes have been estimated to provide over 50% of the daily vitamin C requirement in the USA and approximately 20% of the dietary vitamin C intake in Europe (Love and Pavek 2008).

Vitamin B_6, also called pyridoxine, is a versatile cofactor for key metabolic processes (Hellmann and Mooney 2010) that plays a major role in various cellular reactions and also confers several health benefits for humans, which may be partly attributed to its antioxidant capabilities (Fitzpatrick et al. 2012). It helps in maintaining normal nerve function and plays a crucial role in the synthesis of neurotransmitters such as dopamine and serotonin. Vitamin B_6 also assists normal nerve cell communication and acts as a coenzyme in the breakdown and utilization of carbohydrates, fats and proteins. In plant, it is a potent antioxidant, critical for plant pathogen resistance (Spinneker et al. 2007).

Potatoes are considered to be a good dietary source of vitamin B_6, with concentrations ranging from 0.450 mg/100 g FW to 0.675 mg/100 g FW (Moonney et al. 2013). Physical and chemical factors such as heat, light exposure, and pH also influence vitamin B_6 content, but this vitamin is relatively stable during storage (Fitzpatrick et al. 2012).

The mean concentration of vitamin B_6 in cooked potatoes (0.299 mg/100 g FW) is higher than the mean concentration of other staple crops such as maize (0.139 mg/100 g FW), rice (0.050 mg/100 g FW), cassava (0.051 mg/100 g FW), and wheat (0.034 mg/100 g FW) (Fudge et al. 2017). One hundred grams of cooked potatoes can provide between 17 and 23% of the Recommended Dietary Allowance (RDA) of B_6 from an adult (1.3–1.7 mg per day).

Potato tuber contains also moderate amount of vitamin E (Chitchumroonchokchai et al. 2017). Vitamin E is the collective name for a set of eight related tocopherols and tocotrienols, characterized by a hydrophobic isoprenoid tail and a more hydrophilic chromanol head (Bramley et al. 2000). In potato, significant amount of α-tocopherol has been found in raw tubers of Andean genotypes, ranging from 68 to 517.5 μg/100 g FW (recalculated from Andre et al. 2007), whereas amounts in commercial varieties varied between 15 and 75 μg/100 g FW (recalculated from Andre et al. 2007 and Chun et al. 2006).

In humans, as in plants, vitamin E is located primarily within the phospholipid bilayer of cell membranes. It reacts with and quenches free radicals in cell membranes, preventing polyunsaturated fatty acids from damage by lipid oxidation. Vitamin E deficiency has been associated with an elevated risk of artherosclerosis and other degenerative diseases. It is generally assumed that increases of α-tocopherol in the diet may contribute to a decreased risk of chronic diseases (Andre et al. 2010). The EAR for vitamin E is of 15 mg for women and men (Otten et al. 2010). The consumption of high vitamin E containing potato tubers, such as the Andean varieties, could therefore significantly increase the dietary vitamin E intake.

Antioxidants

Potato is one of the most important sources of antioxidants in the human diet (Lachman and Hamouz 2005). As such, it supports the antioxidant defense that reduces cellular and tissue toxicities that result from free radical-induced protein, lipid, carbohydrate, and DNA damage (Andre et al. 2010). In this way, potato antioxidants may reduce the risk for cancers, cardiovascular diseases, and type 2 diabetes.

Based on metabolic relationships and structural composition, there are three major groups of antioxidants present in potato, as in most plants. The first group consists in the aromatic phenolic compounds, which encompasses flavonoids including anthocyanins and flavonols produced by the flavonoid pathway, hydroxycinnamic acids and their derivatives produced by the phenylpropanoid pathway, and the amino acids tyrosine, phenylalanine, and tryptophan produced by the shikimate pathway. The second group encompasses the isoprenoid antioxidants such as the carotenoids and tocopherols; and the third group includes antioxidants related to ascorbate and glutathione functions in a redox system of compound-recycling that include ascorbic acid (Lovat et al. 2016).

Phenolics

Phenolic compounds, also known as polyphenols, constitute one of the most widely distributed group of dietary antioxidants in the plant kingdom, presenting more than 10,000 different structures, ranging from relatively simple phenols to complex polymers such as lignans and suberins. Phenolic compounds are produced in the cytoplasm and are subsequently transported in the vacuole or deposited in the cell wall. Routes to the major classes of phenolic compounds involve: (1) the core phenylpropanoid pathway from phenylalanine to an activated (hydroxy)cinnamic acid derivative, as well as specific branch pathways for the formation of (2) simple phenolic acids, lignins and lignans, (3) flavonoids, (4) tannins, and (5) stilbenes (Andre et al. 2009). Their aromatic cycles can be further modified through hydroxylations, methylations, glycosylations, acylations, or prenylations, extending their variability and complexity (Winkel-Shirley 2001). Phenolic acids include chlorogenic, caffeic, ferulic, and sinapic acids. Among flavonoids, anthocyanins are natural pigments, responsible for the red-blue color of many fruits and vegetables. Anthocyanins can also impact the organoleptic characteristics of foods, which may influence their technological behavior during food processing and also have implications in the field of human health (Pascual-Teresa and Sanchez-Ballesta 2008). Flavonols represent one of the most widespread flavonoid classes in plant and include compounds like quercetin and kaempferol that are most commonly found in their glycosylated form, i.e., linked with glucose or rutinose. As compared to other phenolic compounds, flavonol concentrations are known to be largely influenced by the environmental conditions during plant growth (Lancaster et al. 2000).

Phenolic compounds are considered to be health-promoting phytochemicals as they have shown in vitro antioxidant activity and have been reported to exhibit beneficial anti-bacterial, hypoglycemic, anti-viral, anti-carcinogenic, anti-inflammatory and vasodilatory properties (Duthie et al. 2000; Mattila and Hellstrom 2006).

Chlorogenic Acid

Chlorogenic acid has been reported as the predominant phenolic acid in raw and boiled potato tubers (Burgos et al. 2013b). The main function of chlorogenic acid in the plant appears to defend against pathogens. Concentrations of chlorogenic acid as well as other hydroxycinnamic acids are significantly induced following pathogen invasion, and deposited to enforce the cell walls to arrest pathogen development (Yogendra et al. 2015). In humans, these compounds consumed through diet are increasingly considered as effective protective agents against reactive oxygen species (ROS), which are known to be involved in aging and many degenerative diseases (Liang and Kitts 2015).

The isomers of chlorogenic acid, neo-chlorogenic acid, and crypto-chlorogenic acid, as well as caffeic acid are also found in potato tubers (Andre et al. 2007). Potatoes contain three isomers of chlorogenic acid depending on whether the hydroxycinnamate is attached to 3-, 4-, or 5-position of the quinic acid moiety with 5-*O*-caffeoylquinic acid as the principal chlorogenic acid. The 5-*O*-caffeoylquinic acid (CQA) isomer is also the principal chlorogenic acid component of coffee and apples (Stalmach et al. 2010; Clifford 1999). In vitro and ex vivo studies have demonstrated a reduction in oxidation of human LDL following the consumption of coffee suggesting that 5-*O*-CQA protects against in vitro oxidation of human LDL, a key step in the formation of atherosclerotic plaques (Natella et al. 2007; Richelle et al. 2001). 5-*O*-CQA has also been shown to exert anti-carcinogenic effects in animal models (Stalmach et al. 2010).

Lachman et al. (2013) have reported chlorogenic acid concentrations of raw potatoes ranging from 7.87 to 60.07 mg/100 g FW in nonpeeled potatoes and from 5.11 to 46.13 mg/100 g FW in their peeled counterparts, while Burgos et al. (2013b) report a wider range of variation in the chlorogenic acid concentration of raw purple potatoes (ranging from 63 to 329.75 mg/100 g FW). Boiling, baking, and microwaving reduce the chlorogenic acid concentration of potatoes with boiled tubers having a higher retention of chlorogenic acid than baked and microwaved ones (Lachman et al. 2013). In a recent study conducted by Piñeros-Niño et al. (2017), the chlorogenic acid concentration of cooked tubers from 193 potato varieties ranged from 19.25 to 399 mg/100 g FW. In a previous study by Burgos (2014), the chlorogenic acid concentration in cooked tubers of purple-fleshed cultivars ranged from 36.17 to 395.73 mg/100 g FW and in red-fleshed cultivars from 14.45 to 48.60 mg/100 g FW.

The highest concentration of chlorogenic acid reported in 100 g of cooked potato tubers is similar to the maximum amount provided by a single cup of coffee (350 mg chlorogenic acid; Clifford 1999) and is tenfold higher than the maximum amount provided by whole apples (38.5 mg/100 g FW, Spanos and Wrolstad 1992).

Chlorogenic acid is only partially bioavailable and its bioactivity may be modulated by the gut microbiota that can generate bioactive secondary microbial phenolic metabolites such as caffeic acid that have much greater bioavailability (Tomas-Barberan et al. 2014; Olthof et al. 2003). Chlorogenic acid may also promote a healthy gut microbiome. In a batch culture fermentation model of the colon, chlorogenic acid was found to promote growth of Bifidobacterium bacterial species that could be beneficial for gut health (Mills et al. 2015).

Anthocyanins

Anthocyanins are a class of water-soluble flavonoids, which show a range of pharmacological effects, such as prevention of cardiovascular disease, obesity control, and anti-tumor activity. Their potential anti-tumor effects are reported to be based on a wide variety of biological activities including antioxidant, anti-inflammation, anti-mutagenesis, induction of differentiation, inhibiting proliferation by modulating signal transduction pathways, inducing cell cycle arrest, and stimulating apoptosis or autophagy of cancer cells; anti-invasion; anti-metastasis; reversing drug resistance of cancer cells and increasing their sensitivity to chemotherapy (Lin et al. 2017).

Anthocyanins are present in the flesh and skin of several purple- and red-fleshed potatoes such as those landraces found in the Andes, which show a wide range of anthocyanin structures and concentrations that are largely cultivar-dependent (Brown et al. 2003) and location-dependent (Ieri et al. 2011). Increased height above sea level, higher annual sum of precipitation, and lower annual average temperatures cause higher anthocyanin concentrations (Lachman et al. 2009).

The total anthocyanin concentration of raw and cooked purple-fleshed potatoes ranges from 63 to 588 mg/100 g FW and from 71 to 453 mg/100 g FW, respectively (Burgos et al. 2013a, b). Total anthocyanin concentration of cooked red-fleshed potatoes ranges from 8.2 to 55.3 mg/100 g FW (Burgos 2014).

Giusti et al. (2014) identified five major anthocyanidins (cyanidin, petunidin, pelargonidin, peonidin, and malvidin) in extract from purple potato and three major anthocyanidins (cyanidin, pelargonidin, and peonidin) in extracts of red potatoes. The extract of purple potatoes contained four major anthocyanins: cyanidin-3-rutinoside-5-glucoside, petunidin-3-rutinoside-5-glucoside, pelargonidin-3-rutinoside-5-glucoside, and peonidin-3-rutinoside-5-glucoside, with petunidin and peonidin glycosides being the most predominant. The extract of red-fleshed potatoes contained four major anthocyanins: cyanidin-3-rutinoside-5-glucoside, pelargonidin-3-rutinoside-5-glucoside, peonidin-3-rutinoside-5-glucoside, and pelargonidin-3-rutinoside, with pelargonidin-3-rutinoside-5-glucoside being the most predominant.

Burgos (2014) characterized the anthocyanin profile of 12 purple-fleshed accessions and 6 red-fleshed accession from CIP's genebank and found that in purple-fleshed accessions the predominant anthocyanin is petunidin-3-(coumaroyl) rutinoside-5-glucoside (petanin), representing from 37 to 78% of the total anthocyanins. It is followed by peonidin-3-(coumaroyl) rutinoside-5-glucoside, cyanidin-3-(coumaroyl) rutinoside-5-glucoside, and minor proportions of malvidin 3-(coumaroyl) rutinoside-5-glucoside and pelargonidin-3-(coumaroyl) rutinoside-5-glucoside. In red-fleshed accessions, the predominant anthocyanin is pelargonidin-3-(coumaryl) rutinoside-5-glucoside, representing 41–75% of the total anthocyanins. It is followed by peonidin-3-(coumaroyl) rutinoside-5-glucoside, pelargonidin-3-rutinoside, and cyanidin-3-(coumaroyl) rutinoside-5-glucoside in various proportions and then by pelargonidin-3-(coumaryl) rutinoside in minor proportions. Figure shows as an example the anthocyanin profile for two purple-fleshed accessions (CIP 705534 and CIP 702363) and two red-fleshed accessions (CIP 703625 and CIP 702453). Pt3(c)R5G represented by the purple bar is dominant in the purple-fleshed accessions while PI3R(c)R5G represented by the pink bar is dominant in the red-fleshed potatoes.

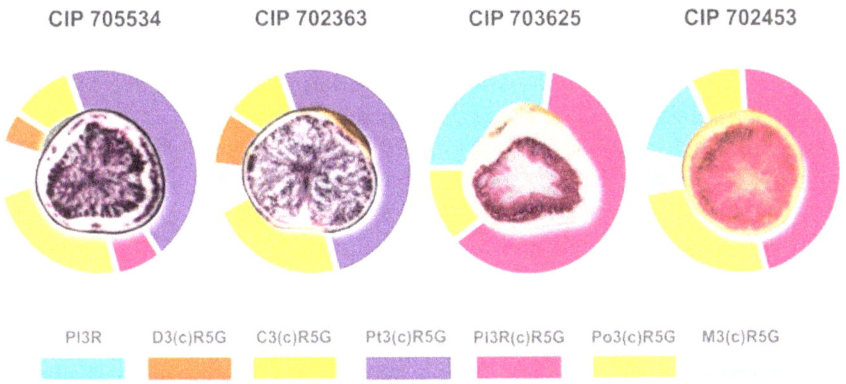

Fig. Anthocyanin profile in purple-fleshed and red-fleshed potatoes. (PI3R: pelargonidin-3-rutinoside, PI3R(c)R5G: pelargonidin-3-(coumaroyl) rutinoside-5-glucoside, Pt3(c)R5G: petunidin-3-(coumaroyl) rutinoside-5-glucoside, Po3(c)R5G: peonidin-3-(coumaroyl) rutinoside-5-glucoside, C3(c)R5G: cyanidin-3-(coumaroyl) rutinoside-5-glucoside, M3(c)R5G: malvidin 3-(coumaroyl) rutinoside-5-glucoside)

The most prominent anthocyanins present in the red- and purple-fleshed accessions are acylated with hydroxycinnamic acid (Fossen and Andersen 2000). Three different cinnamic acids were found acylating the anthocyanins in the extract of purple and red potatoes: caffeic, *p*-coumaric and ferulic acid (Giusti et al. 2014). Acylated anthocyanins are known to be stable and hence can be considered as promising natural colorants for the food industry.

The highest anthocyanin concentration reported in a dark purple-fleshed potato (above 400 mg/100 g FW; Andre et al. 2007) is lower than in blueberries (558 mg/100 g, FW; Hosseinian and Bea 2007), cranberries (589 mg/100 g FW; Wada and Ou 2002), eggplant (750 mg/100 g FW; Wu et al. 2006), and purple corn (1642 mg/100 g FW; Cevallos-Casals and Cisneros Zevallos 2003). However, the contribution of purple-fleshed potatoes to the diet can be considerably higher considering the high mean intake of potatoes in some areas like the Andean highlands where consumption may reach 500 g per day, as compared to the mean intake of blueberries, cranberries, and eggplant (1 g per day in the United States; Wu et al. 2006).

Kubow et al. (2017) studied the biotransformation of anthocyanins from cooked purple-fleshed potatoes using a dynamic human gastrointestinal (GI) model that includes stomach, small intestine, and colonic vessels. After 24 h digestion, liquid chromatography-mass spectrometry identified 15–36 anthocyanin species throughout the GI vessels. Genetic background of the purple potato cultivars led to major variances in the pattern of anthocyanin breakdown and release during digestion composition. Diminished concentrations of several anthocyanin species in the colonic vessels indicated microbial biotransformation which is, in turn, associated to increased bioaccessibility.

The cytotoxicity and cell viability of colonic Caco-2 cancer cells and nontumorigenic colonic CCD-112CoN cells after 24 h exposure to colonic fecal water of purple-fleshed potato digests has been also tested by Kubow et al. (2017). The cultivar Leona showed a significant potency to induce cytotoxicity and decrease viability of Caco-2 cells. The differing microbial anthocyanin metabolite profiles in colonic vessels between cultivars were indicated to play a significant role in the impact of fecal water toxicity on tumor and nontumorigenic cells.

In white- and yellow-fleshed potato tubers, flavonols predominate in the flavonoid profile (Andre et al. 2007). Flavonols have been extensively studied in the past 10 years as they present a range of putative health-promoting effects, including reduced risk of cancer and cardiovascular diseases (Wang et al. 2016). Rutin in particular has shown strong antioxidative and anti-inflammatory effects at the cellular level (Habtemariam and Lentini 2015).

In potato tubers, rutin (quercetin-3-*O*-rutinoside) and kaempferol-3-*O*-rutinoside are the most important compounds, with reported concentrations of 0–4.78 mg and 0–5.68 mg/100 g FW, respectively, in Andean raw potato tubers (Andre et al. 2007). The influence of various cooking methods on potato flavonols has been investigated, which revealed the stability of the concentrations through treatment (Navarre et al. 2010). The bioaccessibility of these compounds was also high (close to 100% on average) when evaluated in a collection of 12 Andean potato genotypes (Andre et al. 2015).

Carotenoids

Potatoes contain lipophilic phytonutrients in the form of carotenoids that have numerous health-promoting properties including decreasing risk of several chronic diseases (Gammone et al. 2015; Wu et al. 2015). Carotenoids have been reported to exhibit chemoprevention by a variety of mechanisms including immune system activation, protection against oxidative stress, promotion of gap junction communication, inhibition of DNA damage, enhanced metabolic detoxification, and tumor suppressor action and inhibition of oncogene expression (Khachik et al. 1999; Fiedor and Burda 2014).

Potato carotenoid concentrations and profiles are related to the flesh color with dark yellow cultivars showing approximately tenfold higher concentrations of total carotenoids than white-fleshed varieties (Brown et al. 2005). Significant and predominant amounts of zeaxanthin and antheraxanthin are found in deep yellow-fleshed potatoes while the carotenoid profile of yellow potatoes is composed of violaxanthin, antheraxanthin, lutein, and zeaxanthin and that of cream-fleshed potatoes of violaxanthin, lutein, and β-carotene (Burgos et al. 2009).

The violaxanthin, antheraxanthin, lutein, and β-carotene concentration of raw tubers of potatoes from the *Tuberosum* group ranged from 1.5 to 87.8 μg/100 g FW, 0.6 to 15.8 μg/100 g FW; 1.6 to 35.1 μg/100 g FW; and 0.1 to 2.1 μg/100 g FW, respectively (Fernandez-Orozco et al. 2013), while the concentration of these carotenoids in tubers from the *Phureja* group ranged from 20.0 to 410 μg/100 g FW; 9.3 to 503 μg/100 g FW; 55 to 211 μg/100 g FW and 4.8 to 27 μg/100 g FW, respectively (Burgos et al. 2009), and in tubers from the Andigenum group from 14.3 to 173 μg/100 g FW, 7 to 16 μg/100 g FW; 43.3 to 442 μg/100 g FW and 10.5 to 54.8 μg/100 g FW, respectively (Andre et al. 2007).

Boiling does not affect the lutein and zeaxanthin concentration of potato; however, violaxanthin and antheraxanthin concentrations of potatoes are significantly reduced after boiling. Lutein and zeaxanthin concentrations of cooked yellow-fleshed potatoes ranged from 73 to 253 μg/100 g FW and from 0 to 1048 μg/100 g FW, respectively (Burgos et al. 2012) with deep yellow-fleshed potatoes being a significant source of zeaxanthin (above 500 μg/100 g FW).

Lutein and zeaxanthin are important dietary carotenoids that are selectively taken up into the macula of the eye, where they protect against development of age-related macular degeneration and cataracts (Wu et al. 2015). Moreover, these compounds have been reported to have other health-promoting effects, including immune-enhancement and reduction of the risk of developing degenerative diseases such as cancer and cardiovascular diseases (Krinsky and Johnson 2005). There is no recommended daily intake for lutein and zeaxanthin, but many studies show a health benefit for lutein supplementation at 10 mg per day and zeaxanthin at 2 mg per day (American Optometric Association 2009).

The highest values of lutein and zeaxanthin reported in 100 g of yellow-fleshed potatoes are lower compared to the amount of lutein provided by 100 g of lettuce (540 μg; Kimura and Rodriguez-Amaya 2003), broccoli (3250 μg; Khachick et al.

1992), parsley (5800 µg; Hart and Scott 1995), or spinach (4180 µg; Tee and Lim 1991); and lower than the amount of zeaxanthin provided by 100 g of maize at its highest zeaxanthin concentration (3800 µg/100 g) (Brenna and Berardo 2004) and of red paprika (2200 µg/100 g) (Müller 1997; Mínguez Mosquera and Hornero-Méndez 1994). However, it is important to consider that potato consumption can be as high as 500 g per day whereas the mean intake of the above-mentioned vegetables is less than 50 g per day; hence, the overall contribution of potato-based carotenoids to the dietary intake can be higher. Furthermore, the contribution of a food source to lutein and zeaxanthin intake depends on their digestive stability, bioaccessibility, and bioavailability in the respective food matrix. Bioaccessibility refers to the proportion of ingested carotenoid that is released from the food matrix and incorporated into micelles in the gastrointestinal tract, and thus available for intestinal absorption (Rodriguez-Amaya 2015). Bioavailability refers to the portion of the carotenoid that is absorbed in the body, enters in systemic circulation and becomes available for utilization in normal physiological functions or for storage in the human body (van Het Hof et al. 2000). The bioavailability of carotenoids from plant foods is influenced by the species and structure of carotenoids present in the food, composition, and release of carotenoids from the food matrix, absorption in the intestinal tract, transportation within the lipoprotein fractions, biochemical conversions, and tissue-specific depositions, as well as by the nutritional status of the ingesting consumer (Bohn 2017). Research on bioavailability of potato carotenoids is required to have more useful and complete information regarding their nutritional and health benefits.

Burgos et al. (2013a) evaluated the in vitro digestive stability and the efficiency of micellarization or bioaccessibility of lutein and zeaxanthin in yellow-fleshed potatoes. The gastric and duodenal digestive stability of lutein and zeaxanthin in boiled tubers ranged from 70 to 95% while the bioaccessibility ranged from 33 to 71% for lutein and from 51 to 71% for zeaxanthin. A more recent study has reported that bioaccessibility of lutein and zeaxanthin in yellow-fleshed clones range from 76 to 82% for lutein and from 24 to 55% for zeaxanthin (Andre et al. 2015).

The maximum bioaccessible lutein concentration reported in yellow-fleshed potatoes is around 300 µg/100 g FW while the maximum bioaccessible zeaxanthin concentration is around 600 µg/100 g FW. Considering the mean potato intake in the Andes of Peru, Ecuador, and Bolivia (500 g per day), potato tubers from the variety with the highest bioaccessible lutein could provide 14% of the suggested level of lutein intake for having health benefit (10 mg per day). Likewise, potato tubers of the variety with the highest bioaccessible zeaxanthin concentration could provide 50% more than the suggested level of zeaxanthin intake (2 mg per day). In Spain, potato was shown to contribute 13–20% towards the total dietary intake of zeaxanthin and was ranked as the third main contributor after citrus fruits and green vegetables (Garcia-Closas et al. 2004). More population studies are needed, however, regarding the nutritional and health contributions of carotenoids as provided by potatoes.

Antioxidant Activity

Antioxidant activity (AA) describes the capacity of redox molecules in foods and biological systems to scavenge free radicals considering the additive and synergistic effects of all antioxidants rather than the effect of single compounds, and may, therefore, be useful to study the potential health benefits of antioxidants on oxidative stress-mediated diseases (Puchau et al. 2010). In that context, mixtures of phytochemicals found in plant foods are more effective in improving antioxidant status than isolated phytochemicals (DeGraft-Johnson et al. 2007). The antioxidant activity of foods as assessed by indices such as ferric reducing ability of plasma (FRAP) has been indicated to be a valid and reproducible determinant of human plasma AA measurements (Rautiainen et al. 2008). Antioxidant capacities of food staples such as potatoes could thus potentially affect the antioxidant status of that population.

The antioxidants in potato are mainly hydrophilic (polyphenols, ascorbic acid, anthocyanins, and flavanols) (Reyes et al. 2005). In white- or yellow-fleshed potatoes, prevalent contributors of AA are chlorogenic acid, gallic acid, caffeic acid, and catechin (Reddivari et al. 2007a), while in purple- and red-fleshed potatoes the major contributors to AA are anthocyanins and chlorogenic acid (Lachman et al. 2009). Potatoes also contain lipophilic antioxidants (carotenoids and vitamin E).

Because antioxidant activity of potato anthocyanins results from the synergistic effect of each anthocyanin pigment (Hayashi et al. 2003), it is important to assess different pigmented potato cultivars for individual anthocyanidin content, as well as the contribution of the anthocyanidin composition to their antioxidant activity. A high degree of hydroxylation and/or methoxylation of individual anthocyanidins could contribute in conjunction with other phenolics to high AA (Lachman et al. 2009).

Burgos et al. (2013b) reported that boiled potatoes of purple-fleshed potato varieties have an AA ranging from 4017 to 17,304 mg Trolox equivalents (TEq)/g, FW as determined by the 2,2-azino-bis-3-ethylbenzthiazoline-6-sulfonic acid (ABTS) antioxidant capacity measure and from 2369 to 9754 mg TEq/g, FW as determined by 1,1-diphenyl-2-picryl-hydrazyl (DPPH) antioxidant capacity assay. Compared to other sources of antioxidants, potato has lower AA than strawberry, blackberry, and blueberry as determined by the ABTS assay (around 25,030–50,000 mg TEq/g, FW, Garcia-Alonso et al. 2004). However, as indicated above the overall contribution of potato to the antioxidant intake of a population will finally depend in the amount of potatoes typically consumed.

Ombra et al. (2015) reported after simulated gastrointestinal digestion, the extracts from purple potato have high in vitro antioxidant, antimicrobial, and anti-proliferative activities against the colon cancer cells Caco-2 and SW48 and the breast cancer cells MCF-7 and MDA-MB-231.

After digestion of cooked tubers from purple-fleshed potatoes using a dynamic human gastrointestinal model, Kubow et al. (2017) found an increased FRAP anti-

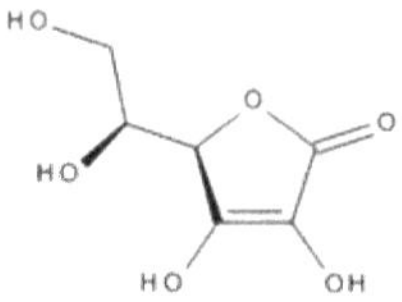

Ascorbic Acid

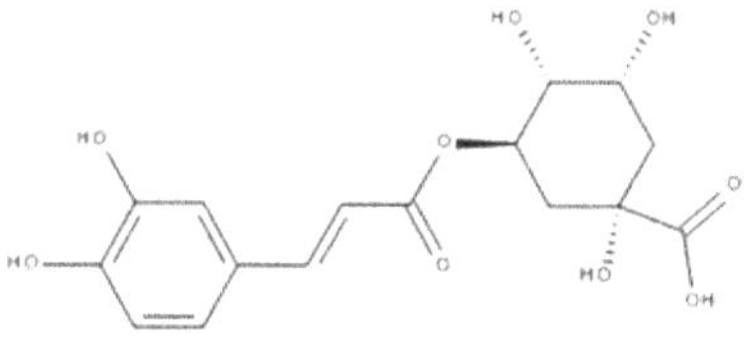

Chlorogenic Acid

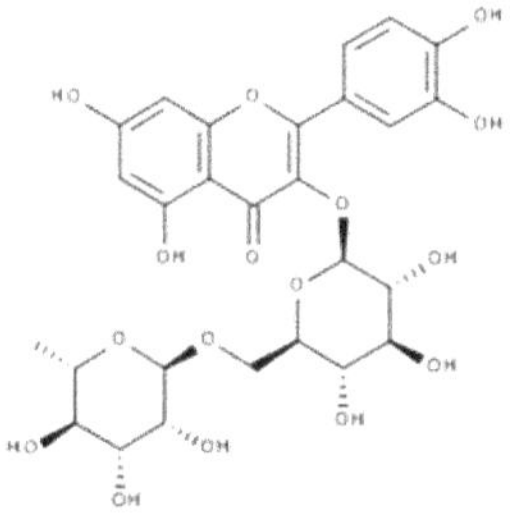

Rutin

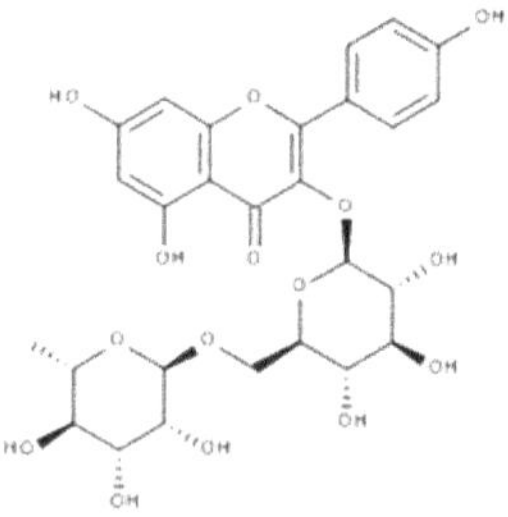

Kaempferol-3-rutinoside

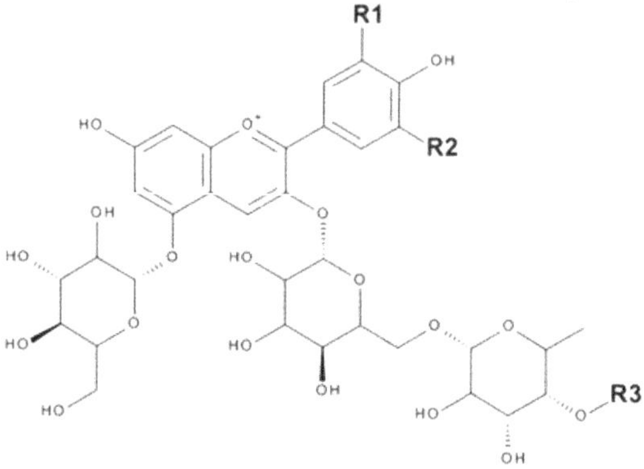

Anthocyanins	R1	R2	R3
pet-3-coum-rut-5-glc	OMe	OH	p-coumaric acid
peo-3-coum-rut-5-glc	OMe	H	p-coumaric acid
cyan-3-coum-rut-5-glc	OH	H	p-coumaric acid
mal-3-coum-rut-5-glc	OMe	OMe	p-coumaric acid
pel-3-coum-rut-5-glc	H	H	p-coumaric acid

Fig. Chemical structure of hydrophilic antioxidants in potato

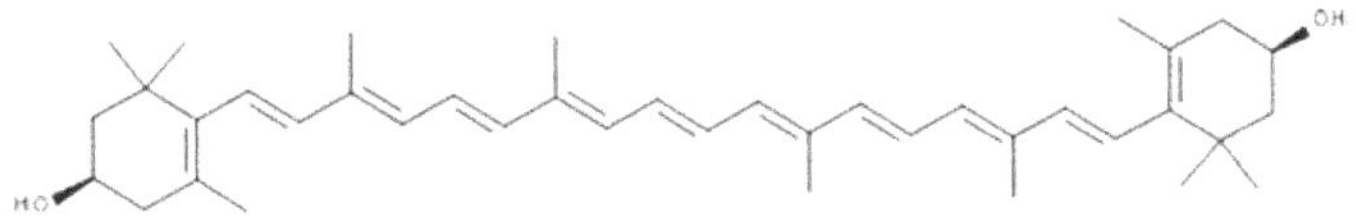

zeaxanthin

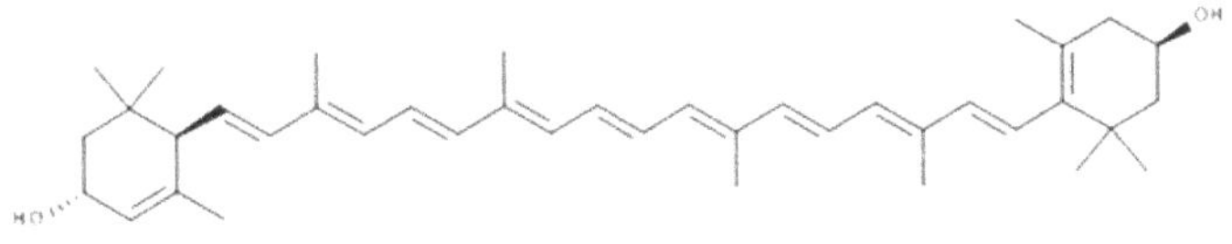

lutein

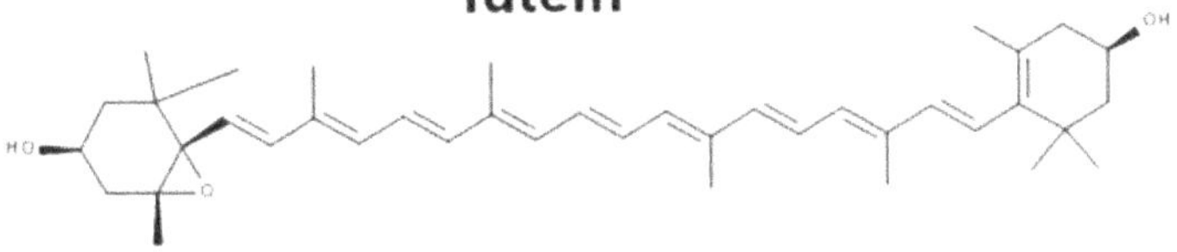

antheraxanthin

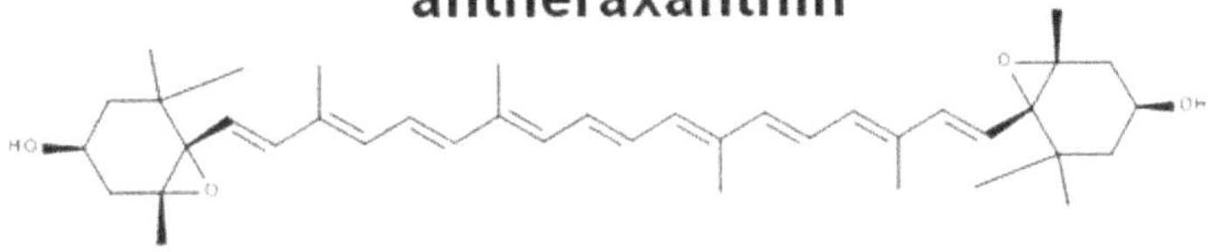

violaxanthin

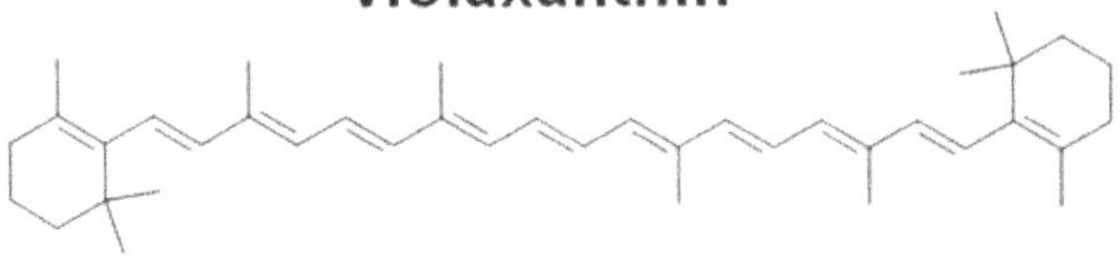

β-carotene

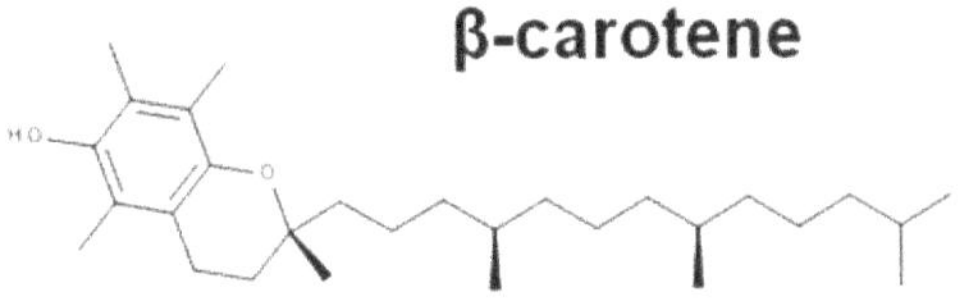

α-tocopherol

Fig. Chemical structure of lipophilic antioxidants in potato

oxidant activity in the colonic reactors. Metabolic microbial breakdown of anthocyanins over a 24 h period appeared to generate sufficient amounts of microbial metabolites to produce an improvement in antioxidant capacity. Anthocyanins and their metabolites can, via antioxidant activity, provide protection for intestinal cells against oxidative stress in the gut, and hence alleviate gut inflammation, protect against colorectal cancer, and generally enhance colorectal health.

Glycoalkaloids

Glycoalkaloids are secondary plant metabolites that serve as natural defenses against bacteria, fungi, viruses, and insects (Friedman 2004). They can be toxic for humans when present in high concentrations, and can impart a bitter taste to potatoes. However, glycoalkaloids and hydrolysis products without the carbohydrate side chain (aglycones) also have beneficial effects that include: lowering of cholesterol (Friedman et al. 2003) and inhibition of the growing of cancer cells in culture as well as tumor growth in vivo (Friedman 2015).

Although there are many glycoalkaloids, α-chaconine and α-solanine make up 95% of the total glycoalkaloids present (Friedman et al. 1997); α-solanine is found in greater concentrations than α-chaconine, and α-solanine has only half as much specific toxic activity as a α-chaconine (Lachman et al. 2001).

Experiments with human taste panels revealed potato varieties with glycoalkaloid levels exceeding 14 mg/100 g FW tasted bitter (Friedman 2006). Those in excess of 22 mg/100 g FW also induced mild to severe burning sensations in the mouths and throats of panel members.

Glycoalkaloid levels vary greatly in different potato varieties and may be influenced by factors such as light, mechanical injury, and storage. They are also influenced by stress such as heat and drought during production. This raises concern for maintaining the quality of potatoes under climate change (Andre et al. 2009), and suggests increased attention may be needed to glycoalkaloid concentrations of potato varieties bred for or grown in warm environments.

Glycoalkaloid concentration of raw potatoes ranges from 0.7 to 18.7 mg/100 g FW (Friedman et al. 2003). Peeling significantly reduced the glycoalkaloid levels in the tubers: solanine to 43.6% and chaconine to 31% (Lachman et al. 2013). Cooking also significantly reduced the levels of glycoalkaloids (Tajner-Czopek et al. 2008), with boiling reducing the levels of glycoalkaloids more than baking and microwaving (Lachman et al. 2013).

Glycoalkaloid content in potato tubers should not exceed 20 mg/100 g FW, because this level is dangerous for human health (Ruprich et al. 2009). The toxicity of glycoalkaloids at appropriate high levels may be due to adverse effects such as anticholinesterase activity on the central nervous system and to disruption of cell membranes adversely affecting the digestive system and general body metabolism (Friedman et al. 2003). The toxicity of glycoalkaloids is associated with the synergistic interaction between two main components of glycoalkaloids: α-solanine and α-chaconine.

However, glycoalkaloids also have anti-carcinogenic properties. Exposure of cancer cells to glycoalkaloids produced potatoes (α-chaconine and α-solanine) or their hydrolysis products (mono-, di-, and trisaccharide derivatives and the aglycones solasodine, solanidine, and tomatidine) inhibits the growth of the tumor cells in culture as well as in vivo tumor growth (Friedman 2015). On the basis of the anticarcinogenic properties of these potato components, it is conceivable that the levels typically noted in commercial potatoes might help to protect against multiple cancers. Epidemiological studies, however, are needed to substantiate this possibility.

Contribution to Health

Population-based epidemiological studies have emphasized the importance of nutrition to combat metabolic disorders emerging worldwide that have been associated with diet, such as diabetes, cancer, and cardiovascular diseases. In that regard, higher intakes of fruits and vegetables have been consistently indicated to exert protective effects against such chronic diseases (Dragsted et al. 2006). Potato has been underappreciated relative to other vegetables as it has been subject to controversy such as being labeled as a contributor to development of obesity and diabetes (Burlingame et al. 2009). On the other hand, potatoes contain relatively high concentrations of key phytonutrients that have shown bioactivities that could counteract chronic disease development (Ezekiel et al. 2013). Potatoes have shown promising health-promoting effects in human cell culture, experimental animals, and human clinical studies, including anti-cancer, hypocholesterolemic, anti-inflammatory, anti-obesity, and anti-diabetic properties. Nutritional compounds of potatoes such as phenolics, anthocyanins, fiber, starch as well as compounds considered anti-nutritional such as glycoalkaloids, lectins, and proteinase inhibitors are believed to contribute to the health benefits of potatoes. As there are many biological activities attributed to the compounds present in potato, some of which could be beneficial or detrimental depending on specific circumstances, long-term studies investigating the association between potato consumption and diabetes, obesity, cardiovascular disease, and cancer while controlling for fat intake are needed (Visvanathan et al. 2016).

Anticancer Effect

Several studies have shown a reduction in proliferation of cancer cells when treated with potato extracts. Potato antioxidants such as phenolic acids and anthocyanins, glycoalkaloids, fiber, and proteinase inhibitors identified in potatoes have been implicated in the suppression of cancer cell proliferation in vitro and in vivo.

Role of Potato Antioxidants

Phenolic acids and anthocyanins are potato antioxidants that have reported anti-carcinogenic activity. Hayashi et al. (2006) reported that anthocyanins in steamed purple and red potatoes suppressed the growth of benzopyrene-induced stomach cancer in mice. Reddivari et al. (2007b) found that the anthocyanin fractions from potato extracts were cytotoxic to prostate cancer cells through activation of caspase-dependent and caspase-independent pathways. Madiwale et al. (2011) reported that purple flesh potatoes rich in anthocyanins suppressed proliferation and elevated

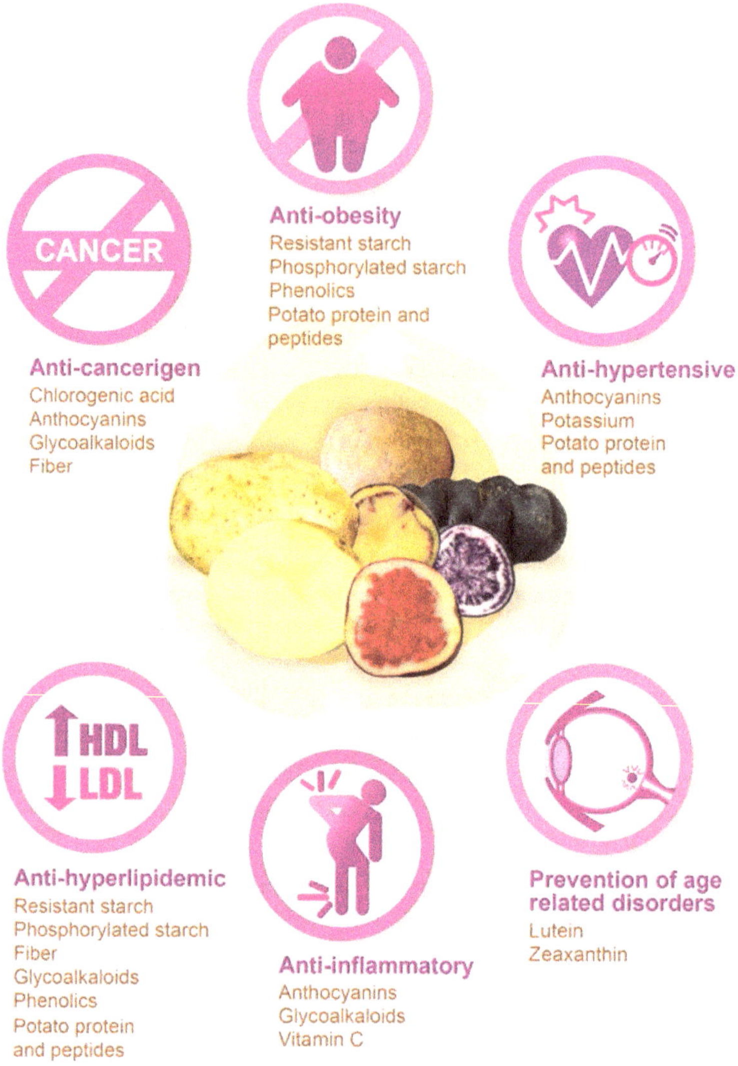

Fig. Health benefits of potatoes

apoptosis of colon cancer cells compared with white and yellow flesh potatoes. In a more recent study, Charepalli et al. (2015) found that extracts of purple-fleshed potatoes suppress colon tumorigenesis via elimination of colon cancer stem cells. Chlorogenic acid, the main phenolic acid of potato, is effective against human liver, colon, and prostate cancer cells (Wang et al. 2011) and inhibits significantly the proliferation of colon cancer and prostate cancer cells.

Role of Potato Glycoalkaloids

α-Solanine and α-chaconine, the main steroidal glycoalkaloids in potatoes, are well studied for their antitumor properties (Friedman 2015). Lee et al. (2004) found that α-solanine exhibited growth inhibition and apoptosis induction in multiple cancer cells such human colon (HT29) and liver (HepG2) cancer cells. Friedman et al. (2005) evaluated the anti-carcinogenic effect of α-solanine and α-chaconine extracted from five fresh potato varieties (Dejima, Jowon, Sumi, Toya, and Vora Valley) and found that glycoalkaloids exerted anti-proliferative effects of the following human tumor cell lines: cervical (HeLa), liver (HepG2), lymphoma (U937), stomach (AGS and KATO III) cells, and on normal liver (Chang) cells. Friedman et al. (2005) also reported that the anti-proliferative effects of the glycoalkaloids were concentration dependent and that α-chaconine was more bioactive than α-solanine. Yang et al. (2006) found that α-chaconine induced the apoptosis of HT-29 human colon cancer cells through caspase-3 activation and inhibition of extracellular signal-regulated kinase phosphorylation.

Reddivari et al. (2010) showed that α-chaconine exhibited potent anti-proliferative properties and increased cyclin-dependent kinase inhibitor p27 levels in two prostate cancer cell lines, LNCaP and PC3. More recently, it has been reported that α-solanine, has a positive effect on the inhibition of pancreatic cancer cell growth in vitro and in vivo. Sun et al. (2014) demonstrated that α-solanine inhibited cancer cell growth through caspase 3-dependent mitochondrial apoptosis and that the expression of tumor metastasis-related proteins, MMP-2 and MMP-9, was also decreased in the cells treated with α-solanine. Lv et al. (2014) reported that α-solanine inhibited proliferation of PANC-1, sw1990, MIA PaCa-2 cells in a dose-dependent manner, as well as cell migration and invasion with a toxic dose and that the administration of α-solanine during 2 weeks in a xenograft model reduces the tumor volume and weight by 43–61%. These studies showed beneficial effects on pancreatic cancer in vitro and in vivo, which may be mediated via suppressing pathways involving proliferation, angiogenesis, and metastasis.

Role of Potato Fiber

Langner et al. (2009) reported that commercially available potato fiber extract (Potex) exhibited anti-proliferative effects in several tumor cell cultures. The fiber extract decreased cancer cell motility, induced apoptosis, and also caused morphological changes in tumor cells.

Anti-diabetic and Anti-obesity Effects

Potato consumption has often been associated in cohort studies with elevated risk of type 2 diabetes (Muraki et al. 2016) and obesity (Borch et al. 2016), which has been attributed to a relatively high glycemic index in some potato varieties and processed potato products containing added saturated and trans fats. A major confounding factor in such studies is typical Western dietary patterns associated with increased disease risk typically include potato consumption along with high intake of red and processed meat, refined grains, high-fat dairy products, fried foods and sugar (Pastorino et al. 2016). More research is needed to adjust association of food items such as potatoes in such dietary patterns (Hu 2002). Moreover, RDS present in cooked potatoes (especially amylose) tends to retrograde upon cooling generating appreciable amounts of slowly digestible starch (SDS) or resistant starch (RS) that contribute to dietary fiber content (Sajilata et al. 2006) and potentially positively impact health by slowing postprandial glucose release from cooked potatoes (King and Slavin 2013).

GI values below 56 are considered as low glycemic index while values above 74 are considered to indicate a high glycemic index. GI values in potato ranged from 56 to 94 for eight British cultivars (Henry et al. 2005) and from 53 to 103 for seven Australian cultivars (Wang et al. 2014). When boiled red potatoes were served hot to volunteers, a GI of 89.4 was found (Fernandes et al. 2005). When cooking is followed by cooling, amylose retrogrades to produce resistant starch. The GI response was only 56.2 when cooking was followed by refrigeration of 12–24 h.

Potato chips and French fries have been implicated by some nutrition researchers as major contributors to obesity risk as these products contain a high fat and caloric content. Potato servings, however, are not likely in themselves to promote obesity as potatoes are considered to have a low energy density as they are a low-fat food with a high-water content (Anderson et al. 2013). Potato-based foods with high calorie fat additions have been considered as a major culprit towards obesity risk (Camire et al. 2009).

Conversely, potatoes may have a role in controlling appetite and therefore weight gain, by contributing to satiety. Satiety is the feeling of fullness and the loss of hunger that occur after eating. Many factors influence satiety, including the rate of gastric emptying and the proportion of macronutrients in the food. Foods that increase satiety are thought to promote weight control by delaying subsequent meals and total calories consumed (Camire et al. 2009). Compared to rice and pasta, adult feeding studies have shown that satiating amounts of potatoes co-ingested with meat resulted in lower energy intake and postprandial insulin concentrations; and higher levels of ghrelin, which is a gastric orexigenic appetite-stimulating hormone that contributes to feeding regulation (Erdmann et al. 2007). Likewise, studies involving children showed that meat co-ingestion with boiled mashed potato resulted in an approximate 40% lower energy intake as compared to meat consumed together with either pasta or rice (Akilen et al. 2016). The stronger satiety of boiled mashed potato for the calories consumed was related to similar suppression of ghrelin postprandially relative to the other carbohydrate-rich foods despite the lower potato meal intake. Short-term intervention studies have generally indicated that high GI meals decrease satiety, and an increase in the return of hunger and energy intake at a later meal as opposed to low glycemic index meals containing potatoes (Roberts 2000).

Anti-hyperlipidemic, Anti-hypertensive and Anti-inflammatory Effects

A variety of animal feeding studies have shown cholesterol-lowering properties from potato intake that have been related to its content of protein, resistant and phosphorylated starch, fiber, glycoalkaloids (Friedman 2006), and phenolic compounds (Friedman 1997). Robert et al. (2006) found that consumption of cooked potatoes (consumed with skin) improved lipid metabolism in cholesterol-fed rats. Rats fed a potato-enriched diet for 3 weeks had lower concentrations of plasma cholesterol and triglycerides and reduced liver cholesterol content. Hashimoto et al. (2006) showed that retrograded starch from two varieties of potato pulp lowered serum total cholesterol and triglyceride concentrations. The authors indicated that the retrograded starch promoted the excretion of bile acids resulting in a low concentration of serum cholesterol; and that retrograded starch inhibited the synthesis of fatty acids at the mRNA levels of fatty acid synthase (FAS) and SREBP-1c, which might be related to the observed reduction of the serum triglyceride concentrations. Kanazawa et al. (2008) reported that gelatinized potato starch containing a high level of phosphate reduced concentrations of serum-free fatty acids and triglycerides and liver triglycerides.

Liyanage et al. (2008) have demonstrated the hypocholesterolemic effect of potato peptides. Rats fed a cholesterol-free diet containing 20% (w/w) potato peptides showed greater concentrations of serum high-density lipoprotein (HDL) cholesterol and increased fecal steroid output and lesser non-HDL cholesterol concentrations than rat fed diets containing 20% casein peptides. The results were attributed to inhibition of cholesterol absorption, possibly via suppression of micellar solubility of cholesterol. In a follow-up study, Liyanage et al. (2009) found that potato peptides reduced the serum non-HDL cholesterol concentrations by stimulating fecal steroid excretion, accelerated by cecal short-chain fatty acids in a hypercholesterolemic rat model. There is a lack of data, however, from randomized controlled trials to demonstrate a relationship between potato consumption and blood lipid parameters in humans.

Vinson et al. (2012) showed a significant lowering of systolic blood pressure in humans after supplementation to hypertensive subjects in a 4-week cross-over trial involving consumption of six to eight small purple potatoes twice daily versus no potato intake. The blood pressure lowering effect was related to high intake of polyphenols associated with the pigmented potatoes. This latter intervention trial is contrasted by an analysis from three large prospective cohort studies indicating increased hypertension risk in association with potato intake of four or more servings per month as opposed to one serving per month (Borgi et al. 2016). A major limitation of such trials is that co-ingestion of salt, high-salt foods, saturated or trans fats with potatoes could have contributed to the hypertension risk as opposed to potato per se, particularly since potatoes are typically eaten in a meal context (Miller and Stanner 2016). In support of this contention, a 3-year longitudinal study of Japanese people showed that adherence to a traditional Japanese dietary pattern exerted favorable effects on blood pressure that was partly associated with potato intake (Niu et al. 2016).

Relatively high intake of potassium is needed to counteract the blood pressure raising effects of a high sodium diet and so protect against hypertension (Camire et al. 2009). An increase in consumption of potassium-rich foods has been promoted to combat hypertension and cardiovascular disease (WHO 2012). In that regard, intake of potassium-rich foods has been indicated to protect against stroke risk (Adebamowo et al. 2015). As potatoes are rich in potassium and are naturally very low in sodium content, this food could counter development of hypertension-associated diseases. Additionally, Makinen et al. (2008) reported that a protein isolated from vascular bundle and inner tuber tissues of potato enhanced the inhibition of the angiotensin converting enzyme I, a biochemical factor affecting blood pressure that contributes to hypertension.

Kaspar et al. (2011) found anti-inflammatory effects in healthy men consuming white and pigmented potatoes with greater effects from pigmented potatoes. Potato phenolics and glycoalkaloids have shown evidence of anti-inflammatory activities (Kenny et al. 2013). Indigestible carbohydrates including resistant starch and fiber have demonstrated the ability to modulate inflammatory markers in both animal models (Vaziri et al. 2014) and human clinical trials (Jiao et al. 2015). Hence, the contribution of resistant starch or fiber from select potato products may have also direct impact on inflammatory stress in humans.

Potato and Its Relationship with Cardiovascular Diseases

As a key dietary source of potassium, vitamin C, and dietary fiber, potatoes contribute significantly to nutrients with defined roles in promoting cardiovascular health (McGill et al. 2013). Boiled potatoes have been shown to have favorable impact on several measures of cardiometabolic health in animals and humans, including lowering blood pressure, improving lipid profiles, and decreasing markers of inflammation (McGill et al. 2013). When eaten as a regular food item and consumed with skin, potato intake can significantly enhance cardioprotective fiber intake that is generally lacking in Western-type diets (Lockyer et al. 2016). Large prospective studies in Sweden involving a 13-year follow-up showed no adverse relationship of higher potato intake with cardiovascular risk for either morbidity or mortality (Larsson and Wolk 2016). Similarly, a systematic review of five observational studies carried out by Borch et al. (2016) showed no convincing evidence to support an adverse association between unprocessed potato intake and the risk of developing metabolic disorders including obesity, type 2 diabetes, and cardiovascular disease. On the other hand, processed potato products like French fries and potato crisps, with high lipid and trans fats content and added sodium can have adverse effect of the heart health and so should be minimized in the diet. In that regard, despite the lack of a relationship between chronic disease risk and potato intake in the above comprehensive review by Borch et al. (2016), they showed that French fries and fried potato were associated with an increased risk for obesity and type 2 diabetes. Likewise, a longitudinal study involving 4440 subjects with an 8-year follow-up showed no association between higher potato intake and mortality risk, whereas participants who consumed fried potatoes two to three times/week had an increased mortality risk (Veronese et al. 2017). Camire et al. (2009) have recommended preparing potatoes with minimum lipid addition and consume potatoes with peels to conserve their cardiovascular health promoting properties.

Concluding Remarks

Potato is an important source of carbohydrates, resistant starch, quality protein, vitamins C and B6 as well as potassium. Potatoes are also a source of antioxidants. Chlorogenic acid and glycoalkaloids are present in all potatoes independently of the flesh color while deep yellow-fleshed potatoes contain high amounts of lutein and zeaxanthin; and purple-fleshed potatoes contain high amounts of anthocyanins. Potatoes glycoalkaloids in high concentrations can be toxic to humans but in low concentrations can have beneficial effects such as inhibition of the growth of cancer cells.

The contribution of potato to the diet is affected by cooking, potato intake, and the bioavailability of potato nutrients. Potato vitamins are significantly reduced after cooking. However, 100 g of cooked potatoes provide around 30% of the requirement of vitamin C and 20% od the requirement of vitamin B6. Potato carotenoids and anthocyanins show high recoveries after cooking due to an improved release of these antioxidants. In vitro studies demonstrate that potato lutein and zeaxanthin have a high bioaccessibility and that potato phenolics undergo microbial transformation in the intestinal tract producing metabolites that may also promote a healthy gut microbiome. Further research in humans is needed to confirm the beneficial effect of potato phenolics in the gut.

In areas where potato is consumed in large quantities like in the highlands of Latin American countries, the potato contribution to the energy, protein, iron, and zinc intake is significant. In those areas, iron and zinc biofortified potatoes are expected to contribute to reduce malnutrition and anemia. Nevertheless, to assess the full potential of the biofortified potatoes, human studies are required to gain insight on how much of the iron from biofortified potatoes are absorbed by the human body.

Regarding its contribution to human health, potatoes have shown promising health-promoting effects in human cell culture, experimental animals, and human clinical studies. Potato compounds such as phenolic acids and anthocyanins, glycoalkaloids, fiber, and proteinase inhibitors have been implicated in the suppression of cancer cell proliferation in vitro and in vivo and are believed to contribute to the hypocholesterolemic, anti-inflammatory, anti-obesity and anti-diabetic properties of potato.

References

Adebamowo SN, Spiegelman D, Willett WC (2015) Association between intakes of magnesium, potassium, and calcium and risk of stroke: 2 cohorts of US women and updated metaanalyses. Am J Clin Nutr 101:1269–1277

Akilen R, Deljoomanesh N, Hunschede S, Smith CE, Arshad MU, Kubant R, Anderson GH (2016) The effects of potatoes and other carbohydrate side dishes consumed with meat on food intake, glycemia and satiety response in children. Nutr Diabetes 6:e195. https://doi.org/10.1038/nutd.2016.1

American Optometric Association (2009) Diet, nutrition and eye health. http://www.aoa.org/documents/nutrition/Diet_Nutrition_Eye_Health_booklet.pdf. Accessed 28 Sept 2017

Amrein T, Bachmann S, Noti A, Biedermann M, Barbosa M, Biedermann-Brem S, Grob K, Keiser A, Realini P, Escher F, Amadó R (2003) Potential of acrylamide formation, sugars, and free asparagine in potatoes: a comparison of cultivars and farming systems. Agric Food Chem 51(18):5556–5560

Anderson GH, Soeandy CD, Smith CE (2013) White vegetables: glycemia and satiety. Adv Nutr 4(Suppl):356S–367S

Andre CM, Oufir M, Guignard C, Hoffmann L, Hausman JF, Evers D, Larondelle Y (2007) Antioxidant profiling of native Andean potato tubers (*Solanum tuberosum* L.) reveals cultivars with high levels of β-carotene, alpha-tocopherol, chlorogenic acid, and petanin. Agric Food Chem 55:10839–10849

Andre CM, Schafleitner R, Legay S, Lefèvre I, Aliaga CA, Nomberto G, Hoffmann L, Hausman JF, Larondelle Y, Evers D (2009) Gene expression changes related to the production of phenolic compounds in potato tubers grown under drought stress. Phytochemistry 70:1107–1116

Andre CM, Larondelle Y, Evers D (2010) Dietary antioxidants and oxidative stress from a human and plant perspective: a review. Curr Nutr Food Sci 6(1):2–12

Andre CM, Evers D, Ziebel J, Guignard C, Hausman JF, Bonierbale M, zum Felde T, Burgos G (2015) In vitro bioaccessibility and bioavailability of iron from potatoes with varying vitamin C, carotenoid, and phenolic concentrations. J Agric Food Chem 63(41):9012–9021. https://doi.org/10.1021/acs.jafc.5b02904

Auestad N, Hurley J, Fulgoni V, Schweitzer CM (2015) Contribution of food groups to energy and nutrient intakes in five developed countries. Nutrients 7:4593–4618

Bach S, Yada RY, Bizimungu B, Fan M, Sullivan JA (2013) Genotype by environment interaction effects on starch content and digestibility in potato (*Solanum tuberosum* L.). J Agric Food Chem 61(16):3941–3948. https://doi.org/10.1021/jf3030216. Epub 2013 Apr 12

Bamberg JB, del Rio A (2005) Conservation of potato genetic resources. In: Razdan MK, Mattoo AK (eds) Genetic improvement of solanaceous crops. Volume I: Potato. Science Publishers, Inc., Plymouth, p 476

Bates C (1997) Vitamin Analysis. Ann Clin Biochem 34(6):599–626. https://doi.org/10.1177/000456329703400604

Bethke P, Jansky S (2018) The Effects of Boiling and Leaching on the Content of Potassium and Other Minerals in Potatoes. J Food Sci 73(5):H80–H85

Bohn T (2017) Bioactivity of carotenoids–chasms of knowledge. Int J Vitam Nutr Res 10:1–5. https://doi.org/10.1024/0300-9831/a000400

Bonierbale M, Burgos G, zum Felde T, Sosa P (2010) Composition nutritionnelle des pommes de terre. Cahiers de nutrition et dietétique 45:S28–S36

Borch D, Juul-Hindsgaul N, Veller M, Astrup A, Jaskolowski J, Raben A (2016) Potatoes and risk of obesity, type 2 diabetes, and cardiovascular disease in apparently healthy adults: a systematic review of clinical intervention and observational studies. Am J Clin Nutr 104:489–498

Borgi L, Rimm EB, Willett WC, Forman JP (2016) Potato intake and incidence of hypertension: results from three prospective US cohort studies. BMJ 353:i2351. https://doi.org/10.1136/bmj.i2351

Bramley PM, Elmadfa I, Kafatos A, Kelly FJ, Manios Y, Roxborough HE, Schuch W, Sheehy PJ, Wagner JH (2000) Vitamin E. J Sci Food Agric 80:913–938

Brenna O, Berardo N (2004) Applications of near-infrared reflectance spectroscopy (NIRS) to the evaluation of carotenoids in maize. J Agri Food Chem 52:5577–5582

Brennan CS (2005) Dietary fibre, glycaemic response, and diabetes. Mol Nutr Food Res 49:560–570

Brown CR, Wrolstad R, Durst R, Yang P, Clevidence B (2003) Breeding studies in potatoes containing high concentrations of anthocyanins. Am J Pot Res 80:241. https://doi.org/10.1007/BF02855360

Brown CR, Culley D, Yang CP, Durst R, Wrolstad R (2005) Variation of anthocyanin and carotenoid contents and associated antioxidant values in potato breeding lines. J Am Soc Hortic Sc 130:174–180

Burgos G (2014) Concentración y bioaccesibilidad de carotenoides y compuestos fenólicos en papas cocidas. Tesis doctoral. Universidad de la Laguna. Departamento de Química Analítica, Nutrición y Bromatología

Burgos G, Amoros W, Morote M, Stangoulis J, Bonierbale M (2007) Iron and zinc concentration of native Andean potato varieties from a human nutrition perspective. J Sci Food Agric 87(4):668–675

Burgos G, Salas E, Amoros W, Auqui M, Munoa L, Kimura M, Bonierbale M (2009) Total and individual carotenoid profiles in the Phureja group of cultivated potatoes: I. concentrations and relationships as determined by spectrophotometry and high performance liquid chromatography (HPLC). J Food Compos Anal 22:503–508

Burgos G, Amoros W, Salas E, Muñoa L, Sosa P, Díaz C, Bonierbale M (2012) Carotenoid concentrations of native Andean potatoes as affected by cooking. Food Chem 133:1131–1137

Burgos G, Muñoa L, Sosa P, Bonierbale M, Zum Felde T, Díaz C (2013a) In vitro bioaccessibility of lutein and zeaxanthin of yellow fleshed boiled potatoes. Plant Foods Hum Nutr 68(4):385–390

Burgos G, Amoros W, Muñoa L, Sosa P, Cayhualla E, Sanchez C, Diaz C, Bonierbale M (2013b) Total phenolic, total anthocyanin and phenolic acid concentrations and antioxidant activity of purple-fleshed potatoes as affected by boiling. J Food Compos Anal 30:6–12

Burlingame B, Mouille B, Charrondiere R (2009) Nutrients, bioactive non-nutrients and antinutrients in potatoes. J Food Compos Anal 22:494–502

Camire ME, Kubow S, Donnelly DJ (2009) Potatoes and human health. Crit Rev Food Sci Nutr 49:823–840

Cevallos-Casals B, Cisneros Zevallos L (2003) Stoichiometric and kinetic studies of phenolic antioxidants from Andean purple corn and red-fleshed sweet potato. J Agric Food Chem 51:3313–3319

Champ MMJ (2004) Physiological aspects of resistant starch and in vivo measurements. J AOAC Int 87:749–755

Charepalli V, Reddivari L, Radhakrishnan S, Vadde R, Agarwal R, Vanamala JK (2015) Anthocyanin-containing purple-fleshed potatoes suppress colon tumorigenesis via elimination of colon cancer stem cells. J Nutr Biochem 26(12):1641–1649. https://doi.org/10.1016/j.jnutbio.2015.08.005

Chitchumroonchokchai C, Diretto G, Parisi B, Giuliano G, Failla ML (2017) Potential of golden potatoes to improve vitamin A and vitamin E status in developing countries. PLoS One 12(11):e0187102

Chun J, Lee J, Ye L, Exler J, Eitenmiller RR (2006) Tocopherol and tocotrienol contents of raw and processed fruits and vegetables in the United States diet. J Food Compos Anal 19:196–204

Clifford M (1999) Chlorogenic acids and other cinnamates – nature, occurrence and dietary burden. J Sci Food Agric 79:362–372

Crapo PA, Reaven G, Olefsky J (1977) Postprandial plasma-glucose and -insulin responses to different complex carbohydrates. Diabetes 26:1178–1183

DeGraft-Johnson J, Kolodziejczyk K, Krol M, Nowak P, Krol B, Nowak D (2007) Ferric-reducing ability power of selected plant polyphenols and their metabolites: implications for clinical studies on the antioxidant effects of fruits and vegetable consumption. Basic Clin Pharmacol Toxicol 100:345–352

De Haan S, Burgos G, Liria R, Rodriguez F, Creed-Kanashiro H, Bonierbale M (2019) The Nutritional Contribution of Potato Varietal Diversity in Andean Food Systems: a Case Study. Am J Potato Res 96:151. https://doi.org/10.1007/s12230-018-09707-2

Dragsted LO, Krath B, Ravn-Haren G, Vogel UB, Vinggaard AM, Jensen PB, Loft S, Rasmussen SE, Sandstrom B, Pedersen A (2006) Biological effects of fruit and vegetables. Proc Nutr Soc 65:61–67

Duarte-Delgado D, Ñústez-López C, Narváez-Cuenca C, Restrepo-Sánchez L, Melo S, Sarmiento F, Kushalappa A, Mosquera-Vásquez T (2016) Natural variation of sucrose, glucose and fructose contents in Colombian genotypes of *Solanum tuberosum* Group Phureja at harvest. J Sci Food Agric 96(12):4288–4294. https://doi.org/10.1002/jsfa.7783. Epub 2016 Jun 21

Duthie G, Duthie S, Kyle J (2000) Plant polyphenols in cancer and heart disease: implications as nutritional antioxidants. Nutr Res Rev 13:79–106

Ek KL, Brand-Miller J, Copeland L (2012) Glycaemic effect of potatoes. Food Chem 133:1230–1240

Englyst HN, Kingsman SM, Cummings JH (1992) Classification and measurement of nutritionally important starch fractions. Eur J Clin Nutr 46:S33–S50

Erdmann J, Hebeisen Y, Lippl F, Wagenpfeil S, Schusdziarra V (2007) Food intake and plasma ghrelin response during potato-, rice- and pasta-rich test meals. Eur J Clin Nutr 46:196–203

Ezekiel R, Singh N, Sharma S, Kaur A (2013) Beneficial phytochemicals in potato—a review. Food Res Int 50:487–496

FAO (2013) FAO statistical databases FAOSTAT. http://www.fao.org/faostat/en/#data/FBS. Accessed October 12th 2017

FAO/OMS/UNU (2004) Human energy requirements. Report of joint FAO/WHO/UNU expert consultation. FAO, Rome

FAO/WHO (2001) Human vitamin and mineral requirements. Report of a joint FAO/WHO expert consultation. Bangkok, Thailand

Fernandes G, Velangi A, Wolever T (2005) Glycemic index of potatoes commonly consumed in North America. J Am Diet Assoc 105:557–562

Fernández-Garcia E, Carvajal-Lérida I, Pérez-Gálvez A (2009) In vitro bioaccessibility assessment as a prediction tool of nutritional efficiency. Nutr Res 29:751–760

Fernandez-Orozco R, Gallardo-Guerrero L, Hornero-Méndez D (2013) Carotenoid profiling in tubers of different potato (*Solanum* sp.) cultivars: accumulation of carotenoids mediated by xanthophyll esterification. Food Chem 141:2864–2872

Fiedor J, Burda K (2014) Potential role of carotenoids as antioxidants in human health and disease. Nutrients 6(2):466–488

Fitzpatrick TB, Basset GJ, Borel P, Carrari F, Della Penna D, Fraser PD, Hellmann H, Osorio S, Rothan C, Valpuesta V (2012) Vitamin deficiencies in humans: can plant science help? Plant Cell 24:395–414

La Frano M, de Moura F, Boy E, Lönnerdal B, Burri B (2014) Bioavailability of iron, zinc, and provitamin A carotenoids in biofortified staple crops. Nutr Rev 72(5):289–307. https://doi.org/10.1111/nure.12108.

Fossen T, Andersen Ø (2000) Anthocyanins from tubers and shoots of the purple potato, Solanum tuberosum. J Hortic Sci Biotechnol 75(3):360–363. https://doi.org/10.1080/14620316.2000.11511251

Friedman M (1997) Chemistry, biochemistry, and dietary role of potato polyphenols. A review. J Agric Food Chem 45:1523–1540

Friedman M (2004) Analysis of biologically active compounds in potatoes (Solanum tuberosum), tomatoes (Lycopersicon esculentum), and jimson weed (Datura stramonium) seeds. J Chromatogr A 1054(1–2):143–155

Friedman M (2006) Potato glycoalkaloids and metabolites: roles in the plant and in the diet. J Agric Food Chem 54:8655–8681

Friedman M (2015) Chemistry and anticarcinogenic mechanisms of glycoalkaloids produced by eggplants, potatoes, and tomatoes. J Agric Food Chem 63:3323–3337
Friedman M, McDonald G, Filadelfi-Keszi M. (1997) Potato glycoalkaloids: chemistry, analysis, safety, and plant physiology. Critical Reviews in Plant Sciences 16(1):55–132. https://doi.org/10.1080/07352689709701946
Friedman M, Roitman JN, Kozukue N (2003) Glycoalkaloid and calystegine contents of eight potato cultivars. J Agric Food Chem 51(10):2964–2973
Friedman M, Lee KR, Kim HJ, Lee IS, Kozukue N (2005) Anticarcinogenic effects of glycoalkaloids from potatoes against human cervical, liver, lymphoma, and stomach cancer cells. J Agric Food Chem 53:6162–6169
Fudge J, Mangel N, Gruissem W, Vanderschuren H, Fitzpatrick T (2017) Rationalising vitamin B_6 biofortification in crop plants. Curr Opin Biotechnol 44:130–137. https://doi.org/10.1016/j.copbio.2016.12.004
Galliard T (1973) Lipids of potato tubers. I. Lipid and fatty acid composition of tubers from different varieties of potato. J Sci Food Agric 24(5):617–622
Gammone MA, Riccioni G, D'Orazio N (2015) Carotenoids: potential allies of cardiovascular health? Food Nutr Res 59:26762. https://doi.org/10.3402/fnr.v59.26762
Garcia-Alonso M, Pascual-Teresa S, Santos-Buelga C, Rivas-Gonzalo J (2004) Evaluation of the antioxidant properties of fruits. Food Chem 84:13–18
Garcia-Closas R, Berenguer A, Tormo MJ, Sanchez MJ, Quiros JR, Navarro C, Arnaud R, Dorronsoro M, Chirlaque MD, Barricarte A, Ardanaz E, Amian P, Martinez C, Agudo A, Gonzalez CA (2004) Dietary sources of vitamin C, vitamin E and specific carotenoids in Spain. Br J Nutr 91:1005–1011
Gibson S, Kurilich A (2013) The nutritional value of potatoes and potato products in the UK diet. Nutr Bull 38:389–399
Giusti MM, Polit MF, Ayvaz H, Tay D, Manrique I (2014) Characterization and quantitation of anthocyanins and other phenolics in native Andean potatoes. J Agric Food Chem 62:4408–4416
Habtemariam S, Lentini G (2015) The therapeutic potential of rutin for diabetes: an update. Mini Rev Med Chem 15(7):524–528
Han J, Kosukue N, Young K, Lee K, Friedman M (2004) Distribution of ascorbic acid in potato tubers and in home-processed and commercial potato foods. J Agric Food Chem 52(21):6516–6521
Hart D, Scott KJ (1995) Development and evaluation of an HPLC method for the analysis of carotenoids in foods and measurement of the carotenoid content of vegetables and fruits commonly consumed in the UK. Food Chem 54:101–111
Hashimoto N, Ito Y, Han KH, Shimada K, Sekkikawa M, Topping DL (2006) Potato pulps lowered the serum cholesterol and triglyceride levels in rats. J Nutr Sci Vitaminol 52:445–450
Hayashi K, Mori M, Matsunani Y, Suzutan T, Ogasawara M, Yoshida I, Hosokawa TA, Azuma M (2003) Anti Influenza Virus Activity of a Red-Fleshed Potato Anthocyanin. Food Sci Technol Res 9(3):242–244
Hayashi K, Hibasami H, Murakami T, Terahara N, Mori M, Tsukui A (2006) Induction of apoptosis in cultured human stomach cancer cells by potato anthocyanins and its inhibitory effects on growth of stomach cancer in mice. Food Sci Technol Res 12:22–26
Hellmann H, Mooney S (2010) Vitamin B6: A Molecule for Human Health? Molecules 15:442–459
Henry C, Lightowler H, Strik C, Storey M (2005) Glycaemic index values for commercially available potatoes in Great Britain. Br J Nutr 94(6):917–921
Holst B, Williamson G (2008) Nutrients and phytochemicals: from bioavailability to bioefficacy beyond antioxidants. Curr Opin Biotechnol 19:73–82
Hosseinian FS, Bea T (2007) Saskatoon and wild blueberries have higher antho-cyanin content than other Manitoba berries. J Agric Food Chem 55(26):10832–10838
Hu FB (2002) Dietary pattern analysis: a new direction in nutritional epidemiology. Curr Opin Lipidol 13(1):3–9
Ieri F, Innocenti M, Andrenelli L, Vecchio V, Mulinacci N (2011) Rapid HPLC/DAD/MS method to determine phenolic acids, glycoalkaloids and anthocyanins in pigmented potatoes (*Solanum tuberosum* L.) and correlations with variety and geographical origin. Food Chem 125:750–759

Jiao J, Xu J-Y, Zhang W, Han S, Qin L-Q (2015) Effect of dietary fiber on circulating C-reactive protein in overweight and obese adults: a meta-analysis of randomized controlled trials. Int J Food Sci Nutr 66:114–119

Kanazawa T, Atsumi M, Mineo H, Fukushima M, Nishimura N, Noda T (2008) Ingestion of gelatinized potato starch containing a high level of phosphorus decreases serum and liver lipids in rats. J Oleo Sci 57:335–343

Kaspar KL, Park JS, Brown CR, Mathison BD, Navarre DA, Chew BP (2011) Pigmented potato consumption alters oxidative stress and inflammatory damage in men. J Nutr 141:108–111

Kenny OM, McCarthy CM, Brunton NP, Hossain MB, Rai DK, Collins SG, Jones PW, Maguire AR, O'Brien NM (2013) Anti-inflammatory properties of potato glycoalkaloids in stimulated Jurkat and Raw 264.7 mouse macrophages. Life Sci 92:775–782

Khachick F, Goli M, Beecher G, Holden J, Lusby W, Tenorio M, Barrera M (1992) Effect of food preparation on qualitative and quantitative distribution of major carotenoid constituents of tomatoes and several green vegetables. J Agric Food Chem 40:390–398

Khachik F, Cohen L, Zhao Z (1999) Metabolism of dietary carotenoidsand their possible role in prevention of cancer and macular de-generation. In: Shibamoto T, Terao J, Osawa T (eds) Functional foods for disease prevention I, American Chemical Society symposium series. Oxford University Press, New York, pp 71–85

Kimura M, Rodriguez-Amaya DB (2003) Carotenoid Composition of Hydroponic Leafy Vegetables. J Agric Food Chem 51:2603–2607. http://dx.doi.org/10.1021/jf020539b

King J, Slavin J (2013) White potatoes, human health, and dietary guidance. Adv Nutr 4:393S–401S. https://doi.org/10.3945/an.112.003525

Krinsky NI, Johnson EJ (2005) Carotenoid actions and their relation to health and disease. Mol Aspects Med 26(6):459–516. Review

Kubow S, Iskandar MM, Melgar-Bermudez E, Sleno L, Sabally K, Azadi B, How E, Prakash S, Burgos G, Felde TZ (2017) Effects of simulated human gastrointestinal digestion of two purple-fleshed potato cultivars on anthocyanin composition and cytotoxicity in colonic cancer and non-tumorigenic cells. Nutrients 9(9):pii: E953. https://doi.org/10.3390/nu9090953

Külen O, Stushnoff C, Holm DG (2012) Effect of cold storage on total phenolics content, antioxidant activity and vitamin C level of selected potato clones. J Sci Food Agric 93:2437–2444

Kumar D, Singh B, Parveen K (2004) An overview of the factors affecting sugar content of potatoes. Ann Appl Biol 145:247–256

Lachman J, Hamouz K (2005) Red and purple coloured potatoes as a significant antioxidant source in human nutrition – a review. Plant Soil Environ 51(11):477–482

Lachman J, Hamouz K, Orsak M, Pivec V (2001) Potato glycoalkaloids and their significance in plant protection and human nutrition – review. Rosstlinna Vyrova 47(4):181–191

Lachman J, Hamouz K, Sulc M, Orsak M, Pivec V, Hejtmankova A (2009) Cultivar differences of total anthocyanins and anthocyanidins in red and purple-fleshed potatoes and their relation to antioxidant activity. Food Chem 114:836–843

Lachman J, Hamouz K, Musilová J, Hejtmánková K, Kotíková Z, Pazderu K, Domkárová J, Pivec V, Cimr J (2013) Effect of peeling and three cooking methods on the content of selected phytochemicals in potato tubers with various colour of flesh. Food Chem 138:1189–1197

Lancaster J, Reay P, Norris J, Butler R (2000) Induction of flavonoids and phenolic acids in apple by UV-B and temperature. J Hortic Sci Biotechnol 75(2):142–148. https://doi.org/10.1080/14620316.2000.11511213

Langner E, Rzeski W, Kaczor J, Kandefer-Szersze M, Pierzynowski SG (2009) Tumour cell growth-inhibiting properties of water extract isolated from heated potato fibre (Potex). J Pre-Clin Clin Res 3:36–41

Larsson SC, Wolk A (2016) Potato consumption and risk of cardiovascular disease: 2 prospective cohort studies. Am J Clin Nutr 104:1245–1252

Lee KR, Kozukue N, Han JS, Park JH, Chang EY, Baek EJ, Chang JS, Friedman M (2004) Glycoalkaloids and metabolites inhibit the growth of human colon (HT29) and liver (HepG2) cancer cells. J Agric Food Chem 52:2832–2839

Leeman M, Östman E, Björck I (2005) Vinegar dressing and cold storage of potatoes lowers postprandial glycaemic and insulinaemic responses in healthy subjects. Eur J Clin Nutr 59:1266–1271

Liang N, Kitts D (2015) Role of chlorogenic acids in controlling oxidative and inflammatory stress conditions. Nutrients 8(1). https://doi.org/10.3390/nu8010016

Lin BW, Gong CC, Song HF, Cui YY (2017) Effects of anthocyanins on the prevention and treatment of cancer. Br J Pharmacol 174:1226–1243

Liu Q, Tarn R, Lynch D, Skjodt N (2007) Physicochemical properties of dry matter and starch from potatoes grown in Canada. Food Chem 105:897–907

Liyanage R, Han KH, Watanabe S, Shimada K, Sekikawa M, Ohba K (2008) Potato and soy peptide diets modulate lipid metabolism in rats. Biosci Biotechnol Biochem 72:943–950

Liyanage R, Han KH, Shimada KI, Sekikawa M, Tokuji Y, Ohba K (2009) Potato and soy peptides alter caecal fermentation and reduce serum non-HDL cholesterol in rats fed cholesterol. Eur J Lipid Sci Technol 111:884–892

Lockyer S, Spiro A, Stanner S (2016) Dietary fibre and the prevention of chronic disease—should health professionals be doing more to raise awareness? Nutr Bull 41:214–231

Lombardo S, Pandino G, Mauromicale G (2013) The influence of growing environment on the antioxidant and mineral content of "early" crop potato. J Food Compos Anal 32(1):28–35

Lovat C, Nassar A, Kubow S, Li X, Donnelly D (2016) Metabolic biosynthesis of potato (*Solanum tuberosum* L.) antioxidants and implications for human health. Crit Rev Food Sci Nutr 56(14):2278–2303

Love SL, Pavek JJ (2008) Positioning the potato as a primary food source of vitamin C. Am J Potato Res 85:277–285

Lv C, Kong H, Dong G, Liu L, Tong K, Sun H, Chen B, Zhang C, Zhou M (2014) Antitumor efficacy of α-solanine against pancreatic cancer in vitro and in vivo. PLoS One 9(2):e87868

Lynch DR, Liu Q, Tarn TR, Bizimungu B, Chen Q, Harris P, Chik CL, Skjodt NM (2007) Glycemic index: a review and implications for the potato industry. Am J Potato Res 84:179–190

Madiwale GP, Reddivari L, Holm DG, Vanamala J (2011) Storage elevates phenolic content and antioxidant activity but suppresses antiproliferative and pro-apoptotic properties of colored-flesh potatoes against human colon cancer cell lines. J Agric Food Chem 59:8155–8166

Makinen S, Kelloniemi J, Pihlanto A, Makinen K, Korhonen H, Hopia A (2008) Inhibition of angiotensin converting enzyme I caused by autolysis of potato proteins by enzymatic activities confined to different parts of the potato tuber. J Agric Food Chem 56:9875–9883

Mattila P, Hellstrom J (2006) Phenolic acids in potatoes, vegetables, and some of their products. J Food Compos Anal 20:152–160

McGill C, Kurilich A, Davignon J (2013) The role of potatoes and potato components in cardiometabolic health: a review. J Ann Med 45(7):467–473

Miao M, Jiang B, Cui S, Zhang T, Jin Z (2015) Slowly digestible starch–a review. Crit Rev Food Sci Nutr 55(12):1642–1657. https://doi.org/10.1080/10408398.2012.704434

Miller R, Stanner S (2016) Baked, mashed, boiled or fried—can potatoes increase the risk of hypertension? Nutr Bull 41:252–256

Mills CE, Tzounis X, Oruna-Concha MJ, Mottram DS, Gibson GR, Spencer JP (2015) In vitro colonic metabolism of coffee and chlorogenic acid results in selective changes in human faecal microbiota growth. Br J Nutr 113:1220–1227

Mínguez Mosquera M, Hornero-Méndez D (1994) Comparative Study of the Effect of Paprika Processing on the Carotenoids in Peppers (Capsicum annuum) of the Bola and Agridulce Varieties. J Agric Food Chem 42(7):1555–1560

Monro J, Mishra S, Blandford E, Anderson J, Genet R (2009) Potato genotype differences in nutritionally distinct starch fractions after cooking, and cooking plus storing cool. J Food Compos Anal 22(6):539–545

Moonney S, Chen L, Kuhn C, Navarre R, Knowles NR, Hellman H (2013) Genotype-specific changes in vitamin B_6 content and the PDX family in potato. Bio Med Res Int. Article ID 389723, 7 pages. https://doi.org/10.1155/2013/389723

Müller HZ (1997) Determination of the carotenoid content in selected vegetables and fruit by HPLC and photodiode array detection. Lebensm Unters Forsch 204:88. https://doi.org/10.1007/s002170050042

Muraki I, Rimm EB, Willett WC, Manson JE, Hu FB, Sun Q (2016) Potato consumption and risk of type 2 diabetes: results from three prospective cohort studies. Diabetes Care 39(3):376–384. https://doi.org/10.2337/dc15-0547

Muttucumaru N, Elmore J, Curtis T, Mottram D, Parry M, Halford N (2008) Reducing acrylamide precursors in raw materials derived from wheat and potato. J Agric Food Chem 56(15):6167–6172. https://doi.org/10.1021/jf800279d. Epub 2008 Jul 15

Naidu K (2003) Vitamin C in human health and disease is still a mystery? An overview. Nutr J 2:7

Nassar AM, Sabally K, Kubow S, Leclerc YN, Donnelly DJ (2012) Some Canadian-grown potato cultivars contribute to a substantial content of essential dietary minerals. J Agric Food Chem 60(18):4688–4696. https://doi.org/10.1021/jf204940t

Natella F, Nardini M, Belelli F, Scaccini C (2007) Coffee drinking induces incorporation of phenolic acids into LDL and increases the resistance of LDL to ex vivo oxidation in humans. Am J Clin Nutr 86:604–609

Navarre DA, Shakya R, Holden J, Kumar S (2010) The effect of different cooking methods on phenolics and vitamin C in developmentally young potato tubers. Am J Potato Res 87(4):350–359

Neilson J, Lagüe M, Thomson S, Aurousseau F, Murphy AM, Bizimungu B, Deveaux V, Bègue Y, JME J, Tai HH (2017) Gene expression profiles predictive of cold-induced sweetening in potato. Funct Integr Genomics 17(4):459–476

Niu K, Momma H, Kobayashi Y, Guan L, Chujo M, Otomo A, Ouchi E, Nagatomi R (2016) The traditional Japanese dietary pattern and longitudinal changes in cardiovascular disease risk factors in apparently healthy Japanese adults. Eur J Nutr 55:267–279

Ombra M, Fratianni F, Granese T, Cardinale F, Cozzolino A, Nazzaro F (2015) In vitro antioxidant, antimicrobial and anti-proliferative activities of purple potato extracts (Solanum tuberosum cv Vitelotte noire) following simulated gastro-intestinal digestion. Nat Prod Res 29(11):1087–1091. https://doi.org/10.1080/14786419.2014.981183

Olthof MR, Hollman PC, Buijsman MN, van Amelsvoort JM, Katan MB (2003) Chlorogenic acid, quercetin-3-rutinoside and black tea phenols are extensively metabolized in humans. J Nutr 133:1806–1814

Otten J, Hellwig P, Meyers L (2010) Dietary reference intakes: the essential guide to nutrient requirements. The National Academies Press, Washington, DC

Pascual-Teresa S, Sanchez-Ballesta MT (2008) Anthocyanins: from plant to health. Phytochem Rev 7:281–299

Pastorino S, Richards M, Pierce M, Ambrosini GL (2016) A high-fat, high-glycaemic index, low-fibre dietary pattern is prospectively associated with type 2 diabetes in a British birth cohort. Br J Nutr 115:1632–1642

Piñeros-Niño C, Narváez-Cuenca C, Kushalappa A, Mosquera T (2017) Hydroxycinnamic acids in cooked potato tubers from Solanum tuberosum group Phureja. Food Sci Nutr 5(3):380–389

Puchau B, Zulet MA, Gonzalez de Echavarri A, Hermsdorff HH, Martinez JA (2010) Dietary total antioxidant capacity is negatively associated with some metabolic syndrome features in healthy young adults. Nutrition 26(5):534–541

Raigond P, Ezekiel R, Raigond B (2014) Resistant starch in food. J Sci Food Agric 95(10):1968–1978. https://doi.org/10.1002/jsfa.6966. Review

Ramadan M, Oraby H (2016) Fatty acids and bioactive lipids of potato cultivars: an overview. J Oleo Sci 65(6):459–470. https://doi.org/10.5650/jos.ess16015

Rautiainen S, Serafini M, Morgenstern R, Prior RL, Wolk A (2008) The validity and reproducibility of food-frequency questionnaire–based total antioxidant capacity estimates in Swedish women. Am J Clin Nutr 87:1247–1253

Reddivari L, Hale AL, Miller JC (2007a) Determination of phenolic content, composition and their contribution to antioxidant activity in specialty potato selections. Am J Potato Res 84:275–282

Reddivari L, Vanamala J, Chintharlapalli S, Safe SH, Miller J (2007b) Anthocyanin fraction from potato extracts is cytotoxic to prostate cancer cells through activation of caspase-dependent and caspase-independent pathways. Carcinogenesis 28(10):2227–2235

Reddivari L, Vanamala J, Safe SH, Miller JC (2010) The bioactive compounds α-chaconine and gallic acid in potato extracts decrease survival and induce apoptosis in LNCaP and PC3 prostate cancer cells. Nutr Cancer 62:601–610

Reyes LF, Miller JC, Cisneros-Zevallos L (2005) Antioxidant capacity, anthocyanins and total phenolics in purple- and red-fleshed potato (*Solanum tuberosum* L.) genotypes. Am J Potato Res 82:271–277

Richelle M, Tavazzi I, Offord E (2001) Comparison of antioxidant activity of commonly consumed polyphenolic beverages (coffee, cocoa, tea) prepared per cup serving. J Agric Food Chem 49:3438–3442

Robert L, Narcy A, Rock E, Demigne C, Mazur A, Rémésy C (2006) Entire potato consumption improves lipid metabolism and antioxidant status in cholesterol-fed rat. Eur J Nutr 45:267–274

Roberts SB (2000) High-glycemic index foods, hunger, and obesity: is there a connection? Nutr Rev 58:163–169

Rodríguez B, Ríos D, Rodríguez E, Díaz C (2010) Influence of the cultivar on the organic acid and sugar composition of potatoes. J Sci Food Agric 90(13):2301–2309. https://doi.org/10.1002/jsfa.4086

Rodriguez-Amaya DB (2015) Status of carotenoid analytical methods and in vitro assays for the assessment of food quality and health effects. Curr Opin Food Sci 1:56–63. https://doi.org/10.1016/j.cofs.2014.11.005

Ruprich J, Rehurkova I, Boon PE, Svensson K, Moussavian S, Van der Voet H, Bosgra S, Van Klaveren JD, Busk L (2009) Probabilistic modeling of exposure doses and implications for health risk characterization: glycoalkaloids from potatoes. Food Chem Toxicol 47:2899–2905

Sajilata MG, Singhal RS, Kulkarni PR (2006) Resistant starch—a review. Compr Rev Food Sci Food Saf 5:1–17

Salmeron J, Manson JE, Stampfer MJ, Colditz GA, Wing AL, Willett WC (1997) Dietary fiber, glycemic load, and risk of non-insulin-dependent diabetes mellitus in women. J Am Med Assoc 277:472–477

Slavin JL (2008) Position of the American Dietetic Association: health implications of dietary fiber. J Am Diet Assoc 108:1716–1731

Soh NL, Brand-Miller J (1999) The glycaemic index of potatoes: the effect of variety, cooking method and maturity. Eur J Clin Nutr 53:249–254

Spanos GA, Wrolstad RE (1992) Phenolics of apple, pear, and white grape juices and their changes with processing and storage. A review. J Agric Food Chem 40:1478–1487

Spinneker A, Sola R, Lemmen V, Castillo MJ, Pietrzik K, González-Gross M (2007) Vitamin B_6 status, deficiency and its consequences—an overview. Nutr Hosp 22(1):7–24

Stalmach A, Steiling H, Williamson G, Crozier A (2010) Bioavailability of chlorogenic acids following acute ingestion of coffee by humans with an ileostomy. Arch Biochem Biophys 501:98–105

Sun H, Lv C, Yang L, Wang Y, Zhang Q, Yu S, Kong H, Wang M, Xie J, Zhang C, Zhou M (2014) Solanine induces mitochondria-mediated apoptosis in human pancreatic cancer cells. BioMed Res Int 2014:805926

Tajner-Czopek M, Jarych-Szyszka M, Lisińska G (2008) Changes in glycoalkaloids content of potatoes destined for consumption. Food Chem 106(2):706–711

Tee E, Lim C (1991) Carotenoid composition and content of Malaysian vegetables and fruits by the AOAC and HPLC methods. Food Chem 41:309–339

Teucher B, Olivares M, Cori H (2004) Enhancers of iron absorption: ascorbic acid and other organic acids. Int J Vitam Nutr Res. 74(6):403–419

Tian J, Chen C, Ye X, Chen S (2016) Health benefits of the potato affected by domestic cooking: a review. Food Chem 202:165–175

Tomas-Barberan F, García-Villalba R, Quartieri A, Raimondi S, Amaretti A, Leonardi A, Rossi M (2014) In vitro transformation of chlorogenic acid by human gut microbiota. Mol Nutr Food Res 58(5):1122–1131. https://doi.org/10.1002/mnfr.201300441. Epub 2013 Dec 23

van Het Hof KH, West CE, Weststrate JA, Hautvast JG (2000) Dietary factors that affect the bioavailability of carotenoids. J Nutr 130(3):503–506

Vaziri ND, Liu SM, Lau WL, Khazaeli M, Nazertehrani S, Farzaneh SH, Kieffer DA, Adams SH, Martin RJ (2014) High amylose resistant starch diet ameliorates oxidative stress, inflammation, and progression of chronic kidney disease. PloS One 9:e114881

Venn B, Green T (2007) Glycemic index and glycemic load: measurement issues and their effect on diet-disease relationships. Eur J Clin Nutr 61(Suppl 1):S122–S131

Veronese N, Stubbs B, Noale M, Solmi M, Vaona A, Demurtas J, Nicetto D, Crepaldi G, Schofield P, Koyanagi A, Maggi S, Fontana L (2017) Fried potato consumption is associated with elevated mortality: an 8-y longitudinal cohort study. Am J Clin Nutr 106:162–167

Vinson J, Demkosky C, Navarre D, Smyda M (2012) High-antioxidant potatoes: acute in vivo antioxidant source and hypotensive agent in humans after supplementation to hypertensive subjects. J Agric Food Chem 60(27):6749–6754. https://doi.org/10.1021/jf2045262

Visvanathan R, Jayathilake C, Chaminda Jayawardana B, Liyanage R (2016) Health-beneficial properties of potato and compounds of interest. J Sci Food Agric 96(15):4850–4860. https://doi.org/10.1002/jsfa.7848. Epub 2016 Jul 7. Review

Wada L, Ou B (2002) Antioxidant activity and phenolic content of Oregon cane-berries. J Agric Food Chem 50(12):3495–3500

Wahyuni Y, Ballester A, Sudarmonowati E, Bino RJ, Bovy AG (2011) Metabolite biodiversity in pepper (Capsicum) fruits of thirty-two diverse accessions: variation in health-related compounds and implications for breeding. Phytochemistry 72(11–12):1358–1370

Wang Q, Chen Q, He M, Mir P, Su J, Yang Q (2011) Inhibitory effect of antioxidant extracts from various potatoes on the proliferation of human colon and liver cancer cells. Nutr Cancer 63:1044–1052

Wang S, Copeland L, Brand-Miller J (2014) Discovery of a low-glycaemic index potato and relationship with starch digestion in vitro. Br J Nutr 111(4):699–705. https://doi.org/10.1017/S0007114513003048

Wang W, Sun C, Mao L, Ma P, Liu F, Yang J, Gao Y (2016) The biological activities, chemical stability, metabolism and delivery systems of quercetin: a review. Trends Food Sci Technol 56:21–38

WHO (World Health Organization) (2012) Guideline: potassium intake for adults and children. World Health Organization, Geneva

Winkel-Shirley B (2001) Flavonoid biosynthesis. A colorful model for genetics, biochemistry, cell biology, and biotechnology. Plant Physiol 126:485–493

Woolfe J, Poats S (1987) The Potato in the Human Diet. Cambridge University Press. 0521326699, 9780521326698

Wu X, Beecher GR, Holden JM, Haytowitz DB, Gebhardt SE, Prior RL (2006) Concentration of anthocyanins in common foods in the United States and estimation of normal consumption. J Agric Food Chem 54(11):4069–4075

Wu J, Cho E, Willett WC, Sastry SM, Schaumberg DA (2015) Intakes of lutein, zeaxanthin, and other carotenoids and age-related macular degeneration during 2 decades of prospective follow-up. JAMA Ophthalmol 133(12):1415–1424. https://doi.org/10.1001/jamaophthalmol.2015.3590

Yang SA, Paek SH, Kozukue N, Lee KR, Kim JA (2006) Alpha-chaconine, a potato glycoalkaloid, induces apoptosis of HT-29 human colon cancer cells through caspase-3 activation and inhibition of ERK 1/2 phosphorylation. Food Chem Toxicol 44:839–846

Yogendra KN, Kushalappa AC, Sarmiento F, Rodriguez E, Mosquera T (2015) Metabolomics deciphers quantitative resistance mechanisms in diploid potato clones against late blight. Funct Plant Biol 42:284–298

Yuan G, Sun B, Yuan J, Wang Q (2009) Effects of different cooking methods on health-promoting compounds of broccoli. J Zhejiang Univ Sci B 10(8):580–588. https://doi.org/10.1631/jzus.B0920051

Zeeman SC, Kossmann J, Smith AM (2010) Starch: its metabolism, evolution, and biotechnological modification in plants. Annu Rev Plant Biol 61:209–234

Zeng C (2013) Effects of different cooking methods on the vitamin C content of selected vegetables. Nutr Food Sci 43(5):438–443. https://doi.org/10.1108/NFS-11-2012-0123

8

Viral Diseases in Potato

J. F. Kreuze, J. A. C. Souza-Dias, A. Jeevalatha, A. R. Figueira, J. P. T. Valkonen, and R. A. C. Jones

Abstract Viruses are among the most significant biotic constraints in potato production. In the century since the discovery of the first potato viruses we have learned more and more about these pathogens, and this has accelerated over the last decade with the advent of high-throughput sequencing in the study of plant virology. Most reviews of potato viruses have focused on temperate potato production systems of Europe and North America. However, potato production is rapidly expanding in tropical and subtropical agro-ecologies of the world in Asia and Africa, which present a unique set of problems for the crop and affect the way viruses can be managed. In this chapter we review the latest discoveries in potato virology as well as the changes in virus populations that have occurred over the last 50 years, with a particular focus on countries in the (sub-)tropics. We also review the different management approaches including use of resistance, seed systems, and cultural approaches that have been employed in different countries and reflect on what can be learnt from past research on potato viruses, and what can be expected in the future facing climate change.

J. F. Kreuze (✉)
International Potato Center, Lima, Peru
e-mail: j.kreuze@cgiar.org

J. A. C. Souza-Dias
Instituto Agronômico de Campinas (IAC)/APTA/SAA-SP, Campinas, São Paulo, Brazil
e-mail: jcaram@iac.sp.gov.br

A. Jeevalatha
ICAR-Central Potato Research Institute, Shimla, Himachal Pradesh, India

A. R. Figueira
Department of Plant Pathology, Lavras Federal University, Lavras, Minas Gerais, Brazil
e-mail: antonia@ufla.br

J. P. T. Valkonen
Department of Agricultural Sciences, University of Helsinki, Helsinki, Finland
e-mail: jari.valkonen@helsinki.fi

R. A. C. Jones
Institute of Agriculture, University of Western Australia, Crawley, Perth, Australia
e-mail: roger.jones@uwa.edu.au

Introduction

In terms of human consumption, potato (*Solanum tuberosum*) is currently the third most important food crop globally after rice and wheat, and over half of its production currently occurs in developing countries (Devaux et al. 2014). Worldwide, over the last few decades, potato production has increased at a much higher rate compared to other major staple crops. This increase has occurred principally in developing countries located in largely tropical and subtropical regions. Due to its ability to produce high amounts of digestible energy per unit time and unit area for home consumption, but at the same time provide income as a cash crop, planting potato tends to be popular with farmers wherever they are able to grow them. In the future, development of new heat tolerant and early maturing potato cultivars will likely lead to further expansion of its production into warmer areas of the tropics. However, as temperatures increase virus vectors often become more abundant and the incidence of virus epidemics increases. This increase in insect vectors and virus disease incidence, combined with the fact that virus-tested seed systems are weak or entirely absent, explains why potato virus diseases are of particular importance in the developing world and estimated to account for 50% or more of the potential total yield being lost (Harahagazwe et al. 2018). In addition, the presence of year-round potato cultivation in some tropical regions and the lack of cool upland areas where insect vector pressure is low enough to produce high-quality seed potatoes, both exacerbate potato virus disease problems in these regions. Several global or regional reviews and intercontinentally focused research papers on potato virus diseases have been written during the last decade, devoted to many different aspect or particular viruses, including economic losses, detection methods, molecular variability, resistance genes, and evolution (Valkonen 2007; Gray et al. 2010; Karasev and Gray 2013; Jones 2014; Gibbs et al. 2017; Lacomme et al. 2017; Santillan et al. 2018) and we refer to them for details on those specific aspects. In this chapter we will review potato viruses with a focus on developing world regions and changes that have occurred over the past 50 years. The reason for this is that they have traditionally received less attention in the literature, but also are generally located in places with warmer climates, and with global temperatures rising, may be representative for what the future holds also for the currently more temperate regions. We will start with a general overview of potato infecting viruses of global importance, the damages they cause to potato production, and where they occur. Next, we describe the viruses found in the center of origin of wild and cultivated potatoes, the Andes. Then we describe the situation in two emerging economies in (sub-)tropics with functional seed systems, one, Brazil which is largely based on imported basic seed tubers and one, India, which largely produces its own potato cultivars and seed. This is followed by a brief review of the situation in Africa, and a description of the situation in two contrasting developed economies, Australia and Europe. Finally, we consider the main control methods for potato viruses, how they are being applied in the different agro-ecologies, and how they might be affected by changing climates.

Viruses of Potato

Whereas more than 50 different viruses and one viroid have been reported infecting potatoes worldwide, only a handful of them cause major losses globally. However, some are locally and/or temporarily relevant, while others

Table List of viruses of potato worldwide

Virus[a]	Genus, family	Transmission	Distribution
Potato virus Y (PVY)	*Potyvirus, Potyviridae*	Aphids	Worldwide
Potato virus A (PVA)			Worldwide
Potato virus V (PVV)			Europe, South America
Wild potato mosaic virus (WPMV)			Andes, only reported in wild potatoes
Potato virus X (PVX)	*Potexvirus, Alphaflexiviridae*	Contact	Worldwide
Potato aucuba mosaic virus (PAMV)			Worldwide, very rare
Potato leaf roll virus (PLRV)	*Polerovirus*	Aphids	Worldwide
Potato virus S (PVS)	*Carlavirus, Betaflexiviridae*	Contact, aphids	Worldwide
Potato latent virus (PotLV)		Aphids	North America, rare
Potato virus M (PVM)		Aphids	Worldwide
Potato virus H (PVH)		unknown	China
Potato virus P (PVP syn. Potato rought dwarf virus: PRDV)		unknown	Brazil & Argentina
Potato virus T (PVT)	*Tepovirus, Betaflexiviridae*	Contact, seed	Southern Andean region
Andean potato mottle virus (APMoV)	*Comovirus, Secoviridae*	Beetles	Andean region, Brazil
Potato black ringspot virus (PBRSV = TRSV-Ca)	*Nepovirus, Secoviridae*	true seed, nematodes	Peru
Potato virus U (PVU)		nematodes	Peru, only reported once
Potato virus B (PVB)		nematodes?	Peru, recently reported, relatively common
Cherry leaf roll virus (CLRV)		Nematodes, TPS, pollen?	Europe, North & South America
Lucerne Australian latent virus (LALV)		Unknown	Australia and New Zealand, rare in potato
Tomato black ring virus (TBRV)		Nematodes	Europe, rare
Cherry rasp leaf virus (CRLV)	*Cheravirus, Secoviridae*		North America, only reported once
Arracacha virus B (AVB)		TPS, pollen	Andes

(continued)

Table (continued)

Virus[a]	Genus, family	Transmission	Distribution
Tomato spotted wilt virus (TSWV)	*Tospovirus, Bunyaviridae*	Thrips	Worldwide
Tomato chlorotic spot virus (TCSV)			South America
Groundnut bud necrosis virus (GBNV)			India
Groundnut ringspot virus (GRSV)			Americas
"Tomato yellow fruit ring virus" (TYFRV)			Reported from potato in Iran
Impatiens necrotic spot virus (INSV)			Worldwide, reported in greenhouse grown potatoes in USA
Andean potato latent virus (APLV)	*Tymovirus, Tymoviridae*	Beetles	Andean region
Andean potato mild mottle virus (APMMV)			Andean region
Potato yellow vein virus (PYVV)	*Crinivirus, Closteroviridae*	Whiteflies	Northern Andean region, Panama
Tomato chlorosis virus (ToCV)			Brazil, Spain, India
"Potato yellowing virus" (PYV)	*Ilarvirus, Bromoviridae*	Unknown	Andean region
Tobacco streak virus (TSV)		Pollen, thrips	Worldwide, reported in potato in Brazil
Cucumber mosaic virus (CMV)	*Cucumovirus, Bromoviridae*	Aphids	Worldwide, sporadic in potato
Alfalfa mosaic virus (AlMV)	*Alfamovirus, Bromoviridae*	Aphids	Worldwide, sporadic in potato
Tomato leaf curl New Delhi virus (ToLCNDV)	*Begomovirus, Geminiviridae*	Whiteflies	India
Tomato severe rugose virus (ToSRV)			Brazil
Tomato yellow vein streak virus (ToYVSV=PDMV)			Brazil, Argentina
Tomato mottle Taino virus (ToMoTV)			Cuba
"Solanum apical leaf curl virus" (SALCV)			Peru, only reported once
Potato yellow mosaic virus (PYMV)			Carribean
Beet curly top virus (BCtV)	*Curtovirus, Geminiviridae*	Leaf hopper	Americas, Europe, Asia under dry conditions

(continued)

Table (continued)

Virus[a]	Genus, family	Transmission	Distribution
Potato mop-top virus (PMTV)	*Pomovirus, Virgaviridae*	Spongospora	Americas, Europe, Asia in cool and humid environments
"Colombian potato soil-borne virus" (CPSbV)		Spongospora?	Colombia, isolated from potato soils; CPSbV could infect potatoes symptomless
"Soil-borne virus 2" (SbV2)		Spongospora?	
Tobacco rattle virus (TRV)	*Tobravirus, Virgaviridae*	Nematodes	Worldwide, common in cool climates, or Australia
Tobacco mosaic virus (TMV)	*Tobamovirus, Virgaviridae*	Contact	Worldwide, rare on potato
Tomato mosaic virus (ToMV)			Europe, Andes, rare on potato
Tobacco necrosis virus (TNV)	*Necrovirus, Tombusviridae*	Fungus	Worldwide, rare on potato
Sowbane mosaic virus (SbMV)	*Sobemovirus*, unassigned	Contact	Worldwide, rare on potato
SB26/29 "potato rugose stunting virus"	Torradovirus-like, *Secoviridae*	Psyllids	Southern Peru
Potato yellow dwarf virus (PYDV)	*Nucleorhabdovirus, Rhabdoviridae*	Leafhoppers	North America, has become rare
Eggplant mottle dwarf virus (EMDV)		Aphids	Europe, Africa, Asia, occasionally infects potatoes
Cauliflower mosaic virus (CaMV)	*Caulimovirus, Caulimoviridae*	Aphids	Intercepted once in potato from South America
Potato spindle tuber viroid (PSTVd)	*Pospiviroid, Pospiviroidae*	Contact, aphids (when co infecting with PLRV)	Worldwide
"Potato stunt virus" (PStV)	?	?	Europe

[a]Officially accepted virus species names italicized, whereas unofficial names are between quotation marks and not in italics

are currently only of minor importance anywhere in the world. PVY (see Table for virus acronyms) and PLRV are now the most damaging viruses of potato worldwide, with PVY having overtaken PLRV as the most important. Tuber yield losses are caused by either of them in single infections and can reach more than 80% in combination with other viruses. PVX occurs commonly worldwide and causes losses of 10–40% in single infections and is particularly damaging in combination with PVY or PVA. This is due to its synergism with both potyviruses leading to tuber yield losses of up to 80%. PVS also occurs commonly worldwide but generally causes only minor tuber yield losses unless severe strains are present or it occurs in mixed infection with PVX. PVA can cause yield losses of up to 40% by itself but is far less prevalent than PVY, PVS, or PLRV. PVM is relatively uncommon in most

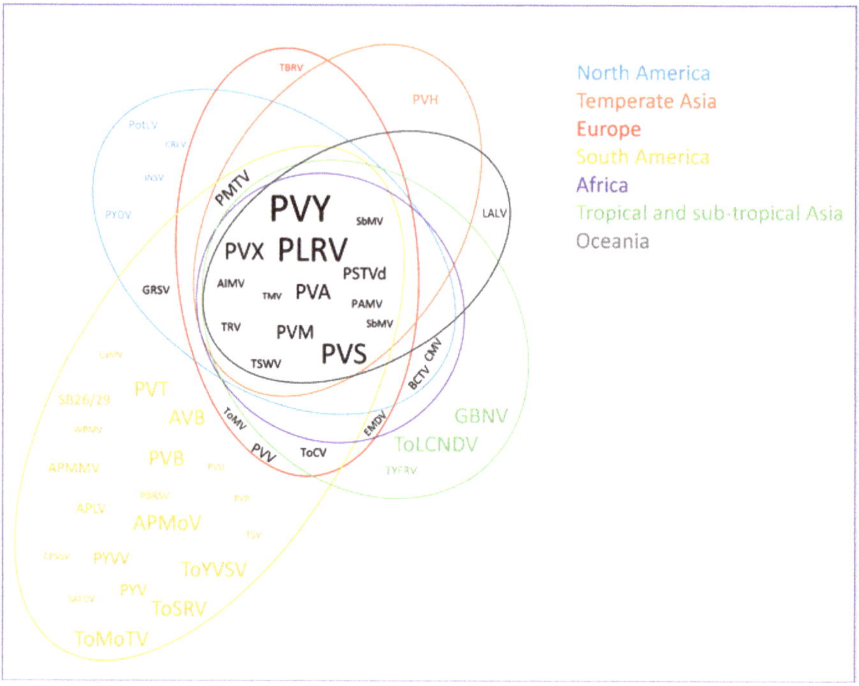

Fig. VENN diagram of geographic occurrence of commoner potato viruses worldwide. Their relative importance/incidence globally, or regionally in case they are geographically restricted, is illustrated by the size of the virus acronym. See Table for virus acronyms

countries and, like PVS, mostly causes only minor tuber yield losses, except in mixed infection with PVX or other viruses.

Besides yield reduction, several viruses cause economic losses by affecting potato quality, particularly by inducing internal and surface tuber necrosis. PLRV sometimes causes necrosis of the tuber vascular system known as "net necrosis." Tuber necrosis, consisting of necrotic rings or arcs in the flesh, sometimes develop with the thrips-transmitted virus TSWV, and with soil-borne viruses like nematode vectored TRV (Sahi et al. 2016) and protist-transmitted PMTV (Abbas and Madadi 2016). TSWV generally infects potato in warmer regions but TRV and PMTV both occur globally in cooler regions where their vectors are established. Certain phylogenetically defined recombinant strains of PVY cause similar necrotic symptoms known as "potato tuber ringspot disease." Over the last three decades, these have caused particularly heavy economic losses to potato industries in Europe and North America as well as in many developing countries in Asia and South America but have not yet reached all parts of the world, e.g. south-west Australia (Kehoe and Jones 2016) or Peru (Fuentes et al. 2019a). Therefore, PVY "strains" have been heavily studied worldwide over the past two decades revealing an exceptional amount of variation and a plethora of genotypes, many of them recombinants. PVY "strains" separate into at least 13 different subgroups defined either biologically or by phylogenetics (Karasev and Gray 2013; Kehoe and Jones 2016; Glais et al. 2017; Gibbs et al. 2017).

Biological strains of PVY are differentiated by the phenotypes they develop when different strain-specific hypersensitive HR resistance genes are present in potato cultivar differentials and whether they introduce necrotic symptoms in tobacco. Strain groups PVY^C, PVY^O and PVY^Z elicit HR phenotypes with hypersensitivity genes *Nc, Ny*, or *Nz*, respectively. Strain groups PVY^N and PVY^E overcome all three hypersensitivity genes, but differ in the phenotypes they induce in tobacco, only PVY^N eliciting veinal necrosis (Chikh-Ali et al. 2014; Karasev and Gray 2013; Rowley et al. 2015; Kehoe and Jones 2016; Jones and Vincent 2018). Such biological strains do not necessarily coincide with the phylogenetic lineages named after them. For example, previously potato biological strain groups PVY^C and PVY^O were thought to coincide with major lineages PVY^C and PVY^O, respectively. However, this proved incorrect as isolates within biological strain group PVY^D fitted within phylogroup PVY^C and those in PVY^Z within phylogroup PVY^O (Kehoe and Jones 2016; Jones and Vincent 2018). As the number of complete PVY genome sequences from different world regions grows, phylogenetic nomenclature based on biological, geographical, and sequence names is becoming increasingly unsustainable so substituting Latinised numerals for current PVY subgroup names was suggested (Jones 2014; Kehoe and Jones 2016; Jones and Kehoe 2016). Biological strains of PVA and PVV are also differentiated by the phenotypes they develop with strain-specific hypersensitive resistance genes present in potato cultivar differentials, but their phylogenetics is little studied. PLRV strains are differentiated biologically based on the severity of symptom expression in potato, but are phylogenetically very homogenous with limited sequence variation between isolates worldwide.

Potato spindle tuber disease caused by a viroid, PSTVd also impairs tuber quality in addition to direct yield loss. Although it led to several disease outbreaks in potato in different parts of the world in the past, through implementation molecular detection and eradication programs its presence in potato has now been significantly diminished in North America and Europe. By contrast, PSTVd is still prevalent in Central-Asia (CIP, unpublished) and China (Qiu et al. 2016). Although PSTVd presence in potato has declined recently globally, the opposite is the case for tomato where outbreaks have been increasing due to its worldwide spread in tomato seed via the international seed trade (Constable et al. 2019). This is of concern for global potato production as tomato PSTVd can cause severe yield and tuber quality losses in potato (Mackie et al. 2019). Additional collateral damage can be caused by virus infections as was demonstrated for PVY infection, which compromises plant defense responses rendering them more vulnerable to Colorado potato beetle (*Leptinotarsa decemlineata*) attack (Petek et al. 2014). Similarly, Lin et al. (2014) found that PVS infection rendered late blight (*Phytophpora infestans*) resistant cultivars more susceptible to late blight.

Among the most important viruses PVY, PLRV, PVA, PVS, and PVM are all aphid-transmitted. All of these except PLRV are transmitted nonpersistently by aphids, whereas PLRV is persistently transmitted. Insecticides have long been known to be effective only against persistently transmitted viruses, which likely explains the decline in prevalence observed in PLRV over the last 50 years in developed and emerging economies. Thrips- and whitefly-transmitted viruses continue to cause outbreaks in potato in warmer climatic regions. These outbreaks are usually only occasional, but TSWV is found commonly infecting potatoes in some countries, e.g. Australia and Argentina such that it is among the common viruses tested for in seed potato production schemes. Such outbreaks of thrips and whitefly-transmitted viruses are becoming steadily more frequent due to a warming climate, and at least one of these viruses, ToLCNDV has recently become a major potato pathogen in India (Jeevalatha et al. 2017a).

Recent phylogenetic studies, that use dating programs to compare the complete genomic sequences of common potato virus isolates, obtained at different times, are providing new insights into their evolution. So far, this has only been done with PVY and PVS (Gibbs et al. 2017; Santillan et al. 2018). Gibbs et al. (2017) inferred the phylogeny of the genomic sequences of 240 PVY isolates collected since 1938, 42% of which were nonrecombinants; sequences from the Andean region were lacking. The nonrecombinants all fitted into major lineages C, O, and N, and recombinants all into lineages R-1 and R-2. The main parents of R-1 and R-2 were PVY^N or PVY^O, respectively, and vice versa for their minor parents. The minor phylogroups within these major lineages [roman numerals in parentheses are from PVY classification system of Kehoe and Jones 2016] were: C with C1(II) and C2(III); O with O (=I) and O5 (=X); N with Eu-N (=IV), XIII and NA-N (=IX); R-1 with NTN-NW + SYR-I (XII), NTN-B (VI), NTN-NW + SYR-II (XI), N-Wi (VII), and N:O (VIII); and R-2 with NTN-A (V). Analysis of the nonrecombinant genomes found the estimated "time to most recent common ancestor" (TMCRA) for PVY to be around 1000 CE which corresponds with the end of the Tiahuanaco and start of the Inca civilizations in the Andes. A more comprehensive study including Andean PVY sequences recently found that PVY-N (=III) could be divided into three phylogroups (N1, N2 and N3), and two of them were unique to the Andes, suggesting PVY-N originated from the Andean region in contrast to PVY-O for which no such evidence could be found (Fuentes et al. 2019a).

Santillan et al. (2018) studied the phylogenetics of PVS genomic sequences collected since 1976, including Andean region sequences. The nonrecombinant genomes belonged to three major PVS lineages, two evenly branched and predominantly South American and a non-South American one with a long basal branch and many distal subdivisions. The South American lineages contained isolate sequences from three cultivated potato species, pepino (*Solanum muricatum*) and arracacha (*Arracacia xanthorrhiza*), whereas only isolates from a single cultivated potato species (*Solanum tuberosum*) were present in the other lineage. The two nodes of the basal PVS trifurcation were dated at 1079 and 1055 CE corresponding with the end of the Tiahuanaco and well before the start of the Inca civilizations, and the basal node to the non-SA lineage at 1837 CE corresponding roughly with the start of the European potato famine caused by late blight (*Phythophora infestans*). The PRDV/PVP cluster diverged from PVS 5–7000 years ago. This suggests a potato-infecting proto-PVS/PRDV/PVP emerged in South America, and spread into a range of local *Solanum* and other species, one early lineage spreading worldwide in potato.

Kutnjak et al. (2014) studied the phylogeny of PVX genomic sequences from the Andes. What they found was similar to the PVS situation with three major lineages, two of which were South American and one non-South American. However, an earlier study with PVX coat protein gene sequences had found several European and North American sequences in the single major South American lineage known at that time, and their presence was confirmed by Kutnjak et al. (2014).

At least 37 of the known potato viruses are found in South America and this number is set to increase further with the application of high-throughput sequencing (HTS) techniques to screen for virus infections (Kreuze 2014; Fuentes et al. 2019b; CIP, http://potpathodiv.org/). Only the above-mentioned viruses and PAMV have so far established themselves globally as potato infecting viruses, whereas most of the other global viruses may be the result of generalist viruses that have managed to become established in potatoes, or more recent newcomers from related crops that have achieved a foothold in potato due to increasing vector populations, principally whiteflies (crini- and begomoviruses) and thrips (tospoviruses). At least 20 potato viruses remain restricted to South America and most of these represent viruses that evolved together with wild and/or cultivated potatoes in the Andean region.

Viruses in the Andean Region

Cultivated potatoes were first domesticated in the Andean region of South America where they show the highest level of genetic diversity including four cultivated potato species with various ploidy levels and many native cultivar groups and wild potato relatives. Their viruses evolved with them and it is therefore not surprising that more viruses are found infecting potatoes in this region than elsewhere in the world. Besides the usual viruses associated with potato production throughout the world and which were distributed globally through infected tubers, the Andean region hosts several unique viruses that do not seem to have established themselves beyond their geographical region of origin. These include the nepoviruses PBRSV, PVU, and PVB, the tymoviruses APLV, APMMV, the Ilarvirus PYV, the crinivirus PYVV, the cheravirus AVB, and the tepovirus PVT. HTS-based approaches have recently also detected the presence of new viruses corresponding to at least 14 different genera (Fuentes et al. 2019b; CIP, http://potpathodiv.org/). Many of the Andean potato viruses have also been reported infecting other root and tuber crops that are grown in the same environments as potatoes, such as Ulluco (PVS, PVT, PLRV, AVA, APLV/APMMV), Oca (PBRSV, PVT, AVB), Mashua (PVT), Aracacha (PVS, AVA, AVB, PBRSV) & Maca (APLV/APMMV), and in the solanaceous bush fruit pepino (PVS). These are crops from such diverse species that it is likely many wild hosts may also be infected constituting a continual environmental reservoir for these viruses present in the absence of cultivated plant hosts. In addition, the viruses commonly found in other parts of the world often show much higher level of variability in the Andean region (Gil et al. 2016a, b; Kalyandurg et al. 2017; Santillan et al. 2018; Gutierrez et al. 2013; Gibbs et al. 2017; Henao-Díaz et al. 2013).

In the pre-ELISA era, surveys to establish the occurrence of potato viruses in Andean countries mostly involved potato germplasm collections, but in the 1970s Peruvian potato crops were widely sampled. They were undertaken using older serological detection assays and inoculation to indicator hosts complemented by electron microscopy. The viruses found in that era included nine also found in Europe and North America (PLRV, PVX, PVY, PVV, PVS, PVA, PVM, PMTV, PAMV) and eight others only found in Andean countries (APLV, APMoV, PVT, AVB, PBRV, PVU, PYVV, WPMV) (Jones 1981, references therein). Since the 1980s, surveys have been conducted using ELISA to detect the most common potato viruses (PVY, PVX, PVS, PLRV, APMoV, APLV) in potatoes growing at higher altitude (>3000 m) in the Peruvian highlands. The most frequently detected viruses have consistently been contact-transmitted with PVX (30–82% incidence) and PVS (20–50%) being the commonest followed by APMoV (4–15%) and APLV (2–6%). Similar viruses and incidences were found in higher altitude plantings in Ecuador where PYVV was also found occurring at low frequency (0–3% incidence). In contrast, PLRV and PVY were usually only detected at 0–5% in these materials and, when included, PMTV was uncommon (Pérez Barrera et al. 2015). When similar surveys were undertaken at lower altitudes in the Andean region (<3000 m), the findings resembled those in other areas of the world, with PVY and PLRV dominating. The differences in PVY and PLRV incidences between potato crops growing at different altitudes likely reflects the greater abundance of their aphid vectors below 3000 m.

Whereas the potyviruses PVY, PVA, and, to a lesser extent, PVV are established worldwide, another potato potyirus, WPMV has never been reported infecting cultivated potato even in the Andean region. So far, it has only been reported from a wild potato species growing in an isolated Lomas ecosystem and the cultivated bush fruit crop pepino both in the coastal desert in Peru (Fribourg et al. 2019). In earlier studies, when Andean potato cultivars were inoculated with PVY isolates belonging to biological strain groups PVY^{C}, PVY^{O}, PVY^{Z}, and PVY^{N}, one developed HR phenotypes consistent with presence of genes *Nc* and *Ny*, one an HR phenotype consistent with gene *Nc* alone, and 1 with neither, so both genes were easily found in commercial cultivars in potato's original center of domestication. With European potato cultivars inoculated at the same time, the corresponding figures were six with *Nc*, six with *Ny* and one with neither. An HR phenotype consistent with *Nz* presence and an ER phenotype consistent with *Ry* presence developed in two cultivars each. There is still a need to determine how common resistance genes *Ry, Nc, Nz, Ny*, and putative *Nd* are among Andean potato cultivars and the degree of protection they provide against PVY in Andean potato crops. This applies not only to commercial plantings grown with or without access to healthy seed programs but also to Andean native potato landraces belonging to the four potato species that grow in Andean subsistence plantings (Jones 1981; Jones and Vincent 2018).

The potexvirus PVX consists of four biological strains differentiated by their phenotypic reactions when they infect potato cultivars with strain-specific hypersensitive resistance genes *Nx* and *Nb*. Group 1 strains fail to overcome either gene, group 2 strains overcome *Nx*, group 3 strains overcome *Nb*, and group 4 strains overcome both genes. All four strain groups occur in the Andean region, along with an additional strain (HB) that has not been reported elsewhere in the world, which overcomes not only these two genes but also extreme PVX resistance gene *Rx*. PVX^{HB} caused a mild or symptomless infection in eight native potato landraces, systemic necrotic symptoms in cultivar "Mi Peru," and bright yellow leaf markings in "Renacimiento." Phylogenetic analysis of coat protein gene sequences placed HB in the major PVX lineage that contained group 2 and 4 isolates from South America, North America, or Europe, whereas strain group 1, 3, and 4 sequences, none of which were from South America, were in the main lineage that lacked any South American sequences. Thus, only strain group 4 sequences were in both lineages (Kutnjak et al. 2014).

Three carlaviruses have been reported infecting potatoes in South America (PVS, PVM, and PRDV). Early PVS isolates from the Andean region caused systemic infection in *Chenopodium quinoa*, but isolates from elsewhere only infected its inoculated leaves. The former were therefore called Andean strain (PVS^{A}) and the latter ordinary strain (PVS^{O}). However, PVS^{A} was found subsequently in many countries (Jones 2014, references therein), and, more recently, PVS^{O} was found in the Andean region (Santillan et al. 2018). These biological defined strains are not coincident with phylogenetically defined PVS strains as PVS^{O} occurs in both South American and non-South American lineages (Santillan et al. 2018). A new strain of PVS was recently found infecting Arracacha (Santillan et al. 2018; de Souza et al. 2018). The carlavirus PVM has been reported from Bolivia, Chile, and Peru and in the Andean region of northern Argentina, but in recent surveys is conspicuously absent from Peru. The carlavirus PRDV on the other hand has been reported only from Argentina and Brazil (PVP isolate) and probably infected potato from indigenous hosts as they have not been reported from the Andean region, although in evolutionary terms PRDV and PVP are considered likely ancestral parents of PVS from this region (Santillan et al. 2018).

Several nepoviruses infect potatoes in the Andes all of which were originally isolated from potato plants showing calico symptoms, although none of them have been linked to this syndrome by reproducing the symptoms in experimentally infected potato plants. The first was discovered simultaneously by two research groups who separately named it PBRV and the calico strain of TRSV. Subsequently, however, the virus was shown not to be TRSV (Souza Richards et al. 2013) so only the name PBRV remains in use. PVB was recently reported infecting potatoes in Peru (de Souza et al. 2017) where it is now known to be relatively common (Fuentes et al. 2019b; CIP, http://potpathodiv.org/). Although PVB was first identified from plants showing calico symptoms, these symptoms were not necessarily caused by it as other viruses, such as PVX, were also present. The virus has not yet been characterized biologically and is impact on tuber yield is unknown. Another subgroup C nepovirus, PVU, was isolated previously from potatoes with calico symptoms in central Peru, but sequence comparison distinguished it from PVB (de Souza et al. 2017).

Fig. Field with typical calico and yellowing symptoms, typically associated with nepoviruses, in the highlands of Peru. Photo credits: CIP

The comovirus APMoV (family *Secoviridae*) was identified infecting potato in Peru, Argentina, and in Brazil where it was also found infecting eggplants. In Peru, it was relatively common and widespread in the past based on ELISA results, but in a recent survey using HTS, was not commonly encountered (Fuentes et al. 2019b; http://potpathodiv.org/); it is transmitted by beetles of the genus Diabrotica, as well as by seed and contact. It also occurs outside South America having been found infecting tabasco peppers in Honduras in Central America.

The tymoviruses APLV and APMMV (family *Tymoviridae*), which was recently separated from APLV (Kreuze et al. 2013; Koenig and Ziebell 2013), have been identified in potato germplasm from Colombia, Ecuador, Peru, and Bolivia. Nevertheless, they seem to be becoming less common in field grown potatoes in the Andean region than when they were first found in the 1970s.

PVT is the only known member of the genus *Tepovirus* (family *Betaflexiviridae*). It has been detected not only in Peruvian, Bolivian, and Chilean potato germlasm, but also ulluco (*Ullucus tuberosus*), oca (*Oxalis tuberosa*), and mashua (*Tropaeolum tuberosum*) plants growing in the field in these countries. It is transmitted through contact and potato true seed, but causes only mild mosaic or no symptoms in potato plants and seems relatively uncommon (Lizárraga et al. 2000).

Two ilarviruses (Family *Bromoviridae*), AlMV and PYV, both sometimes infect potatoes in the Andean region. Whereas AMV is found worldwide and normally causes calico symptoms, including in the highlands of Peru, PYV is largely symptomless and restricted to the Andean region where it has been identified in germplasm from Ecuador, Peru, Bolivia, and Chile.

PYVV (genus *Crinivirus*, family *Closteroviridae*) causes obvious veinal yellowing symptoms, which were first seen in an early Andean potato germplasm collection in Europe. It has been known in Colombia for many years (Jones 1981, references therein; Franco-Lara et al. 2013). PYVV is thought to have originated from the Andean region of Northern Ecuador and Central West Colombia and causes up to 50% yield losses. It is transmitted in a semi-persistent manner by the greenhouse whitefly (*Trialeurodes vaporariorum* Westwood; *Hemiptera*: *Aleyrodidae*) (Salazar et al. 2000), through tuber seed and underground stem-grafts. It also infects tomatoes and various weed species. The virus also occurs in potato-producing areas of Northern Peru, and in the Venezuelan Andes, and recently spread to Panama in Central America (CIP, unpublished). Its whitefly vector occurs globally so it could spread to other continents. Its prevalence in plantings at lower altitudes in the Andes reflects the restriction of its whitefly vector to warmer conditions (Jones 2016, references therein).

A begomovirus PALCV was reported infecting potatoes and wild solanaceaous hosts from the highland jungle region of central Peru in the late 1980s (Hooker and Salazar 1983). Although apparently relatively common at the time, the original isolates were lost and the virus was never found again in the same region. On the other hand, during the late 90s a novel virus coded SB-26/29 and transmitted by brown leafhoppers (*Russelliana solanicola*) was associated with a novel and rapidly spreading rugose stunting disease in Southern Peru. The disease caused yield reductions of 20–90% (Tenorio et al. 2003). The disease has now become rare, likely due to changes in cropping patterns that led to reduction in leafhopper populations. Partial sequence determination identified this virus as related to torradoviruses (CIP, unpublished), and recent surveys (Fuentes et al. 2019b; http://potpathodiv.org/) in Peru indicate it, and related viruses, can still be found with some frequency in potatoes. In Colombia, two new Pomoviruses related to PMTV were identified in soil samples from potato fields (Gil et al. 2016b). At least one of them (CPSbV) was shown to infect potatoes and transmitted through tubers, but the virus could only be detected in roots, and the plants were without symptoms.

Brazil

In Brazil, the potato is grown throughout the year, in three successive crops: rainy season, with harvest from December to March, with more than 50% of the total production; dry season, with harvest from April to August, representing about 30% of production, and winter season, with harvest from September to November, with lower production volume (IBGE 2017). The positive aspect of three harvesting seasons is a supply of fresh potatoes on the market throughout the year. However, this year-long field production also means a greater opportunity for uninterrupted spread of insect-transmitted viruses. This is a major factor why Brazil imports virus-tested seed-potato stocks from abroad annually, especially from Netherlands, Germany, France, Canada and Chile. This frequent importation has, historically, allowed the introduction of new pathogens into the country.

A first report of potato cultivation in Brazil dates from the 1920s in São Paulo, and since extended to other neighboring states of the South and Southeast, and later further expanded to the Northeast and Central-West regions. However, the largest production still occurs in the South and Southeast regions, with the State of Minas Gerais ranking as the first, with about 1.3 million tons/year (Agrianual 2016). Since the beginning of potato cultivation in Brazil, *Potato leafroll virus* (PLRV) had always been the most important viral agent associated with seed-potato tuber degeneration (Souza-Dias et al. 2013). The predominance of PLRV among seed-potato viruses in Brazil lasted until the mid-1990s, when two new strains of PVY were introduced, nearly simultaneously, through seeds imported from countries where their incidence was already known. After their introduction, this virus became a major cause of rapid seed-potato degeneration, overtaking the historical importance of PLRV as main cause for rejecting early field generations (G-1 or G-2), based on tolerance limits for viruses of the Federal Brazilian seed-potato tuber production-certification program.

As mentioned, PLRV has shown high incidences in Brazilian potato fields in the past. Until the mid 1990s it was normal for qualified and traditional seed-potato producers to face high incidences of PLRV (over 20%) in the very first (G-1) field multiplication of imported seed-potato stocks, which are officially considered, in Brazil, as G-0 (Souza-Dias et al. 2016a). This traditional and prevalent virus problem started to decline, coincidently in space and time with the introduction and fast outbreak of new PVY strains detailed below.

A possible reason for the shifting from PLRV to PVY was that seed-potato producers had over the years become more conscious regarding PLRV infected plants, learning to recognize its symptoms such as interveinal yellowing and rolled leaves, but did not recognize or understand the relevance of new symptoms such as mosaic, chlorosis and leaf deformation characteristic of PVY. As PVY expanded in association with the introduction of imported potato seeds, this virus soon became the major cause of seed-potato tuber degeneration (Barrocas et al. 2000). To counter this, a more intensive and efficient action toward controlling aphids, including the use of new insecticides such as neonicotinoids, took place. The vector transmission mechanism of PLRV (persistent circulatory, requiring significant time for acquisition and transmission) and PVY (nonpersistent, almost immediate transmission) would be more affected by the insecticidal effect, which could explain the noticeable reduction in the field incidence and spread of PLRV, whereas it is well know that insecticides have limited effect on nonpersistently transmitted viruses such as PVY. Nowadays, the detection of PLRV in seeds within official certification programs is extremely rare. In the State of Minas Gerais, as well as in most of the potato producing states, PLRV has not been associated with the rejection of seed lots since the beginning of the new millennium. However, imported seed potatoes are a big concern; Villela et al. (2017), testing national and imported potato seeds, found virus incidences as high as 10% in seed potatoes imported by Brazil.

The probable first report of PVY necrotic strain (PVY^N) in South America was in the beginning of 1940s (Nóbrega and Silberschmidt 1944). They studied some PVY isolates from Peru, inducing vein necrosis in leaves of tobacco. It took 20 years before a first official scientific report on PVY in Brazil was published in Sao Paulo State (Kitajima et al. 1962) where the authors describe the virus particle morphology and the histological symptoms. During the following years, although PVY was not considered a major problem for Brazilian potato producers, attempts to discover new methods to avoid PVY infection and to find new indicator plants were carried out in Brazil.

Surveying for PVY isolates in experimental fields, Andrade and Figueira (1992) detected five different strains, based on the reactions induced in the tobacco cultivar "Turkish NN." Although PVY^N was present, the PVY^O strain had a much higher incidence and was identified in almost all cultivars planted during 1980–1990. The investigation of PVY strains, done between 1983 and 1988, showed the presence of PVY^O in 80% of samples infected with PVY collected in experimental fields, planted with "Achat," "Baraka," "Baronesa," "Bintje," "Granola," and "Mona Lisa" cultivars. On the other hand, the incidence of PVY^N ranged from 0 to 12.4%. Even if present, PVY was never found in high incidences in fields of either ware or seed potatoes. Possibly the strains present in the country at that time were not as easily spread as the PVY strains introduced in Brazil in mid-1990s.

It is considered that the first significant introduction of a new PVY strain into Brazil came with seed potatoes of the cultivars "Achat" and "Baraka" imported from Germany, where it encountered optimal conditions for dissemination. This strain was later recognized as being the Wilga strain of PVY ($PVY\text{-}N^{wi}$; Galvino-Costa et al. 2012). All over the country, where seeds of these cultivars were used, large outbreaks of PVY were soon observed, reaching incidence rates above 70% in plants produced from infected seeds. A second introduction is suggested to have occurred through imported seed-potato of cv. Atlantic from Canada. It was later confirmed to be PVY^{NTN}, which causes necrotic rings in tubers of susceptible cultivars, such as "Monalisa". This strain also adapted itself very well to the Brazilian conditions, spreading rapidly to all potato growing regions (Galvino Costa et al. 2012). The introduction of these two strains brought a major problem to the potato growers who were used to plant cvs "Achat", "Baraka", and "Bintje", and easily recognized PLRV infected plants as showing, symptoms of leaf roll and yellowing of the lower leaves, but not chlorosis and mosaic that started appear as new PVY strains expanded. In 1995, in Minas Gerais, the seed-potato areas under certification were covered by more than 50% with cvs "Achat" and "Baraka," and a little less than 18% with "Bintje". Due to the increasing incidence of new PVY strains, these cultivars were abandoned, and only 3 years later, in 1998, little more than 15% of the potato seed production area was planted with those three potato cultivars. Conversely, where "Monalisa", a ware potato, used to occupy only around 5% of the certified area, it rose to about 29%. By 2000, "Monalisa" reached more than 50% of the production area, after which PVY^{NTN} began to spread rapidly, mainly associated to regions were "Atlantic" was grown, implicating it as the source of introduction of PVY^{NTN}. Due to a high susceptibility to PVY^{NTN} and sensitivity to the typical superficial tuber necrotic rings, "Monalisa" rapidly became unmarketable. Soon after in 2001, the Dutch cv "Agata" began to be tested and gained broad acceptance among the producers due to its high productivity, marketable phenotype, soil adaptability, and resistance to different PVY strains (Ramalho et al. 2012). The lack of expression of PVY^{NTN} symptoms in tubers of "Agata", in striking contrast to "Monalisa", contributed much to the fast replacement of "Monalisa" by "Agata". Thus, 4 years later, "Agata" was occupying over 30% of the area planted in Minas Gerais and nowadays is the main cultivar planted in Brazil (Silva et al. 2015).

Early characterization of PVY strains in Brazil was based on host symptoms, and the serological tests employed were DAS-ELISA which did not identify specific strains. The first serological tests using monoclonal antibodies and molecular studies started in the 1990s, confirming the presence of $PVY^{N\text{-}Wi}$ and PVY^{NTN} strains Brazil. Based on the reaction of PVY isolates to monoclonal antibodies, and on the symptoms shown by host plants, a great variability among them became evident (Galvino-Costa et al. 2012). More recent investigations (authors, unpublished data) have shown that there is a large abundance of PVY isolates with uncommon serology and that apparently N and O strains have disappeared from the Brazilian fields. Galvino-Costa et al. (2014) found that the incidence of $PVY^{N:O/N\text{-}Wi}$ was either equal to or greater than that of PVY^{NTN}, depending on the region in which the tubers have been produced and that mixed infections with both strains occur often, although this is sometimes only detectable with more sensitive techniques, such as RT-qPCR.

"Agata" has reached about 80% of the potato producing areas in Brazil and being a symptomless carrier of PVY, it acts as an efficient silent disseminator of the virus, particularly among "home saved" (not certified) seed-potato producers. This scenario has been a serious sanitary problem for potato production in Brazil, as it is strongly correlated with the successive increase of PVY reservoir, favoring spread of this virus into certified seed-potato fields.

Two other viruses that have been monitored in the field by the official certification programs are PVS and PVX. The incidence of these viruses in the field is sporadic and has never been associated with crop losses in Brazil. However, their monitoring is because some countries that export seed potatoes to Brazil have high incidences of these two viruses in the field (Souza-Dias et al. 2016a). If potato seeds with high PVS/PVX incidence reach Brazil, where high incidence of PVY usually occurs, the consequences could be disastrous. Recent surveys have shown incidences as high as 10% of PVX and 20% of PVS in imported potato seeds (Villela et al. 2017).

An increasing number of potato fields showing over 50% of typical PLRV-like symptoms, brought concerns about a possible PLRV outbreak, associated with whitefly instead of with aphids as virus vector. However, later on it was identified as ToCV (Souza-Dias et al. 2013; Lima 2016). The field symptoms of ToCV are characterized by internerval chlorosis and slight curling of the leaf edges, which begin in the apical leaves. ToCV can be transmitted by at least five species of whitefly (Orfanidou et al. 2016). In recent years an outbreak of the whitefly *Bemisia tabaci*, has been noticed in potato crops in Brazil (Moraes et al. 2017), favoring the spread of ToCV, probably from infected tomato plants in the vicinity. Despite of continued reports of whitefly infestation in Brazilian potato fields over the past 5 years, high occurrences of ToCV have been associated with tomato but not with potato crops (Orfanidou et al. 2016). Some outbreaks in potato have been reported in certain areas such as in the states of Goiás and São Paulo (Souza-Dias et al. 2013). However, at least for the time being, it does not appear to be a recurring problem for this crop. Thus, special care is being taken to monitor this disease in the field, as well as other whitefly-transmitted viruses, also reported in commercial Brazilian potato fields (see below).

The overlapping cropping cycles of tomato and potato, combined with favorable climatic conditions and the frequent proximity between the areas where they are planted, has caused other tomato viruses to occasionally migrate into potato crops. Two species of *Begomovirus* (family *Geminiviridae*) have been described in potato in Brazil: *Tomato yellow vein streak virus* (ToYVSV) and *Tomato severe rugose virus* (ToRSV). ToYVSV and ToRSV, the latter which has been prevailing in tomato crops, seems to be also prevalent in potato fields and both inducing similar deforming yellow mosaic symptoms (Souza-Dias et al. 2016a). The vector of both viruses is also the whitefly *Bemisia tabaci* (Pantoja et al. 2014). In contrast to what has been normal for tomato producing areas in Brazil, so far, there has not been any record of widespread begomovirus outbreaks in Brazilian potato producing areas. However, considering they are whitefly-transmitted viruses, and the high populations of whiteflies observed in potato crops in recent times (Moraes et al. 2017), careful monitoring of begomoviruses in potato should take place, as recommended for ToCV.

Tospoviruses, whose type member is *Tomato Spotted wilt virus* (TSWV), are transmitted by several species of thrips in a persistent manner. The viruses are acquired only at larvae stage, replicate in the insect vector and persist through the several stages of its life cycle (Rotenberg et al. 2015). They have always had a sporadic occurrence in potato crops in Brazil; but, in general, they are recorded as current season infection, and not as seed-tuber perpetuated virus. Therefore, they were never considered important as causing damage to this crop. However, from 2010 to 2015 there was a long period of drought in the Southeast of Brazil, with a significant increase in temperature, contributing to the increase of viral diseases in several crops, including potato, clearly associated to the same favorable conditions for increase in vector population. Field surveys by Souza-Dias (data not published) showed a high virus incidence of tospoviruses with a rare and isolated observation of virus perpetuation between 2010 and 2015. The more common species are *Tomato spotted wilt virus* (TSWV), *Groundnut ringspot virus* (GRSV) and *Tomato chlorotic spot virus* (TCSV) (Lima and Michereff 2016). Similar outbreaks have been described elsewhere, such as in Argentina (Salvalaggio et al. 2017) and United States of America (Abad et al. 2005).

As a norm, usually not all stems of a plant-hill show tospovirus symptoms: potato tubers are not only symptomless but also tospovirus-free, even when produced from infected plants. However, as a rare event for potato tospoviruses in Brazil, necrotic rings, both on the surface and penetrating the tuber flesh were observed in some of the tuber progeny of a tospovirus-infected plant (cv "Agata") (Souza-Dias et al. 2016b).

The plants that emerged from these tospovirus symptomatic tubers were however free of the virus. In other countries, there has been evidences of tospovirus species perpetuating via tubers produced by infected plants. These observations cause concern for seed-potato production (Abad et al. 2005; Salvalaggio et al. 2017). Therefore, a close monitoring of the incidence of tospoviruses in Brazilian potato fields is recommended in order to control not only its dissemination in the current season but also its perpetuation by tuber seed transmission.

India

PLRV, PVY, PVX, PVA, PVS, PVM, GBNV, and PAMV are known to occur in India. ToLCNDV-[potato], a begomovirus is reported to infect potato only in India. Mosaics and leafroll are the most common and severe symptoms in the subtropical and tropical climates of India. PLRV is important and occurs widely in almost all varieties. The mosaic causing viruses, PVY, PVA, and PVM as well as severe strains of PVX occur either singly and/or in different combinations. PVA and PVM are not common. PVY^N is almost not known in India, but recent study indicates the possible presence of PVY^N and PVY^{NTN}. ToLCNDV-[potato] has emerged as a serious threat to potato production during recent times. Its incidence is reported in almost all major potato growing states (Jeevalatha et al. 2017a). GBNV is reported in the early planted crop in the central and western parts of India (Jain et al. 2004). However, its occurrence in Pant nagar (Pundhir et al. 2012) and northwestern hills of India (Raigond et al. 2017) indicates the adaptation and spread of the virus to new areas. Recently, mixed infection of CMV with other potato viruses was reported (Sharma et al. 2016). It was found mostly in association with PVX, PVY^n followed by PVA, $PVY^{o/c}$, and PVM. Rarely, it was found associated with PAMV (Ghorai et al. 2017).

PVY is an important potato virus, which occurs widely in almost all the potato cultivars in India. Severe strains of PVY have the potential to reduce yield up to 80%. In India, PVY^O is most common and PVY^C strain has also been reported earlier based on the reactions on biological indicator host. Recently, based on host reactions, serology with monoclonal antibodies and complete genome sequence, the evidence of occurrence of a recombinant strain (N:O type) of PVY (isolate PVY-Del-66) was provided for the first time (Jailani et al. 2017). Isolate PVY-Del-66 shared closest sequences identity of 97.7–99.9% and a close phylogenetic relatedness with the N:O strains reported from USA and Germany. Del-66 isolate caused necrosis in tobacco and reacted positively with the MAb to common strain PVY^O but not with necrotic or chlorotic strains of PVY (Jailani et al. 2017). PCR analysis with strain-specific primers showed the possible presence of PVY^N and PVY^{NTN} in India and is being further confirmed by biological assays (CPRI, unpublished). PVA causes mild mosaic symptoms and not common in India. It reduces yield up to 30–40% and higher in combination with PVY or PVX.

PLRV is one of the most prevalent viral diseases of potato in India. All Indian potato varieties are susceptible to this virus. Yield loss normally ranges from 20 to

50% in India but in extreme cases may be as high as 50–80%, and infected plants produce only a few, small to medium tubers in severe secondary infections. At genome level, Indian isolates are closer to European and Canadian isolates than to an Australian isolate (Jeevalatha et al. 2013a).

PVX is one of the mosaic-causing viruses in almost all varieties of potato. In India, PVX infection may depress yield up to 10–30% and in the presence of PVA or PVY reduces yield up to 40% in potato. Indian PVX isolates were characterized for their biological properties, host range and transmission. Molecular analysis of complete and partial genomes of PVX found that all Indian isolates cluster in clade 1 with isolates from Europe and Asia, and none of them with clade II from south America (Jeevalatha et al. 2016c). Amino acid analysis suggested that these isolates cannot overcome *Rx1* gene or *Nx* gene mediated resistance (Jeevalatha et al. 2016c).

Like most parts of the world, PVS and PVM also infect potato in India. The Andean strain of PVS is reported in India (Garg and Hegde 2000). Complete genome sequence of one isolate of PVM, PVM-Del-144 has been sequenced. PVM isolates from northern plains showed considerable diversity in coat protein gene region (Jebasingh and Makeshkumar 2017).

Tomato leaf curl New Delhi virus-[potato] (ToLCNDV), a species of the genus begomovirus (family *Geminiviridae*) causes apical leaf curl disease of potato in India. Infection leads to severe seed degeneration particularly in susceptible varieties. Primary symptoms appear as curling/crinkling of apical leaves with distinct mosaic symptoms and in case of secondary infection, the entire plant shows severe leaf curling and stunting symptoms (Jeevalatha et al. 2013b; Sohrab et al. 2013). The association of a geminivirus with potato apical leaf curl disease was first reported in northern India and the virus was named tentatively as Potato apical leaf curl virus. However, later it was confirmed that this virus is a strain of

Fig. Symptoms of primary (**a**) and secondary (**b**) infection by ToLCNDV in potato. Photo credits: CIP

ToLCNDV. ToLCNDV-[potato] is a bipartite begomovirus with two genomic components referred as DNA-A and DNA-B. The DNA A components of the ToLCNDV-[potato] isolates shared more than 90.0% similarity to ToLCNDV isolates from vegetable crops such as tomato and okra, 89.0–90.0% to papaya isolates and 70.4–74.0% to other ToLCVs (Jeevalatha et al. 2017a).

Initially, sporadic incidence of the disease was reported in 1996 at Hisar in Haryana, later severe infections were observed in western Uthar Pradesh and other parts of northern India (Saha et al. 2014). It now occurs in almost all the major potato growing states in India and is reported in all cultivated varieties with varying severity levels (Jeevalatha et al. 2013b, 2017a). All the Indian potato varieties are susceptible to this disease except Kufri Bahar, which shows lowest seed degeneration and no/mild leaf curl symptoms even under favorable field (Kumar et al. 2015) and glass house conditions (Jeevalatha et al. 2017b). The virus is transmitted by whiteflies and the infection is more common in crops planted during October than in November because of the large whitefly population. Between 40 and 75% of incidence was recorded in the cultivars grown in Indo-Gangetic plains of India, up to 100% of incidence from the Hisar (Haryana) in susceptible varieties and recently, up to 40% incidence is reported from West Bengal (Saha et al. 2014). Infection results in significant decrease in size and number of tubers. Losses in marketable yield were reported to be as high as 50% in early planted susceptible cultivars. Currently, it is one of the most important viral diseases of potato in India. Repeated use of the same seed stock for 5 years led to 44.83–60.78% yield reduction in susceptible cultivar in Hisar and seed tubers of these cultivars cannot be reused profitably for more than 2 years. Since the virus spreads through seed tubers, it is critical to ensure quality of seed tubers through effective diagnostic tools. Diagnostic protocols like nucleic acid spot hybridization (NASH), polymerase chain reaction (Jeevalatha et al. 2013b), RCA-PCR (Jeevalatha et al. 2014), qPCR (Jeevalatha et al. 2016a) and LAMP assays (Jeevalatha et al. 2018) are available for the detection of ToLCNDV-[potato] in potato. PCR is being used to screen mother plants meant for tissue culture based seed production and also stage I plants in healthy potato seed production. So far, the infection of potato by ToLCNDV is known to occur only in India.

A tospovirus, GBNV causing severe stem/leaf necrosis disease in plains/plateaux of central/western India heavily infects the early crop of potato. It was first reported through morphological and serological studies by Jain et al. (2004). Stem necrosis incidence was recorded up to 90% in some parts of Madhya Pradesh and Rajasthan and up to 50% in Pant nagar. Its occurrence in northwestern hills of India despite of unfavorable conditions indicates possible adaption of the pathogen to new climatic conditions (Raigond et al. 2017).

In northwestern India leaf samples from potato plants with yellow mosaic or flecking symptoms showed positive reaction with PVX and CMV subgroup II in DAS-ELISA and the mixed infection with these viruses was further confirmed by PCR assay using specific primers and sequencing (Sharma et al. 2016). Ghorai et al. (2017) reported above 10% incidence of CMV in potato grown in Punjab. CMV infection in potato occurred mostly in association with PVX (60%), PVY^{n} (60%)

followed by PVA (40%), PVY$^{o/c}$ (30%), and PVM (30%). Rarely, it was found associated with PAMV (10%). Severe symptoms like malformation of leaves, blistering, stunting and reduced leaf size of potato were observed when CMV was present in potato in association with other potato viruses like PVX, PVY^{n}, PVY$^{o/c}$, PVA, PAMV, and PVM (Ghorai et al. 2017). Since the cropping pattern in Punjab corresponds to potato during October to February followed by cucurbits from February to May, the potato serves as an over wintering host of CMV when preferred host plants are not available and CMV is transmitted from potato to cucurbits through aphids (Ghorai et al. 2017).

Africa

Relatively little has been published regarding the viruses infecting potatoes in Africa and the few studies performed have focused on the globally common viruses using antisera. However the same viruses, most of them commonly found elsewhere in the world, have also been found throughout the continent where surveys have been performed. Thus, in Kenya Gildemacher and coworkers (2009) tested over 1000 tubers from 11 markets in seven districts for PLRV, PVY, PVA, and PVX and found average incidences of 71%, 57%, 75%, and 41%, respectively. Mixed infections were common and only 2.4% of tubers were free of any of these viruses. In Tanzania, Chiunga and Valkonen (2013) surveyed for the occurrence of the same viruses, but also PVM and PVS in plants from 16 fields in the south western highlands and found incidences of 55%, 39%, 14% and 5% for PVS, PLRV, PVX, and PVM whereas PVY and PVM were only detected in two locations. In a survey performed in South-West Uganda during 2014, PVX and PLRV were most frequent, followed by PVY and PVM, whereas PVA was not detected (CIP/IITA, unpublished). AlMV and *Beet curly top virus* (BCTV) were found to be locally frequent in Sudan around Kartoum (Baldo et al. 2010).

Although there have been no reports of whitefly-transmitted viruses, whiteflies can be abundant in potato crops in some locations during some seasons and because potatoes are often grown in close proximity to other vegetables there is a clear risk of transfer and possibly emergence of whitefly-transmitted viruses as has already been observed in India and Brazil.

Europe

There are some ten viruses infecting potatoes and causing significant yield losses in Europe . The main viral pathogens include PLRV, PVY, PVA,

PLRV, PVM, and PVS which are all transmitted by aphids. PVV is also aphid-transmitted but occurs only in a few cultivars. They also include PSTVd and PVX which are solely contact-transmitted, and PMTV and TRV both of which are soil-borne and mainly cause problems in the northern more cooler countries of the continent. PLRV was formerly the most important potato virus but has been on the decline for many decades. Conversely, PVY has become the most important, especially its new necrogenic strains which often cause mild foliar symptoms that are often difficult to see in field inspections and tend to be more efficiently aphid transmitted. Their importance resides in the necrotic symptoms they induce in tubers.

One of the viruses with potential to become a serious problem in potato production in Europe could be *Tomato spotted wilt virus* (TSWV). This virus has an unusually large host range of over 1000 plant species, including potato (Bulajic et al. 2014). TSWV is transmitted by thrips preferring climates warmer than those typical for potato growing areas in Europe, which may be a reason why damage caused by TSWV in potato crops has remained mostly modest or negligible and the yield losses affect mainly greenhouse production. However, climate change is predicted to increase temperatures in Europe (Lamichhane et al. 2015), which would provide more favorable conditions for thrips to thrive outdoors in more diversified living environments.

Geographical location and climate seem to create the conditions where different potato viruses get established and spread over the years. Therefore, while the potato viruses transmitted persistently by their vectors can spread over long distances, it is not self-evident that the virus gets established in the new area. For example, PVY is found in potato crops in all potato production areas of Europe, whereas PLRV is rare in northern Europe, i.e., in Finland and the northern parts of Sweden and Norway, despite the fact that both viruses are transmitted by aphids, and transmission of PLRV in the vector aphids continues much longer than PVY. Winged viruliferous aphids carrying PLRV cross the Baltic Sea during warm weather and suitable wind, and transfer PLRV from the potato fields in northern Germany and Poland to southern Finland. Nevertheless, PLRV has not become a significant pathogen in Finland and the infection rate of PLRV in potato crops has remained negligible. Why the abundance of potato viruses is different in different parts in Europe is not fully understood, but the climatic conditions are anticipated to play a role.

Another example of differences in geographical distribution of potato viruses in Europe is PMTV. It is common in potato crops in Scotland, Northern Ireland, all Nordic countries including Denmark, and in Czech Republic and Austria, but rare in the other countries at the southern side of Baltic Sea. It was only recently detected in Poland (Santala et al. 2010). The likely means for spread of PMTV over long distances are seed potatoes produced in an area where soils are contaminated with PMTV and its vector. However, PMTV can spread over long distances also in the resting spores of *S. subterranea* adhering to tools, equipment or vehicles, and in traded materials containing soil, e.g., ornamental plants. Taking the seed potato trade between the European countries to consideration, it seems that factors which are not well-known limit establishment of PMTV and/or development of the necrotic symptoms in tubers in many areas in Europe. Learning more about those factors might also help to design means for control of PMTV.

The advanced seed potato producers in Europe base production on multiplication of pathogen-free in vitro plants of potato cultivars. In practice, the propagation material is tested only for selected viruses considered to be the most harmful and included in the phytosanitary regulations. There are new methods to ensure that the plants are free of those viruses that are not among those routinely tested. First, cryotherapy is an efficient approach to ensure that the promising potato cultivars and breeding lines are virus-free and free from phytoplasma before they are introduced to long-term maintenance in vitro (Wang and Valkonen 2009). Secondly, all known plant viruses and also related, unknown viruses can be detected by analyzing the small RNAs generated by RNA silencing, the main antiviral defense mechanism in plants. The small RNAs (21–24 nucleotides) are extracted from plant tissue, sequenced and used to assemble longer sequences (contigs) using methods of bioinformatics. Viruses in the sample are identified by comparing the contigs with sequences available in databases (Kreuze 2014). This new method called small RNA sequencing and assembly (sRSA) is as sensitive for virus detection as the widely used PCR-based methods (Santala and Valkonen 2018). The advantage of sRSA is that it detects all types of plant viruses in the same assay without need for virus-specific primers or probes.

Australia

The viruses so far found infecting potato in the Australian continent are PLRV, PVY, PVA, PVS, PVM, PAMV, AMV, CMV, TSWV, and *Lucerne Australian latent virus* (LALV), and the viroid PSTVd has also been found (Buchen-Osmond et al. 1988). *Beet western yellows* virus was reported infecting potato in Tasmania but later shown to be confused with PLRV. Of the viruses infecting the potato crop, the most prevalent are PLRV, PVY, PVX, PVS, and TSWV, and these five viruses are the ones tested for routinely in Australian seed potato production schemes. For many years, PVM, PAMV, CMV, and LALV have not been recorded infecting Australian potato crops, but AlMV infection typified by bright yellow calico symptoms still occurs sporadically. Soil-borne viruses, such as TRV and PMTV, that cause problems in other world regions have not yet been recorded infecting potato in Australia, although the PMTV vector *Spongospora subterranea* and TRV vectors *Trichodorus* and *Paratrichodorus* are present.

The most important potato viruses in Australia are PLRV and PVY. The PVY recombinant PVY^{NTN} has been found infecting potato crops in four eastern Australian states (Queensland, New South Wales, Victoria, South Australia) where it is causing similar problems in seed potato production to the ones it causes in Europe. However, it has not, as yet, been found infecting potatoes in Tasmania or Western Australia (Kehoe and Jones 2016). PLRV remains the most prevalent and important potato virus in south-west Australia. PVX and PVS are common contaminants detected during Australian seed potato production but their incidence is now much lower than in the past when roguing was focused on PLRV rather than viruses causing mild foliar symptoms and no routine virus testing was done. TSWV is another common contaminant detected during seed potato production and tuber necrosis due to TSWV infection (Wilson 2001) continues to be found in ware potatoes in some states due to the common occurrence of this virus in weed hosts growing in or near to potato fields and its spread to potato plants by its thrips vector. In the last decade (2000–2010) relaxation of seed potato regulations concerning isolation from commercial potato crops in two Australian states (Victoria, Tasmania) led to a temporary upsurge in the incidence of common potato viruses in high grade seed potatoes. More thorough seed production regulations had to be reintroduced to counteract this situation.

Recent studies on potato viruses in Australia have focused mainly on PVX, PVS, PLRV and PVY. Nyalugwe et al. (2012) inoculated PVX isolates belonging to two strain groups to 38 cultivars grown in Australia to identify phenotypic responses and presence or absence of different PVX resistance genes. They also found that infection with PVX and PVS increased the titer of PVS and enhanced expression of foliar symptoms in potato plants. In a similar study with PVY, Jones and Vincent (2018) inoculated PVY^{O} and PVY^{D} to 39 cultivars to identify phenotypic responses and presence of absence of different PVY resistance genes. Coutts and Jones (2015) investigated PVY^{O}'s contact transmissibility, stability on surfaces, and inactivation with disinfectants. It was contact-transmitted to potato foliage but not to tubers, remained infective for up 24 h on some surfaces, and both bleach and the less caustic nonfat milk were useful PVY disinfectants.

When Cox and Jones (2010a) studied the CP nucleotide sequences of 13 PVS isolates from mainland Australia, all isolates were in phylogroup PVS^{O}. None of them invaded *C. quinoa* systemically so they were all in biological strain PVS^{O}. However, when Lambert et al. (2012) studied 42 PVS isolates from the Island of Tasmania, based on abiity to invade *C. quinoa* systemically three of them belonged to PVS^{A}, while the others belonged to PVS^{O}. When their CP genes were sequenced, they all belonged to phylogroup PVS^{O}. Santillan et al. (2018) included two complete PVS genomes from Australia in their evolutionary study on PVS, and both belonged to the main non-South American grouping, i.e. PVS^{O}. When Cox and Jones (2010b) studied the CP nucleotide sequences of 11 PVX isolates from Australia, all 11 belonged to the main non-South American grouping, i.e. phylogroup I. There was no relationship between biological strain and phylogroup as phylogroup I contained PVX isolates in biological strain groups 1, 3 and 4, whereas minor phylogroups II-1 and II-2 both contained isolates in strain groups 2 and 4. Kehoe and Jones (2014) compared the biological and genomic properties of eight historical European (1943–1984) and five Australian (2003–2012) PVY isolates from potato. Based on eliciting hypersensitivity genes *Nc*, *Ny*, or *Nz*, the European isolates belonged to biological strain groups PVY^{C}, PVY^{O} or PVY^{Z}, whereas the Australian isolates belonged to PVY^{O}, PVY^{Z} or new strain group PVY^{D} which elicited putative hypersensitivity gene *Nd*. The Australian and historical European isolates all fitted in phylogroups Y^{O} or Y^{C}. Moreover, biologically defined PVY^{O} and PVY^{Z} isolates were both within phylogroup Y^{O} while biologically defined Y^{C} and Y^{D} isolates were both phylogroup Y^C revealing disagreement between the current biological and phylogenetic PVY nomenclature systems.

Control of Potato Viruses

Potato is clonally propagated by planting tubers, which increases the risk of virus accumulation in the next crop and tuber generations. Apart from semi-persistently or persistently vector-transmitted viruses, such as PLRV, for which insecticide application as seed tuber dressings or foliar sprays are effective during seed potato production, such treatments are generally ineffective at controlling nonpersistently vector-borne viruses like PVY (Jones 2014). Thus, most potato viruses are controlled by three principal methods: host plant resistance, clean seed systems and cultural practices. Nowadays, in developed countries potato viruses are by and large controlled through formal certified clean seed production systems and to some extent through virus resistance. On the other hand, despite many years of intensive investment, formal certified seed systems have had only very limited, if any, penetration in many developing countries, where farmers mostly obtain their seed from their previous crop or through informal trade involving low-quality planting material. High cost of seed production, lack of adequate infrastructure and economic resources of small scale family farms are some of the reasons contributing to this situation.

In the past, simple seed potato schemes that, for example, relied solely on visual inspection and roguing combined with flooding and livestock to remove any tubers left behind after harvest proved effective at removing PLRV and other viruses causing obvious foliar symptoms, but ineffective at removing viruses causing mild symptoms e.g. PVS and PVX. Formal certified seed systems are expensive to implement in most developing countries as they require rigorous visual inspections and diagnostic testing. Relying solely on visual inspections is cheaper but leads to selection of viral strains that show few foliar symptoms, as occurred with some strains of PVY. Diagnostic testing often requires laboratories. While well-established method such as ELISA (Enzyme Linked Immunosorption Assay) are relatively cheap, they may lack sensitivity. Various PCR (Polymerase Chain Reaction), reverse transcription PCR and real-time PCR, protocols have been developed and multiplexed (e.g. Raigond et al. 2013; Meena et al. 2017; Jeevalatha et al. 2016a) which can provide ultrasensitive detection of viruses in samples. Field diagnostics with viruses is also possible using lateral flow devices that are commercialized by several companies globally but suffer from similar sensitivity issues as regular ELISA and are not available for all viruses. On the other hand, Loop Mediated Isothermal Amplification (LAMP) has recently emerged as a technology that can provide highly sensitive in field detection of potato viruses, with assays developed for PVY (Treder et al. 2018), PLRV (Ahmadi et al. 2013; Almasi et al. 2013), PVX (Jeong et al. 2015), PSTVd (Learcic et al. 2013) and ToLCNDV (Jeevalatha et al. 2018). LAMP assays can rapidly be designed to detect newly identified viruses and can be multiplexed, making it a flexible technology. LAMP is also compatible with crude nucleic acid extractions, can achieve high sensitivity, and be combined with the availability of relatively cheap battery powered real-time devices, such as Bioranger or real-time Genie series of devices, so may soon see more routine use in determining virus infections.

Potato Viruses and Seed Systems

Because of the previously mentioned factors, implementation of healthy seed systems in tropical countries is challenging, especially if there is a lack of cool areas or growing seasons with low aphid vector pressure available to reduce rate of reinfection during seed production. Due to this, as well as lack of appropriate infrastructure, investment, and commercial opportunity for smallholders to recover their investment in expensive seed potatoes, the amount of certified seeds used is generally negligible in most developing countries (Thomas-Sharma et al. 2016). Nevertheless, emerging economies, such as India and Brazil, have implemented seed systems with a level of success.

In **India**, a conventional seed tuber production system based on the "seed plot technique (SPT)" has successfully been used for the last five decades. Since its introduction, the SPT revolutionized the indigenous quality seed production system in the subtropical plains of India by extending it from the hills to the plains. The principle of SPT is growing the seed potato crop using healthy seed during a period with low aphid prevalence from October to the first week of January, coupled with IPM, roguing and dehaulming the seed crop during January before aphids reach critical threshold numbers. Today, 90% of seeds are being produced in northern (Punjab), north central (Gwalior), northwestern (Modipuram), and eastern plains (Patna) of the country. This seed is being supplied to the north east, Deccan plateau, and southern parts of the country which are not suitable for quality seed production. The seed production system in India includes tuber indexing for all major viruses and clonal multiplication of virus free mother tubers in four cycles for breeders seed production. The breeder seed produced by ICAR-CPRI is supplied to various State Government Organizations for further multiplication in three more cycles, viz. Foundation Seed 1 (FS-1), Foundation Seed 2 (FS-2) and Certified Seed (CS) under strict health standards. However, the current situation of breeder seed multiplication by the State Governments is not following the desired seed multiplication chain and breeder seed supplied by ICAR-CPRI is often being multiplied only up to FS-1 stage. Therefore, there is a shortage of certified seed in the country (ICAR-CPRI, Shimla). Incorporation of hi-tech seed production systems coupled with advanced virus detection techniques is the only way out in fulfilling the very large demand of quality seed potatoes in the country.

Continuous monitoring of aphid vector dynamics revealed that aphids cross critical limits 1 week earlier in Punjab and 1–2 weeks earlier in western UP in the recent past. In general terms, vector pressure has increased many folds as compared to the

1980s which is a cause of concern. Therefore, "SPT" is being refined to cope with the changing climate and vector pressure. There is also an urgent need to explore possibilities of seed production in nontraditional areas using modern techniques (Singh et al. 2014).

The major problem for potato production in **Brazil** is related to the low availability of virus-free seed tubers. As mentioned earlier, Brazil has three potato growing seasons per year in a climate where high population density of virus vectors occurs. Therefore, one of the most important measures that must be taken in Brazil is the use of virus-free seed tubers, because if there is a source of virus inoculum in the field, there will be a rapid spread during the potato production cycle. As a consequence, the tubers produced suffer rapid degeneration during their multiplication in the field. To address this problem, Brazil has adopted and periodically revised a seed certification system that establishes norms for seed-potato production in diverse categories: (1) G-0, which is the first generation derived from in-vitro plants, although imported basic classes can also be considered as G-0; (2) the basic and certified seeds, usually going up to G-4, but having virus incidence thresholds mandatorily respected. This is currently regulated under the Federal MAPA IN-32 (as of 20/11/2012, http://www.agricultura.gov.br/assuntos/insumos-agropecuarios/insumos-agricolas/sementes-e-mudas/publicacoes-sementes-e-mudas/INN32de20denovembrode2012.pdf). Producers have a laboratory support system accredited by INMETRO and certified by the Ministry of Agriculture for analysis and diagnosis of viruses in seed material. Currently a small part of the employed seed originates from tissue culture and another part from the production of minitubers in the greenhouse.

An effective low cost alternative system to produce high grade, virus-free, minituber/seed-potato lots involves the sprout/seed-potato technology (Virmond et al. 2017; Souza Dias et al. 2018). This may be accomplished either by the seed-potato exporting or importing country, and the research has shown that sprouts are durable, easy to handle and economic, when compared with potato tubers. The sprouts have to be detached from virus-free seed potatoes when they reach at least 3–5 cm tall (0.5–0.8 g), making sure that they will not be removed by cutting or sectioning, but just by hand-removing at the stolon, and with primordial root formation on the base . After removal, they can easily be packaged into sealed polyethylene bags (100 units) for transport. Storage, before or after export and upon arrival in the country has to be done in under proper environmental conditions, avoiding insect vectors, and having favorable conditions for sprout growth, such as a dark room, 20 °C and 70–80% RU. Upon arrival at the final destination, the sprouts can be directly planted under greenhouse conditions, using horticultural soil-substrate or in a hydro- or aeroponics system to produce minituber/seed-potatoes. Details about the methodology can be found in Virmond et al. (2017). The sprout/seed-potato technology is a novel method to increase the multiplication rate of high-grade (G-0, basic classes), national or imported seed-potato stocks. It has been applied by small and large-scale potato producers in Brazil to obtain minitubers/seed-potato, free of viruses and true-to-type. It has been officially accepted as of 20-11-2012 (MAPA IN 32) to use sprouts, originally from G-0 basic seed-potato lots (national laboratory-

Fig. Producing potato seed from tuber sprouts in Brazil. Photo credits: J. A. C. Souza-Dias

greenhouse minituber/seed-potato, or the annually imported tuber/seed-potato basic classes), as propagating material to produce certified minituber/seed-potato stocks. If properly handled, under the same conditions as used for minituber production, they can have the same phytosanitary health status. Moreover, minitubers/seed-potatoes obtained from sprouts through successive generations (G-3) are safer when compared to tissue culture, because they do not run the risk of presenting the common mutations seen with that technique (Souza-Dias et al. 2017).

Due to the difficulty of potato seed production in the country, producers often import seeds from European countries, Canada and the United States. This exposes the country to the entry of new pathogens, as happened with strains of PVY, that, once in Brazil, underwent a rapid adaptation and changed the epidemiology of the virus in Brazilian potato crops.

Another problem that has occurred with the production of seed potatoes in Brazil was the start in 2011 of the Brazilian legislation for production of potato seeds for personal use. This released the farmer from testing the seed planted, allowing him to plant them without laboratory analysis. Based on visual evaluation only, the producer is at risk of planting seeds with high incidences of virus, especially when dealing with cv. Agata. Besides compromising the yield, it can act as symptomless PVY carrier, especially of PVY^{NTN} as tuber necrotic rings are rarely visible in cv Agata, thus providing a generous source of virus inoculum for potato and other Solanaceous crops nearby.

Resistance

Potato is clonally propagated by planting tubers, which increases the risk of accumulation of viruses in the next crop and tuber generations. Apart from semi-persistent or persistently transmitted viruses such as PLRV, viruses cannot be controlled readily with pesticides, so chemical control of virus vectors provides only partial protection at best or is ineffective. On the other hand, production of healthy seed tubers is an expensive process. Therefore, resistance to viruses in potato cultivars is the most efficient and cost-effective means to control virus diseases in potato when effective seed production systems are absent, as in most developing countries. In developed countries with sophisticated and effective seed tuber production schemes, virus resistance becomes less important than other cultivar characteristics, such as high yield, tuber quality, and adaptation to the local environment.

PVY is now the most widespread viral pathogen in potatoes in most countries. Fortunately, breeders have introduced resistance genes that control PVY to many potato cultivars. Many of them, however, recognize only certain PVY strains. These strain-specific resistance genes can act quickly upon recognition of PVY and kill most of the PVY-infected cells at an early stage of infection leading to localized necrotic lesions, although they are sometimes slower acting resulting in systemic movement followed by single shoot or complete plant death. Therefore, they are called "hypersensitivity resistance" (HR) genes and contrast to "extreme resistance" (ER) genes which do not lead to any visible lesions during the resistance reaction. Furthermore, plasmodesmata connecting the plant cells and used by viruses for movement from cell to cell are sometimes blocked, preventing further spread of the virus. However, mutations in the viral genome can overcome resistance. For example, a few mutations in the helper component protein (HCpro) overcome resistance to PVY^O conferred by the resistance gene *Ny* in potato (Tian and Valkonen 2013). Consequently, the mutant of PVY^O can multiply, spread, and cause leaf mosaic and growth reduction in the potato plants carrying *Ny*.

Jones and Vincent (2018) studied strain-specific HR and ER phenotypes elicited in potato plants by PVY isolates in strain groups PVY^O and PVY^D. These isolates were inoculated to 39 Australasian, European, or North American potato cultivars. HR elicited by infection with strain group PVY^D occurred in 34 of the 39 cultivars, including 2 released as early as 1893–1894 in North America. Since PVY^D elicits putative gene *Nd*, this had apparently been present but unrecognized since the earliest epoch of potato breeding in the second half of the nineteenth century. Systemic hypersensitive resistance (SHR) elicited by strain group PVY^O in presence of hypersensitivity gene *Ny* was present in 23 of the same 34 potato cultivars with putative *Nd*, occurring widely amongst cultivars released in each of the three world regions. Two European cultivars always developed ER following sap and graft inoculation so carried comprehensive PVY resistance gene *Ry*, but no Australasian or North American cultivars carried it. One Australasian and two European cultivars always developed susceptible phenotypes so lacked genes *Ry*, *Ny*, and putative *Nd*. When breeding new PVY-resistant potato cultivars for countries lacking healthy seed potato stocks, or where subsistence farmers cannot afford them, the next best option to gene *Ry* inclusion is incorporating as many strain-specific PVY resistance genes as possible (Jones and Vincent 2018).

The degree of PVY^{O} resistance conferred by *Ny* varies between potato cultivars depending on the extent of localized hypersensitive resistance (LHR) and/or severe SHR versus weak SHR that develops. With LHR the source of infection for further virus spread is removed. When SHR involves death of all systemically infected shoots or entire plant death, foci of PVY infection are eliminated from within the crop so they are unavailable to become infection sources for secondary spread. By contrast, weak SHR that allows PVY-infected plants to persist means they can act as virus sources for secondary spread (Jones and Vincent 2018). A recent example of its effectiveness against PVY^{O} in the field was provided by an investigation in a potato growing region of North America following widespread planting of cultivars with *Ny*. Over a 5-year period, incidence of PVY^{O} dropped from 63 to 7% of the PVY population (Funke et al. 2017). Another example comes from potato cv. Yukon Gold, which is grown in Australia, Canada, Europe and USA, and carries genes *Ny*, *Nz*, putative *Nd* (Rowley et al. 2015; Kehoe and Jones 2016; Jones and Vincent 2018). The quick elimination of PVY-infected plants by the SHR response is beneficial, since it occurs before any tubers of useful size have developed. For example, in Finland it has been possible to grow Yukon Gold over 10 years from farm-owned seed.

Although the SHR response to PVY or PVX infection is a frequently observed phenotypic reaction in breeding populations (of e.g. CIP), breeders have traditionally ignored them as this reaction usually kills the plants or severely stunts them. As the above example of Yukon Gold exemplifies, this phenotype is nevertheless effective in controlling virus infection under field conditions. Growing cultivars with *Ny* and *Nz* is likely to be most helpful in potato growing regions where the recombinant PVY strains that overcome it are still rare or absent.

Nyalugwe et al. (2012) studied strain-specific HR and ER phenotypes elicited in potato plants by isolates in PVX strain groups 1 and 3. They inoculated these isolates to 38 potato cultivars. Presence of extreme PVX resistance gene *Rx* was identified in four Australian, two European cultivars, and one North American cultivar. PVX hypersensitivity gene *Nx* was identified two Australian, four European, and one North American cultivar. PVX hypersensitivity gene *Nb* was identified in one Australian, five European, and one North American cultivars. When breeding new PVX-resistant cultivars potato cultivars for developing countries, incorporation of gene *Rx* is the best option. However, Andean PVX resistance breaking strain XHB not only overcomes *Rx*, but also overcomes *Nx* and *Nb*, so *Rx* is likely to be less effective in potato cultivars growing in the center of origin of the crop.

Breeders have during the past decades focused on the use of ER genes, that are usually strain unspecific and cause no or only microscopic HR reactions in the plants. Thus, the ER genes Ry_{adg}, Ry_{sto}, and Ry_{chc} have been used to introduce resistance for PVY and Rx_1 and Rx_2 for resistance to PVX. To facilitate introgression of PVY resistance, molecular markers have been developed and used, e.g. Bhardwaj et al. (2015) screened potato germplasms and varieties employing SCAR and SSR marker linked to Ry_{adg} and Ry_{sto} genes and identified some elite parental lines that can be exploited for transferring the virus resistance into new potato cultivars. On the other hand, triplex parental potato lines containing three copies of the Ry_{adg} gene have been developed in various breeding programs ensuring 96% of progeny contain at least one copy of the resistance gene (Kaushik et al. 2013; Kneib et al. 2017). High resolution melting markers developed for Ry_{sto} (Nie et al. 2016), Ry_{adg} (Del Rosario et al. 2018) and *Rx1* and *2* (Nie et al. 2018) can accurately predict allele dosage and significantly aide in developing such parental lines. Nevertheless, even ER genes may be sensitive to changes in environment, as exemplified in recent study showing reduced efficiency of resistance to PVY by Ry_{chc} in response to increasing temperatures observed in Japan (Ohki et al. 2018). In contrast to PVY and PVX, a good source of resistance to PLRV has long evaded breeders, but a dominant gene Rl_{adg} conferring high levels of resistance was identified about a decade ago in a potato accession LOP-868 and the subsequently developed SCAR marker (Mihovilovich et al. 2014) has enabled rapid introgression into elite germplasm (Carneiro et al. 2017). Markers have also been developed for another dominant resistance gene, Rlr_{etb} originating from the non-tuber bearing wild species *S. etuberosum* (Kuhl et al. 2016), but its introgression into advanced breeding populations may still take time due to linkage drag from its wild progenitor. Screening of germplasm lines for ToLCNDV-[potato] resistance in field or glass house conditions showed possible presence of resistance source (Kumar et al. 2015; Maan et al. 2017; Jeevalatha et al. 2016b).

Additional resistance genes to PVA, PVV, PVS and PVM have also been identified (Palukaitis 2012) and mapped in potatoes but have to date not been widely utilized due to the considered limited importance of these viruses. Naderpour and Sadeghi (2018) developed a multiplex PCR assay including markers for resistance to PVY, PVS, and PLRV to facilitate introgression of multiple resistances into new varieties. Nevertheless, due to the complex genetics of potato it has not been easy to combine virus resistance with the myriad of other necessary traits needed for a successful variety. Transgenic approaches can readily incorporate resistance to multiple viruses into specific potato varieties (Chung et al. 2013), but considering current controversies surrounding transgenic crops, such products will not likely be released for cultivation in the near future.

Cultural Approaches

Landraces that survive for many years tend to be ones that possess multiple virus resistances as evidenced by the frequent occurrence of virus resistance genes in Andean potato landraces in germplasm collections (Jones 1981, and references therein). However, cultural approaches (such as roguing out plants with obvious virus symptoms, removing volunteer potato plants or weeds likely to harbor potato viruses, deploying reflective mulches to deter insect vector landings, manipulating the planting date to avoid peak flights of insect vectors, and early haulm destruction to avoid late virus infections) are rarely used by developing country farmers unless they are seed producers. In fact, the common habit of small holder farmers of selling and or consuming large tubers and keeping the small ones as seed for a next crop probably maintains virus loads in the seed high, as virus-infected plants often are the ones producing the smallest tubers. Gildemacher et al. (2011)and Schulte-Geldermann et al. (2012) showed how positive selection of healthy looking mother plants to provide seed tubers could reduce virus incidences in subsequent crops by 35–40% and a corresponding yield increase of 30%. Modeling approaches have similarly indicated that the approach of selecting healthy plants for seed production can be as effective as certified seed in maintaining seed quality (Thomas-Sharma et al. 2017). However, this would only apply where viruses causing little or no symptoms are being discounted. Also, the penetration of positive selection techniques among regular farmers has until now been limited.

The seed plot technique as practiced in India (whereas it starts out with certified virus free seed) is largely based on cultural practices to keep tuber seed healthy, growing during seasons and areas with low vector pressure coupled with IPM (Integrated Pest Management), rouging (negative selection), and dehaulming the seed crop before vectors reach a critical threshold limit. The use of straw mulch (Kirchner et al. 2014), mineral oil sprays and intercropping has been shown to enable control of PVY infection, particularly when used in combination (Dupuis et al. 2017a, b), although the economics of it would only justify their application for seed potato production (Dupuis 2017). Insecticide application to prevent PLRV spread in seed potato crops is also routinely used where seed potato stocks are multiplied in more aphid vector prone areas, especially in developed countries.

A unique practice is performed in the Andean region (and also the Himalayas) where farmers traditionally grow their potatoes at higher altitudes to reinvigorate them after several years of cultivation at lower altitudes (De Haan and Thiele 2003). Bertschinger et al. (2017) recently demonstrated that growing potatoes at higher altitude significantly reduced the number of virus-infected tubers produced from infected mother plants. Together with the absence of insect vector populations that could reinfect healthy plants at high altitude, this helps explain how this practice can reduce virus infections in the crop and resulting in subsequent higher yields. The mechanism of this phenomenon (reduced infection of tubers of infected mother plants at higher altitude), however, still remains unknown and should be an interesting topic of further study. Possibly RNA silencing mechanisms as affected by environmental conditions may be involved and understanding them may lead to new approaches for cleaning virus-infected plants.

Other mechanisms may also play a role in reducing losses by virus infections in the Andean region. Anecdotal evidence suggest that healthy planting material repatriated to farming communities often rapidly succumbs to new and more severe virus infections than the original material. An explanation for this may be that farmers have over the generations selected for plants which are infected with mild strains of the viruses, causing only limited yield losses, but protecting from more severe virus strains. This protection is lost when plants are cleaned from viruses.

Whereas this has not been researched until now, it may be worthwhile to investigate this phenomenon as it could lead to identification of new methods to control yield losses by viruses in potato using mild strains.

Final Remarks

Viruses remain a problem for global potato production, even though, over the years, the importance of certain viruses has increased or decreased globally. These changes in relative importance result from a range of factors including not only increased global trade but also regional changes in cultivar usage, cropping patterns, implemented seed systems and diagnostic testing regimes, appearance and evolution of new viruses and virus strains, and vector populations. All of these factors interact with each other and are further affected by climate change, making it difficult to predict what the future will hold. The two examples of potato-infecting geminivirus and torradovirus in Peru represent interesting cases where viruses have rapidly emerged as a significant local problem, only to subsequently disappear again into the background. This may be something that, in the past, has frequently occurred unnoticed in the potato's center of domestication in the Andes, or elsewhere in the world (e.g. PYDV was a devastating virus in the US during the early twentieth century, but has now all but disappeared). Many factors may influence whether a virus eventually manages to get a foothold in a region and become a permanent threat to the potato crop, but it has obviously happened on several occasions during the past 500 years since potato was introduced worldwide, as evidenced by viruses infecting potato uniquely in geographical areas outside of the Andes. The latter represent viruses that the potato did not encounter until it was moved away from the Andean region. Whitefly- and thrips-transmitted viruses should form a particular concern as witnessed by recurrent outbreaks occurring in (sub-)tropical regions, the increasing geographical spread of crinivirus PYVV and the establishment of the begomovirus ToYLCNDV as a major potato pathogen in India. Besides in Brazil, ToCV has also been reported from potatoes in Spain (Fortes and Navas-Castillo 2012) and identified in India associated with leaf-roll disease (CIP, unpublished), and thus the virus does seem to have recurring opportunities to infect potatoes worldwide where conditions are appropriate and it may only be a matter of time until an adaptive mutation appears for it to establish as a significant potato pathogen.

With a warming climate, producing high-quality virus-free seed is set to become more difficult, as opportunities to move to cooler areas with low vector pressure have become fewer in warmer countries, especially those that lack mountainous regions. Some cooler countries have the opportunity to move their seed potato production towards more extreme latitudes, though this rarely applies to developing countries. Thus, there is an increased requirement for new technologies for rapid multiplication of healthy plants under controlled conditions to be able to supply high-quality seeds at an affordable level. Despite the availability of effective resistance genes to, for example, PVY, due to the complex genetics of potatoes, it is difficult to recombine these with other critical traits necessary for a successful cultivar, and as a result only a fraction of potatoes grown globally possess non-strain-specific host resistance. The use of more efficient molecular markers and the promise of diploid potato breeding (Taylor 2018) may change that in the future, but for now, a combined approach for degeneration control adjusted to the local socio-economical and climatic context, as suggested by (Thomas-Sharma et al. 2016), may be the best way to go in developing countries where sophisticated seed production schemes are not currently a viable option.

References

Abad JA, Moyer JW, Kennedy GG, Holmes GA, Cubeta MA (2005) Tomato spotted wilt virus on potato in eastern North Carolina. Am J Potato Res 82(3):255

Abbas A, Madadi M (2016) A review paper on Potato Mop-Top Virus (PMTV): occurrence, properties and management. World J Biology and Biotechnology 1(3):129–134

AGRIANUAL (2016) Anuário da Agricultura Brasileira. FNP Consultoria, São Paulo

Ahmadi S, Almasi MA, Fatehi F, Struik PC, Moradi A (2013) Visual detection of *potato leafroll virus* by one-step reverse transcription loop-mediated isothermal amplification of DNA with hydroxynaphthol blue dye. J Phytopathol 161(2):120–124

Almasi MA, Manesh ME, Jafary H, Dehabadi SMH (2013) Visual detection of *Potato Leafroll* virus by loop-mediated isothermal amplification of DNA with the GeneFinder™ dye. J Virol Methods 192(1–2):51–54

Andrade ER, Figueira AR (1992) Incidência e sintomatologia de estirpes do vírus Y(PVY) nas regiões produtoras de batata do sul de Minas Gerais. Ciência e Prática 16:371–376

Baldo NH, Elhassan SM, Elballa MM (2010) Occurrence of viruses affecting potato crops in Khartoum State-Sudan. Potato Res 53(1):61–67

Barrocas EN, Figueira AR, Morais FR, Santos RC (2000) PVY and PLRV occurrence in lots of potato seeds coming from different regions of Minas Gerais State-Brazil. Fitopatol Bras 25(S):437

Bertschinger L, Bühler L, Dupuis B, Duffy B, Gessler C, Forbes GA, Keller ER, Scheidegger UC, Struik PC (2017) Incomplete infection of secondarily infected potato plants–an environment dependent underestimated mechanism in plant virology. Front Plant Sci 8:74

Bhardwaj V, Sharma R, Dalamu SAK, Baswaraj R, Singh R, Singh BP (2015) Molecular characterization of potato virus Y resistance in potato (*Solanum tuberosum* L.). Ind J Genet 75(3):389–392. https://doi.org/10.5958/0975-6906.2015.00062.0

Buchen-Osmond C, Crabtree K, Gibbs A, McLean G (1988) Viruses of plants in Australia: descriptions and lists from the VIDE database (No. 632.8 V821v). Australian National University, Canberra

Bulajic AR, Stankovic IM, Vucurovic AB, Ristic DT, Milojevic KN, Ivanovic MS, Krstic BB (2014) *Tomato spotted wilt virus* - potato cultivar susceptibility and tuber transmission. Am J Potato Res 91:186–194

Bertschinger L, Bühler L, Dupuis B, Duffy B, Gessler C, Forbes GA, Keller ER, Scheidegger UC, Struik PC (2017) Incomplete infection of secondarily infected potato plants – an environment dependent underestimated mechanism in plant virology. Front Plant Sci (8):74. https://doi.org/10.3389/fpls.2017.00074

Carneiro OL, Ribeiro SR, Moreira CM, Guedes ML, Lyra DH, Pinto CA (2017) Introgression of the Rl adg allele of resistance to potato leafroll virus in Solanum tuberosum L. Crop Breed Appl Biotechnol 17(3):242–249

Chiunga E, Valkonen JPT (2013) First report on viruses infecting potato in Tanzania. Plant Dis 97:1260

Chikh-Ali M, Rowley JS, Kuhl J, Gray S, Karasev AV (2014) Evidence of a monogenic nature of the Nz gene against *Potato virus Y* strain Z (PVYZ) in potato. Am J Potato Res 91:649–654

Chung BN, Yoon JY, Palukaitis P (2013) Engineered resistance in potato against potato leafroll virus, potato virus A and potato virus Y. Virus Genes 47(1):86–92

Constable F, Chambers G, Penrose L, Daly A, Mackie J, Davis K, Rodoni B, Gibbs M (2019) Viroid-infected Tomato and Capsicum Seed Shipments to Australia. Viruses 11(2):98

Cox BA, Jones RAC (2010a) Genetic variability of the coat protein gene of potato virus S and distinguishing its biologically distinct strains. Arch Virol 155:1163–1169

Cox BA, Jones RAC (2010b) Genetic variability of the coat protein gene of potato virus X, and the current relationship between phylogenetic placement and resistance groupings. Arch Virol 155:1349–1356

Coutts BA, Jones RAC (2015) Potato virus Y: contact transmission, stability on surfaces, and inactivation with disinfectants. Plant Dis 99:387–394

De Haan S, Thiele G (2003) In situ conservation and potato seed systems in the Andes. In: Devra I, Sevilla-Panizo JR, Chávez-Servia JL, Hodgkin T (eds) Seed systems and crop genetic diversity on-farm. The International Plant Genetic Resources Institute (IPGRI), Pucallpa, p 126

De Souza J, Gamarra H, Muller G, Kreuze J (2018) First report of potato virus S naturally infecting Arracacha (Arracacia xanthorrhiza) in Peru. Plant Dis 102:1. https://doi.org/10.1094/PDIS-07-17-0945-PDN

De Souza J, Muller G, Perez W, Cuellar W, Kreuze JF (2017) Complete sequence and variability of a new subgroup B nepovirus infecting potato in central Peru. Arch Virol 62:885–889. https://doi.org/10.1007/s00705-016-3147-6

Del Rosario HM, Vidalon LJ, Montenegro JD, Riccio C, Guzman F, Bartolini I, Ghislain M (2018) Molecular and genetic characterization of the Ry adg locus on chromosome XI from Andigena potatoes conferring extreme resistance to potato virus Y. Theor Appl Genet 131(9):1925–1938

Devaux A, Kromann P, Ortiz O (2014) Potatoes for sustainable global food security. Potato Res 57:185–199

Dupuis B (2017) Development of a crop management method to control the spread of Potato Virus Y (PVY). Doctoral dissertation, UCL-Université Catholique de Louvain

Dupuis B, Bragard C, Carnegie S, Kerr J, Glais L, Singh M, Nolte P, Rolot JL, Demeulemeester K, Lacomme C (2017a) Potato virus Y: control, management and seed certification programmes, Potato virus Y: biodiversity, pathogenicity, epidemiology and management. Springer, Cham, pp 177–206

Dupuis B, Cadby J, Goy G, Tallant M, Derron J, Schwaerzel R, Steinger T (2017b) Control of potato virus Y (PVY) in seed potatoes by oil spraying, straw mulching and intercropping. Plant Pathol 66(6):960–969

Fortes IM, Navas-Castillo J (2012) Potato, an experimental and natural host of the crinivirus Tomato chlorosis virus. Eur J Plant Pathol 134:81–86

Franco-Lara L, Rodríguez D, Guzmán-Barney M (2013) Prevalence of *Potato yellow vein virus* (PYVV) in *Solanum tuberosum* group Phureja fields in three states of Colombia. Am J Potato Res 90:324–330. https://doi.org/10.1007/s12230-013-9308-1

Fribourg CE, Gibbs AJ, Adams IP, Boonham N, Jones RAC (2019) Biological and molecular properties of isolates from pepino. Plant Dis 103(7):1746–1756

Funke CN, Nikolaeva OV, Green KJ, Tran LT, Chikh-Ali M, Quintero-Ferrer A, Cating RA, Frost KE, Hamm PB, Olsen N, Pavek MJ, Gray SM, Crosslin JM, Karasev AV (2017) Strain-specific resistance to Potato virus Y (PVY) in potato and its effect on the relative abundance of PVY strains in commercial potato fields. Plant Dis 101:20–28

Fuentes S, Jones RAC, Matsuoka H, Ohshima K, Kreuze JF, Gibbs A (2019a) Potato virus Y; the Andean connection. Virus Evolution, in press https://doi.org/10.13140/RG.2.2.15998.33602

Fuentes S, Perez A, Kreuze JF (2019b) Dataset for: The peruvian potato virome. International Potato Center, V1. https://doi.org/10.21223/P3/YFHLQU

Galvino-Costa SBF, Figueira AR, Rabelo-Filho FAC, Moraes FHR, Nikolaeva OV, Karasev AV (2012) Molecular and serological typing of potato virus Y Isolates from Brazil reveals a diverse set of recombinant strains. Plant Dis 96:1451–1458. https://doi.org/10.1094/PDIS-02-12-0163-RE

Galvino-Costa SBF, Santos BA, Figueira AR, Geraldino-Duarte PS (2014) Detection of potato virus Y (PVY) strains in plants with single and mixed infection. Virus Rev Res 19:215–216

Garg ID, Hegde V (2000) Biological characterization, preservation and ultrastructural studies of Andean strain of Potato virus S. Ind Phytopathol 53:256–260

Ghorai AK, Kang SS, Sharma A, Sharma S (2017) Occurrence of *Cucumber mosaic virus* on potato and its transmission to muskmelon under potato-cucurbit cropping pattern followed in Punjab, India. Int J Curr Microbiol App Sci 6(11):2947–2965

Gibbs AJ, Ohshima K, Yasaka R, Mohammad M, Gibbs MJ, Jones RAC (2017) The phylogenetics of the global population of potato virus Y and its necrogenic recombinants. Virus Evol 3(1):vex002. https://doi.org/10.1093/ve/vex002

Gil JF, Adams I, Boonham N, Nielsen SL, Nicolaisen M (2016a) Molecular and biological characterisation of Potato mop-top virus (PMTV, Pomovirus) isolates from potato growing regions in Colombia. Plant Pathol 65:1210–1220

Gil JF, Adams I, Boonham N, Nielsen SL, Nicolaisen M (2016b) Molecular and biological characterisation of two novel pomo-like viruses associated with potato (*Solanum tuberosum*) fields in Colombia. Arch Virol 161(6):1601–1610

Gildemacher P, Demo P, Barker I, Kaguongo W, Woldegiorgis G, Wagoire W, Wakahiu M, Leeuwis C, Struik P (2009) A description of seed potato systems in Kenya, Uganda and Ethiopia. Am J Potato Res 86:373–382

Gildemacher PR, Schulte-Geldermann E, Borus D, Demo P, Kinyae P, Mundia P, Struik PC (2011) Seed potato quality improvement through positive selection by smallholder farmers in Kenya. Potato Res 54(3):253

Glais L, Bellstedt DU, Lacomme C (2017) Diversity, characterisation and classification of PVY, Potato virus Y: biodiversity, pathogenicity, epidemiology and management. Springer, Cham, pp 43–76

Gray S, De Boer S, Lorenzen J, Karasev A, Whitworth J, Nolte P, Singh R, Boucher A, Xu H (2010) Potato virus Y: an evolving concern for potato crops in the United States and Canada. Plant Dis 94(12):1384–1397

Gutierrez PA, Alzate JF, Marín-Montoya MA (2013) Complete genome sequence of a novel potato virus S strain infecting *Solanum phureja* in Colombia. Arch Virol 158:2205–2208. https://doi.org/10.1007/s00705-013-1730-7

Harahagazwe D, Condor B, Barreda C et al (2018) How big is the potato (Solanum tuberosum L.) yield gap in Sub-Saharan Africa and why? A participatory approach. Open Agriculture 3(1):180–189. https://doi.org/10.1515/opag-2018-0019

Henao-Díaz E, Gutiérrez-Sánchez P, Marín-Montoya M (2013) Phylogenetic analysis of potato virus y (PVY) isolates obtained in potato (*Solanum tuberosum*) and tamarillo (*Solanum betaceum*) crops from Colombia. Actual Biol 35(99):219–232

Hooker WJ, Salazar LF (1983) A new plant virus from the high jungle of the eastern Andes; Solanum apical leaf curling virus (SALCV). Ann Appl Biol 103:449–454

IBGE (Instituto Brasileiro de Geografia e Estatística) (2017) Levantamento Sistemático de Produção Agrícola. Grupo de Coordenação Estatística Agropecuárias/GCEA/DPE/COAGRO, Disponibleat. http:www.ibge.gov.br. Accessed 10 April 2018

Jailani AAK, Shilpi S, Mandal B (2017) Rapid demonstration of infectivity of a hybrid strain of potato virus Y occurring in India through overlapping extension PCR. Physiol Mol Plant Pathol 98:62–68

Jain RK, Khurana SMP, Bhat AI, Chaudhary V (2004) Nucleocapsid protein gene sequence studies confirm that potato stem necrosis is caused by a strain of groundnut bud necrosis virus. Ind Phytopath 57:169–173

Jebasingh T, Makeshkumar T (2017) Characterisation of carlaviruses occurring in India. In: Mandal B, Rao G, Baranwal V, Jain R (eds) A century of plant virology in India. Springer, Singapore

Jeevalatha A, Chakrabarti SK, Sharma S, Sagar V, Malik K, Raigond B, Singh BP (2017a) Diversity analysis of *Tomato leaf curl* New Delhi virus–[potato], causing apical leaf curl disease of potato in India. Phytoparasitica 45:33–43. https://doi.org/10.1007/s12600-017-0563-4

Jeevalatha A, Kaundal P, Kumar A, Guleria A, Sundaresha S, Pant RP, Sridhar J, Venkateswarlu V, Singh BP (2016a) SYBR green based duplex RT-qPCR detection of a begomovirus, *Tomato leaf curl* New Delhi virus-[potato] along with Potato virus X and Potato leaf roll virus in potato. Potato J 43:125–137

Jeevalatha A, Kaundal P, Kumar R, Raigond B, Kumar R, Sharma S, Chakrabarti SK (2018) Optimized loop-mediated isothermal amplification assay for *Tomato leaf curl* New Delhi virus-[potato] detection in potato leaves and tubers. Eur J Plant Pathol 150(3):565–573

Jeevalatha A, Kaundal P, Kumar R, Raigond B, Gupta M, Kumar A, Sharma S, Sagar V, Nagesh M, Singh BP (2016c) Analysis of the coat protein gene of Indian Potato virus X isolates for identification of strain groups and determination of the complete genome sequence of two isolates. Eur J Plant Pathol 145:447–458

Jeevalatha A, Kaundal P, Shandil RK, Sharma NN, Chakrabarti SK, Singh BP (2013a) Complete genome sequence of Potato leafroll virus isolates infecting potato in the different geographical areas of India shows low level genetic diversity. Ind J Virol 24(2):199–204. https://doi.org/10.1007/s13337-013-0138-z

Jeevalatha A, Kaundal P, Venkatasalam EP, Chakrabarti SK, Singh BP (2013b) Uniplex and duplex PCR detection of geminivirus associated with potato apical leaf curl disease in India. J Virol Methods 193:62–67

Jeevalatha A, Singh BP, Kaundal P, Kumar R, Raigond B (2014) RCA-PCR: a robust technique for the detection of *Tomato leaf curl* New Delhi virus-[potato] at ultra-low virus titre. Potato J 41:76–80

Jeevalatha A, Sundaresha S, Kumar A, Kaundal P, Guleria A, Sharma S, Nagesh M, Singh BP (2017b) An insight into differentially regulated genes in resistant and susceptible genotypes of potato in response to tomato leaf curl New Delhi virus-[potato] infection. Virus Res 232:22–33

Jeevalatha A, Vanishree G, Bhardwaj V, Singh R, Nagesh M, Singh BP (2016b) Screening of potato germplasms for apical leaf curl disease resistance. CPRI Newslett 66:1

Jeong J, Cho SY, Lee WH, Lee KJ, Ju HJ (2015) Development of a rapid detection method for Potato virus X by reverse transcription loop-mediated isothermal amplification. Plant Pathol J 31(3):219

Jones RAC (1981) The ecology of viruses infecting wild and cultivated potatoes in the Andean Region of South America. In: Thresh JM (ed) Pests pathogens and vegetation. Pitman, London, pp 89–107

Jones RAC (2014) Virus disease problems facing potato industries worldwide: viruses found, climate change implications, rationalizing virus strain nomenclature and addressing the Potato virus Y issue. The potato: botany, production and uses. CABI, Wallingford, pp 202–224

Jones RAC (2016) Future scenarios for plant virus pathogens as climate change progresses. Adv Virus Res 95:57–147

Jones RAC, Kehoe MA (2016) A proposal to rationalize within-species plant virus nomenclature: benefits and implications of inaction. Arch Virol 161:2051–2057

Jones RA, Vincent SJ (2018) Strain-specific hypersensitive and extreme resistance phenotypes elicited by potato virus Y among 39 potato cultivars released in three world regions over a 117-year period. Plant Dis 102(1):185–196

Kalyandurg P, Gil JF, Lukhovitskaya NI, Flores B, Müller G, Chuquillanqui C, Palomino L, Monjane A, Barker I, Kreuze J, Savenkov E (2017) Molecular and pathobiological characterization of 61 Potato mop-top virus full-length cDNAs reveals great variability of the virus in the centre of potato domestication, novel genotypes and evidence for recombination. Mol Plant Pathol. https://doi.org/10.1111/mpp.12552

Karasev AV, Gray SM (2013) Continuous and emerging challenges of Potato virus Y in potato. Annu Rev Phytopathol 51:571–586

Kaushik SK, Sharma R, Garg ID, Singh BP, Chakrabarti SK, Bhardwaj V, Pandey SK (2013) Development of a triplex (YYYy) parental potato line with extreme resistance to potato virus Y (PVY) using marker assisted selection (MAS). J Hortic Sci Biotechnol 88:580–584

Kehoe MA, Jones RAC (2016) Improving Potato virus Y strain nomenclature: lessons from comparing isolates obtained over a 73-year period. Plant Pathol 65:322–333

Kirchner SM, Hiltunen LH, Santala J et al (2014) Comparison of straw mulch, insecticides, mineral oil, and birch extract for control of transmission of potato virus Y in seed potato crops. Potato Res 57:59–75

Kitajima EW, Carvalho AMB, Costa AS (1962) Microscopia eletrônica de estirpes do vírus Y da batatinha que ocorrem em São Paulo. Bragantia 21:755–763

Kneib RB, Kneib RB, Pereira ADS, Castro CM (2017) Allele dosage of PVY resistance genes in potato clones using molecular markers. Crop Breed Appl Biotechnol 17(4):306–312

Koenig R, Ziebell H (2013) Sequence-modified primers for the differential RT-PCR detection of Andean potato latent and Andean potato mild mosaic viruses in quarantine tests. Arch Virol. https://doi.org/10.1007/s00705-013-1859-4

Kreuze J (2014) siRNA deep sequencing and assembly: piecing together viral infections, Detection and diagnostics of plant pathogens. Springer, Dordrecht, pp 21–38

Kreuze J, Koenig R, De Souza J, Vetten HJ, Müller G, Flores B, Ziebell H, Cuellar W (2013) The complete genome sequences of a Peruvian and a Colombian isolate of Andean potato latent virus and partial sequences of further isolates suggest the existence of two distinct potato-infecting tymovirus species. Virus Res 173:431–435

Kuhl JC, Novy RG, Whitworth JL, Dibble S, Schneider B, Hall D (2016) Development of molecular markers closely linked to the Potato leafroll virus resistance gene, Rlretb, for use in marker-assisted selection. Am J Potato Res 93(3):203–212

Kumar M, Gupta A, Singh J, Singh F (2015) Screening of germplasm lines against Potato apical leaf curl virus disease in potato crop. Int J Trop Agric 33(2):1283–1285

Kutnjak D, Silvestre R, Cuellar W, Perez W, Müller G, Ravnikar M, Kreuze JF (2014) Complete genome sequences of new divergent potato virus X isolates and discrimination between strains in a mixed infection using small RNAs sequencing approach. Virus Res 191:45

Lacomme C, Glais L, Bellstedt D, Dupuis B, Karasev A, Jacquot E (eds) (2017) Potato virus Y: biodiversity, pathogenicity, epidemiology and management. Springer, Cham

Lambert SJ, Scott JB, Pethybridge SJ, Hay FS (2012) Strain characterization of Potato virus S isolates from Tasmania, Australia. Plant Dis 96(6):813–819

Lamichhane JR, Barzman M, Booij K, Boonekamp P, Desneux N, Huber L, Kudsk P, Langrell SRH, Ratnadass A, Ricci P, Sarah JL, Messean A (2015) Robust cropping systems to tackle pests under climate change. A review. Agron Sustain Dev 35:443–459

Learcic R, Morisset D, Mehle N, Ravnikar M (2013) Fast real-time detection of Potato spindle tuber viroid by RT-LAMP. Plant Pathol 62(5):1147–1156

Lima MF (2016) Crinivírus e geminivírus transmitidos por mosca branca: ameaça à produção nacional de batata. Batata Show 44:8–10

Lima MF, Michereff FM (2016) Vira-cabeça: ameaça à bataticultura no Brasil. Batata Show 46:22–25

Lin YH, Johnson DA, Pappu HR (2014) Effect of Potato virus S infection on late blight resistance in potato. Am J Potato Res 91(6):642–648

Lizárraga C, Querci M, Santa-Cruz M, Bartolini I, Salazar LF (2000) Other natural hosts of potato virus T. Plant Dis 84:736–738

Maan DS, Bhatia AK, Rathee M (2017) Screening and evaluation of potato (*Solanum tuberosum*) genotypes to identify the sources of resistance to Potato apical leaf-Curl disease. Int J Pure App Biosci 5(3):53–61. https://doi.org/10.18782/2320-7051.2658

Mackie A, Barbetti MJ, Rodoni B, McKirdy S, Anthony R, Jones C (2019) Effects of a tomato strain on the symptoms, biomass and yields of classical indicator and currently grown potato and tomato cultivars. Plant Disease 3009–3017

Meena PN, Kumar R, Baswaraj R, Jeevalatha A (2017) Simultaneous detection of potato viruses A and M using CP gene specific primers in an optimized duplex RT-PCR. J Pharmacogn Phytochem 6(4):1635–1640

Mihovilovich E, Aponte M, Lindqvist-Kreuze H, Bonierbale M (2014) An RGA-derived SCAR marker linked to PLRV resistance from Solanum tuberosum ssp. andigena. Plant Mol Biol Report 32(1):117–128

Moraes LA, Marchi BR, Bello VH, Watanabe LFM, Yuki VA, Marubayashi JM, Pavan MA, Krause-Sakate R (2017) Uma nova ameaça no Brasil: bemisiatabaci espécie mediterranean (Biótipo Q). Batata show 47:20–21

Naderpour M, Sadeghi L (2018) Multiple DNA markers for evaluation of resistance against potato virus Y, potato virus S and potato leafroll virus. Czech J Genet Plant Breed 54(1):30–33

Nie X, Dickison VL, Brooks S, Nie B, Singh M, De Koeyer DL, Murphy AM (2018) High resolution DNA melting assays for detection of Rx1 and Rx2 for high-throughput marker-assisted selection for extreme resistance to potato virus X in tetraploid potato. Plant Dis 102(2):382–390

Nie X, Sutherland D, Dickison V, Singh M, Murphy A. M, De Koeyer D (2016) Development and validation of high-resolution melting markers derived from Ry sto STS markers for high-throughput marker-assisted selection of potato carrying Ry sto. Phytopathology, 106(11), 1366–1375

Nóbrega NR, Silberschmidt K (1944) Sobre uma provável variante do vírus "Y" da batatinha (Solanum vírus 2, Orton) que tem peculiaridade de provocar necroses em plantas de fumo. Arq Inst Biol 15:307–330

Nyalugwe EP, Wilson CR, Coutts BA, Jones RAC (2012) Biological properties of potato virus X in potato: effects of mixed infection with Potato virus S and resistance phenotypes in cultivars from three continents. Plant Dis 96:43–54

Ohki T, Sano M, Asano K, Nakayama T, Maoka T (2018) Effect of temperature on resistance to potato virus Y in potato cultivars carrying the resistance gene Rychc. Plant Pathol 67:1629–1635. https://doi.org/10.1111/ppa.12862

Orfanidou CG, Pappi PG, Efthimiou KE, Katis NI, Maliogka VI (2016) Transmission of Tomato chlorosis virus (ToCV) by Bemisia tabaci biotype Q and evaluation of four weed species as viral sources. Plant Dis 10:2043–2049

Palukaitis P (2012) Resistance to viruses of potato and their vectors. Plant Pathol J 28(3):248–258

Pantoja KFC, Rocha KCG, ELL B, Pavan MA, Krause-Sakate R (2014) Evaluation of the primary and secondary dispersal of tomato severe rugose virus to Capsicum spp. genotypes by *Bemisia tabaci* MEAM1. Summa Phytopathol 40:375–377. https://doi.org/10.1590/0100-5405/2001

Pérez Barrera W, Valverde Miraval M, Barreto Bravo M, Andrade-Piedra J, Forbes GA (2015) Pests and diseases affecting potato landraces and bred varieties grown in Peru under indigenous farming system. Rev Latinoam Papa 19:29–41

Petek M, Rotter A, Kogovšek P, Baebler Š, Mithöfer A, Gruden K (2014) Potato virus Y infection hinders potato defence response and renders plants more vulnerable to Colorado potato beetle attack. Mol Ecol 23(21):5378–5391

Pundhir VS, Akram M, Ansar M, Rajshekhara H (2012) Occurence of stem necrosis disease in potato caused by groundnut bud Necrosis virus in Uttarakhand. Potato J 39:81–83

Qiu CL, Zhang ZX, Li SF, Bai YJ, Liu SW, Fan GQ, Gao YL, Zhang W, Zhang S, Lu WH, Lü DQ (2016) Occurrence and molecular characterization of Potato spindle tuber viroid (PSTVd) isolates from potato plants in North China. J Integr Agric 15(2):349–363

Raigond B, Sharma M, Chauhan Y, Jeevalatha A, Singh BP, Sharma S (2013) Optimization of duplex RT-PCR for simultaneous detection of Potato virus Y and S. Potato J 40(1):22–28

Raigond B, Sharma P, Kochhar T, Roach S, Verma A, Jeevalatha A, Verma G, Sharma S, Chakrabarti SK (2017) Occurrence of groundnut bud necrosis virus on potato in north western hills of India. Ind Phytopathol. https://doi.org/10.24838/ip.2017.v70.i4.76993

Ramalho TO, Galvino-Costa SBF, Figueira AR (2012) Predominance of potato cultivar Agata in Brazil and its effect in the dissemination and variability of Potato virus Y. Phytopathology 102(S):97–97

Rowley JS, Gray SM, Karasev AV (2015) Screening potato cultivars for new sources of resistance to Potato virus Y. Am J Potato Res 92:38–48

Rotenberg D, Jacobson AL, Schneweis DJ, Whitfield AE (2015) Thrips transmission of tospoviruses. Curr Opin Virol 15:80–89. https://doi.org/10.1016/j.coviro.2015.08.003

Saha A, Saha B, Saha D (2014) Molecular detection and partial characterization of a begomovirus causing leaf curl disease of potato in sub-Himalayan West Bengal, India. J Environ Biol 35:601–606

Sahi G, Hedley PE, Morris J, Loake GJ, MacFarlane SA (2016) Molecular and biochemical examination of spraing disease in potato tuber in response to Tobacco rattle virus infection. Mol Plant-Microbe Interact 29(10):822–828

Salazar L, Muller G, Querci M, Zapata J, Owens R (2000) *Potato yellow vein virus*: its host range, distribution in South America and identification as a Crinivirus transmitted by Trialeurodes vaporariorum. Ann Appl Biol 137(1):7–19

Salvalaggio AE, Lambertini PML, Cendoya G, Huarte MA (2017) Temporal and spatial dynamics of Tomato spotted wilt virus and its vector in a potato crop in Argentina. Ann. Appl. Biol. 171:5–14. https://doi.org/10.1111/aab.12357

Santala J, Valkonen JPT (2018) Sensitivity of small RNA–based detection of plant viruses. Front Microbiol 9:939. https://doi.org/10.3389/fmicb.2018.00939

Santala J, Samuilova O, Hannukkala A, Latvala S, Kortemaa H, Beuch U, Kvarnheden A, Persson P, Topp K, Ørstad K, Spetz C, Nielsen SL, Kirk HG, Uth JG, Budziszewska M, Wieczorek P, Obrepalska-Steplowska A, Pospieszny H, Kryszczuk A, Sztangret-Wisniewska J, Yin Z, Chrzanowska M, Zimnoch-Guzowska E, Jackeviciene E, Taluntytė L, Pūpola N, Mihailova J, Lielmane I, Järvekülg L, Kotkas K, Rogozina E, Sozonov A, Tikhonovich I, Horn P, Broer I, Kuusiene S, Staniulis J, Adam G, Valkonen JPT (2010) Detection, distribution and control of Potato mop-top virus, a soil-borne virus, in northern Europe. Ann Appl Biol 157:163–178

Santillan FW, Fribourg CE, Adams IP, Gibbs AJ, Boonham N, Kehoe MA, Maina S, Jones RAC (2018) The biology and phylogenetics of potato virus S isolates from the Andean region of South America. Plant Dis 102:869–885. https://doi.org/10.1094/PDIS-09-17-1414-RE

Schulte-Geldermann E, Gildemacher PR, Struik PC (2012) Improving seed health and seed performance by positive selection in three Kenyan potato varieties. Am J Potato Res 89(6):429–437

Sharma S, Kang SS, Sharma A, Kaur S (2016) Mixed infection by cucumber mosaic virus and potato virus x in potato with yellow mosaic in India. J Plant Pathol 98:693

da Silva GO, Carvalho ADF, Ponijaleki RS, Bortoletto AC (2015) Desempenho de cultivares de batata para produtividade de tubérculos. Batata Show 42:21–23

Singh BP, Raigond B, Sridhar J, Jeevalatha A, Ravinder K, Venkateswarlu V, Sharma S (2014) Potato seed production systems in India. Conference: national seminar on emerging problems of potato, 1–2 Nov 2014. ICAR-CPRI, Shimla

Sohrab SS, Karim S, Varma A, Abuzenadah AM, Chaudhary AG, Mandal B (2013) Role of sponge gourd in apical leaf curl disease of potato in Northern India. Phytoparasitica 41:403–410

Souza-Dias JAC, Eiras M, Fernandes CRF, Beloni C, Charkowski A (2017) Occurrence of spindle tuber like malformation in seed-potato lots associated with a somatic mutation in tissue culture. In: 101st-Ann Meeting-Potato American Association, Fargo, ND. Book of Abstracts, 78

Souza-Dias JAC, Iamauti MT, Fischer IH (2016a) Doenças da Batateira, Capt 16. In: Amorin L, Rezende JAM, Bergamin Fo A, Camargo LEA (eds) Manual de Fitopatologia v.2. Editora Agron Ceres Ltda, Ouro Fino, pp 125, 772p–147

Souza-Dias JAC, Jefrries C, Rezende JAM, Lima MF (2013) A new virus threat to seed-potato certification in Brazil: the whitefly-transmitted tomato chlorosis virus (Genus: Crinivirus). In: Gaba V, Tsror L (eds) EAPR-pathology section meeting. EAPR, Jerusalem, p 45. http://www.agri.gov.il/download/files/Caram_Souza_Dias_1.pdf

Souza-Dias JAC, Menarim E, Rentz R, Kitajima EW, Sawasaki HE, Duarte M (2016b) Alerta Necessário. Cultivar HF, Junho/Julho ano XIV 98:16–19
Souza Dias, JAC, Trevisan-junior O, Fernandes CRF, Oliveira-Preto TF, Trevisan LH (2018) The sprout/seed-potato technology: establishing as an example of resource usage and alternative income to citrus nurseries, in Brazil. In: 102nd Potato American Association Meeting (PAA 2018) Boise, ID. Book of Proceedings & Abstracts, vol 120, pp 72–73
Souza Richards R, Adams IP, Kreuze JF, De Souza J, Cuellar W, Dulleman AM, RAA VDV, Glover R, Hany U, Dickinson M, Boonham N (2013) The complete genome sequence of two isolates of potato black ringspot virus and its relationship to other isolates and nepoviruses. Arch Virol 159:811–815. https://doi.org/10.1007/s00705-013-1871-8
Taylor M (2018) Routes to genetic gain in potato. Nat Plants 4:631–632
Tenorio J, Chuquillanqui C, Garcia A, Guillen M, Chavez R, Salazar LF (2003) Symptomatology and effect on potato yield of achaparramieto rugoso. Fitopatologia 38(1):32–36
Thomas-Sharma S, Abdurahman A, Ali S, Andrade-Piedra JL, Bao S, Charkowski AO, Crook D, Kadian M, Kromann P, Struik PC, Torrance L (2016) Seed degeneration in potato: the need for an integrated seed health strategy to mitigate the problem in developing countries. Plant Pathol 65(1):3–16
Thomas-Sharma S, Andrade-Piedra J, Carvajal Yepes M, Hernandez Nopsa JF, Jeger MJ, Jones RAC, Kromann P, Legg JP, Yuen J, Forbes GA, Garrett KA (2017) A risk assessment framework for seed degeneration: Informing an integrated seed health strategy for vegetatively propagated crops. Phytopathology 107(10):1123–1135
Tian YP, Valkonen JPT (2013) Genetic determinants of Potato virus Y required to overcome or trigger hypersensitive resistance to PVY strain group O controlled by the gene Ny in potato. Mol Plant-Microbe Interact 26:297–305
Treder K, Chołuj J, Zacharzewska B, Babujee L, Mielczarek M, Burzyński A, Rakotondrafara AM (2018) Optimization of a magnetic capture RT-LAMP assay for fast and real-time detection of potato virus Y and differentiation of N and O serotypes. Arch Virol 163(2):447–458
Valkonen JPT (2007) Viruses: economical losses and biotechnological potential. In: Potato biology and biotechnology. Elsevier Science BV, pp 619–641
Villela JA, Pap T, Salas FJS (2017) Levantamento de viroses presentes em batatas-semente da produção nacional e Importadas. Batata Show 49:20–24
Virmond EP, Kawakami J, Souza-Dias JAC (2017) Seed-potato production through sprouts and field multiplication and cultivar performance in organic system. Hortic Bras 35:335–342. https://doi.org/10.1590/s0102-053620170304
Wang QC, Valkonen JPT (2009) Cryotherapy of shoot tips: novel pathogen eradication method. Trends Plant Sci 14:119–122
Wilson CR (2001) Resistance to infection and translocation of Tomato spotted wilt virus in potatoes. Plant Pathol 50:402–410

Permissions

All chapters in this book were first published in TPC, by Springer Nature; hereby published with permission under the Creative Commons Attribution License or equivalent. Every chapter published in this book has been scrutinized by our experts. Their significance has been extensively debated. The topics covered herein carry significant findings which will fuel the growth of the discipline. They may even be implemented as practical applications or may be referred to as a beginning point for another development.

The contributors of this book come from diverse backgrounds, making this book a truly international effort. This book will bring forth new frontiers with its revolutionizing research information and detailed analysis of the nascent developments around the world.

We would like to thank all the contributing authors for lending their expertise to make the book truly unique. They have played a crucial role in the development of this book. Without their invaluable contributions this book wouldn't have been possible. They have made vital efforts to compile up to date information on the varied aspects of this subject to make this book a valuable addition to the collection of many professionals and students.

This book was conceptualized with the vision of imparting up-to-date information and advanced data in this field. To ensure the same, a matchless editorial board was set up. Every individual on the board went through rigorous rounds of assessment to prove their worth. After which they invested a large part of their time researching and compiling the most relevant data for our readers.

The editorial board has been involved in producing this book since its inception. They have spent rigorous hours researching and exploring the diverse topics which have resulted in the successful publishing of this book. They have passed on their knowledge of decades through this book. To expedite this challenging task, the publisher supported the team at every step. A small team of assistant editors was also appointed to further simplify the editing procedure and attain best results for the readers.

Apart from the editorial board, the designing team has also invested a significant amount of their time in understanding the subject and creating the most relevant covers. They scrutinized every image to scout for the most suitable representation of the subject and create an appropriate cover for the book.

The publishing team has been an ardent support to the editorial, designing and production team. Their endless efforts to recruit the best for this project, has resulted in the accomplishment of this book. They are a veteran in the field of academics and their pool of knowledge is as vast as their experience in printing. Their expertise and guidance has proved useful at every step. Their uncompromising quality standards have made this book an exceptional effort. Their encouragement from time to time has been an inspiration for everyone.

The publisher and the editorial board hope that this book will prove to be a valuable piece of knowledge for researchers, students, practitioners and scholars across the globe.

List of Contributors

R. Ortiz
Swedish University of Agricultural Sciences (SLU), Alnarp, Sweden

E. Mihovilovich
Independent Consultant, Lima, Peru

D. S. Douches
Michigan State University, East Lansing, MI, USA

N. N. Mudege, M. L. Parker P. Kromann and M. Ghislain
International Potato Center, Nairobi, Kenya

S. Sarapura Escobar
Royal Tropical Institute, Amsterdam, The Netherlands

V. Polar and G. Thiele
CGIAR Research Program on Roots, Tubers and Bananas (RTB), Lima, Peru

A. Devaux
International Potato Center, Quito, Ecuador

J.-P. Goffart
Walloon Agricultural Research Center, Gembloux, Belgium

A. Petsakos
Formerly CIP, Seville, Spain

M. Gatto
International Potato Center, Hanoi, Vietnam

J. Okello
International Potato Center, Kampala, Uganda

V. Suarez, G. Hareau, J. Andrade-Piedra and O. Ortiz
International Potato Center, Lima, Peru

G. A. Forbes
Independent Consultant, Servas, Gard, France

A. Charkowski
Colorado State University, Fort Collins, CO, USA

E. Schulte-Geldermann
International Potato Center, Nairobi, Kenya
TH Bingen, University of Applied Sciences, Bingen, Germany

R. Nelson
Cornell University, Ithaca, NY, USA

J. W. Bentley
Independent Consultant, Cochabamba, Bolivia

G. Burgos and T. Zum Felde
International Potato Center, Lima, Peru

C. Andre
The New Zealand Institute for Plant and Food Research Limited/ Luxembourg Institute of Science and Technology, Auckland, New Zealand

S. Kubow
McGill University, Quebec, Canada

J. F. Kreuze
International Potato Center, Lima, Peru

J. A. C. Souza-Dias
Instituto Agronômico de Campinas (IAC)/APTA/SAA-SP, Campinas, São Paulo, Brazil

A. Jeevalatha
ICAR-Central Potato Research Institute, Shimla, Himachal Pradesh, India

A. R. Figueira
Department of Plant Pathology, Lavras Federal University, Lavras, Minas Gerais, Brazil

J. P. T. Valkonen
Department of Agricultural Sciences, University of Helsinki, Helsinki, Finland

R. A. C. Jones
Institute of Agriculture, University of Western Australia, Crawley, Perth, Australia

Index